A Concise Introduction to Functional Analysis

A Concise Introduction to Functional Analysis is designed to serve a one-semester introductory graduate (or advanced undergraduate) course in functional analysis.

The text is pragmatically structured so that each unit corresponds to one class, with the hope of being helpful for both students and teachers. It is expected that this text will provide students with a strong general understanding of the subject, and that they should feel well equipped to take on the more advanced texts and courses covering topics not treated here.

Features

- Numerous examples and counterexamples to illustrate such abstract concepts.
- Over 430 exercises, with partial solutions included in the book itself.
- Minimal pre-requisites beyond linear algebra and general topology.

César R. de Oliveira earned his Ph.D. in Physics from the University of São Paulo in 1987. He has been a visiting professor at the Università degli Studi di Milano (1991–1992) and the University of British Columbia (2008–2009). He is currently a Full Professor in the Department of Mathematics at the Federal University of São Carlos.

His research lies in the field of Mathematical Physics, with publications in both mathematics and physics journals. He has supervised twelve Ph.D. students, and his main areas of interest include Schrödinger and Dirac operators, the Aharonov–Bohm effect, mathematical models of graphene, quantum (in)stability, and dynamical localization.

He also enjoys writing textbooks and has authored four of them. In 2000, he published an introductory mechanics book (in Portuguese, with animations and interactive content on CD-ROM) through his University. In 2010, he released a graduate-level book on functional analysis (in Portuguese) with IMPA (Rio de Janeiro). In 2009, the book *Spectral Theory and Quantum Dynamics* was published by Birkhäuser (Switzerland). Most recently, in 2023, he co-authored *Spectral Measures and Dynamics: Typical Behaviors* with M. Aloisio and S. Carvalho and published by Springer Nature (Switzerland).

A Concise Introduction to Functional Analysis

César R. de Oliveira

CRC Press
Taylor & Francis Group
Boca Raton London New York

CRC Press is an imprint of the
Taylor & Francis Group, an **informa** business

A CHAPMAN & HALL BOOK

Front cover image: Susii/Shutterstock

First edition published 2026
by CRC Press
2385 NW Executive Center Drive, Suite 320, Boca Raton FL 33431

and by CRC Press
4 Park Square, Milton Park, Abingdon, Oxon, OX14 4RN

CRC Press is an imprint of Taylor & Francis Group, LLC

© 2026 **César R. de Oliveira**

ISBN: 978-1-041-10328-8 (hbk)
ISBN: 978-1-041-10650-0 (pbk)
ISBN: 978-1-003-65616-6 (ebk)

DOI: 10.1201/9781003656166

Typeset in Latin Modern font
by KnowledgeWorks Global Ltd.

Publisher's note: This book has been prepared from camera-ready copy provided by the authors.

To
Ana, Natália and Daniel

Contents

Preface

Functional Analysis is a traditional course for mathematicians and students of related areas. This text is planned for a one-semester introductory graduate (or advanced undergraduate) course in Functional Analysis, and it was written with a pragmatic proposal. The idea is that each unit corresponds to one class, with the hope of being helpful for both students and teachers. This is an important difference between this text and others on similar subjects. In order to follow the original plan, it was necessary to restrict the contents to be covered in each chapter, with selection of topics and omission of some interesting applications. Maybe the main omission is the theory of topological vector spaces. The emphasis is on Linear Functional Analysis, and some topics naturally follow the personal taste of the author.

It is expected that this text will give students a general view of the subject and after going through the material, they should be equipped to take on more advanced texts that cover topics not treated here.

Most chapters have *Notes*, which aim at presenting comments that normally are restricted to the classroom, historical data, as well as mentioning extensions and applications not covered in the text. The Notes are very informal, and do not intend to be exhaustive in any sense.

Although the traditional abstract approach to the subject has been followed, there was an attempt to provide a large number of examples and counterexamples to illustrate such abstract concepts. Over 430 exercises are proposed throughout the text. At the end of the book, there are solutions to some of them, and, for sake of completeness, in particular to those exercises whose conclusions are used somewhere in the text.

Readers are supposed to be familiar with Linear Algebra (many results on finite-dimensional spaces are proposed as exercises) and General Topology; apart from a few theorems, basic results of Functions of One Complex Variable and Measure and Lebesgue Integration are usually restricted to examples.

It is worth mentioning some notations. The symbols $\mathcal{N}, \mathcal{B}, \mathcal{H}$ denote normed, Banach and Hilbert spaces, respectively, and they will usually be used without explicitly saying so. The symbol $\mathbb{F}$ denotes either the field of real numbers $\mathbb{R}$ or the field of complex numbers $\mathbb{C}$. The term "non-$\mathcal{P}$" says that the object in question "does not satisfy property $\mathcal{P}$," avoiding some unnecessary accumulation of definitions; for instance, *nonempty set* means a set that is *not empty*. The term *enumerable* (as well as *denumerable*) indicates the cardinality $\aleph_0$ of the set of natural numbers $\mathbb{N}$, while *countable* refers to finite (including zero) or enumerable.

When a subset $\{\xi_j\}$ is also a sequence, usually this is emphasized with the notation (ξ_j). As usual, a vector space is called trivial if it contains only the null element. The symbol := indicates the introduction of a new notation.

This text is a translation of the third edition of a textbook originally published in Portuguese by IMPA, Rio de Janeiro. The good acceptance of these Brazilian editions stimulated me to translate the book into English (please, take into account that English is not the native language of the author!). I would like to express my gratitude to some friends who have corrected many English mistakes in my original translation, particularly Prof. Michael O'Carroll who was very kind and patient.

The Errata for this book will be available on the internet

http://www.dm.ufscar.br/profs/oliveira/FAIErrata.html

I am especially grateful to Dr. Pedro Malagutti for (courageously!) assigning this text in his Functional Analysis classes and for his meticulous reading and insightful critique. I also appreciate the valuable suggestions from my colleagues and students, in particular, by Francisco Caramello, who was very attentive.

São Carlos, SP, Brazil   César R. de Oliveira
                                                                    April 2025

Selected Notation

- The set of natural numbers: $\mathbb{N} = \{1, 2, 3, \cdots\}$.

- The term "enumerable" refers to the cardinality of $\mathbb{N}$, whereas "countable" refers to enumerable or finitely many (including zero).

- $\mathcal{N}, \mathcal{B}, \mathcal{H}$ denote normed, Banach, and Hilbert spaces, respectively.

- (ξ_n) indicates a sequence.

- $w \cdot \lim$ and $s \cdot \lim$ indicate weak and strong convergences, respectively.

- The identity operator is denoted by $\mathbf{1}$.

- $\text{Lin}(A)$ is the set of (finite) linear combinations of elements of A.

- The range, domain and kernel (i.e., null set) of a map T will be denoted by $\text{rng } T, \text{dom } T$ and $\text{N}(T)$, respectively.

- $\text{B}(\mathcal{N}_1, \mathcal{N}_2)$ is the set of bounded linear operators from $\mathcal{N}_1$ to $\mathcal{N}_2$.

- $\text{B}_0(\mathcal{N}_1, \mathcal{N}_2)$ is the set of compact linear operators.

- T^* denotes the Hilbert adjoint of a linear transformation T, whereas T^a denotes the adjoint of T defined on general normed spaces.

- An action "T in X" means that $\text{dom } T \subset X$, whereas "T on X" means that $\text{dom } T = X$. They are abbreviations of "T acting in X" and "T acting on X," respectively.

- The symbol • indicates the end of an example.

- The symbol □ indicates the end of a proof.

CHAPTER 1

Normed Spaces

Broadly speaking, Functional Analysis is recognized as a synthesis of concepts from Linear Algebra, Analysis, and Topology, with a particular emphasis placed on infinite-dimensional vector spaces. This field is considered fundamental in many areas of mathematics and physics. A key concept in Functional Analysis is the norm, which is defined as an abstract notion of length in a vector space. Each norm is associated with a metric, through which a distance between vectors is determined, allowing the vector space to be transformed into a topological space consistent with its linear structure. In this first chapter, the definition of a norm is introduced, fundamental examples are presented, and essential notations are established for use throughout the text.

The vector spaces will be over the fields of real numbers $\mathbb{R}$ as well as over the complex numbers $\mathbb{C}$; the real part of a complex number z will be denoted by Re z and its imaginary part by Im z. In some settings it is not necessary to specify the field, so it is convenient to indicate by $\mathbb{F}$ either $\mathbb{C}$ or $\mathbb{R}$. In general the vector spaces will be denoted by $X, Y, Z, \cdots$, while its elements by $\xi, \eta, \zeta, \cdots$; the *scalars*, that is, elements of $\mathbb{F}$, by $\alpha, \beta, \lambda, \cdots$, or $a, b, c, \cdots$. Recall that a subset A of a vector space X is *linearly independent* if any finite linear combination $\sum_{j=1}^{n} \alpha_j \xi_j = 0$ of elements $\xi_j \in A$, which results in the null vector, implies $\alpha_j = 0$, for all j ($\emptyset$ is linearly independent; verify!). If $\alpha \in \mathbb{F}$, the notation $\alpha > 0$ indicates that $\alpha \in \mathbb{R}$ and it is *strictly positive*; also, $\alpha \geq 0$ will be referred to as "α is *positive*" (similar for *negative* and *strictly negative*). Sometimes it will be used that the set of rational numbers $\mathbb{Q}$ is dense in $\mathbb{F}$ (with the usual topology), and in the case $\mathbb{F} = \mathbb{C}$, it will be understood by *rational* any complex number of the form $r + is$, with $r, s \in \mathbb{Q}$.

Definition 1.1. A *norm* in a vector space X (real or complex) is a mapping $\|\cdot\| : X \to \mathbb{R}$ satisfying:

i. $\|\xi\| \geq 0$ for all $\xi \in X$, and $\|\xi\| = 0 \iff \xi = 0$ (positive length).

ii. $\|\alpha\xi\| = |\alpha| \, \|\xi\|$, for all $\xi \in X$ and any $\alpha \in \mathbb{F}$ (dilation).

iii. $\|\xi + \eta\| \leq \|\xi\| + \|\eta\|$, for all $\xi, \eta \in X$ (triangle inequality).

If in the definition of a norm the condition $\|\xi\| = 0 \Rightarrow \xi = 0$ is not imposed, then $\|\cdot\|$ is said to be a *seminorm*. It is a simple exercise to verify that each norm defines, or induces, a metric d on X by $d(\xi, \eta) = \|\xi - \eta\|$. The pair $(X, \|\cdot\|)$ is called a *normed space*; here, $\mathcal{N}, \mathcal{N}_1, \mathcal{N}_2, \cdots$, will *always* denote normed spaces (with $(\mathcal{N}, \|\cdot\|)$ if one wishes to specify the norm). In case no other topology is specified, it is understood that on $\mathcal{N}$ the topology is the one induced by the norm.

It is convenient to introduce additional notations. If (X, d) is a metric space and $r > 0$, then $B(\xi_0; r) = B_X(\xi_0; r) = \{\xi \in X : d(\xi_0, \xi) < r\}$, $\overline{B}(\xi_0; r) = \overline{B}_X(\xi_0; r) = \{\xi \in X : d(\xi_0, \xi) \leq r\}$, and $S(\xi_0; r) = S_X(\xi_0; r) = \{\xi \in X : d(\xi_0, \xi) = r\}$ indicate the open ball, the closed ball, and the sphere of radius r centered at $\xi_0 \in X$. Finally, a set is *enumerable* (or *denumerable*) if it has the cardinality $\aleph_0$ of $\mathbb{N} = \{1, 2, 3, \cdots\}$, and it is *countable* if it is finite (including zero) or enumerable.

EXERCISE 1.1. (a) Check that if a metric d on X is induced by a norm, then $d(\xi + \zeta, \eta + \zeta) = d(\xi, \eta)$, for all $\xi, \eta, \zeta \in X$. Give a geometric interpretation.

(b) Show that $|\, \|\xi\| - \|\eta\|\, | \leq \|\xi - \eta\|$, for all $\xi, \eta \in X$; conclude then that the norm $\|\cdot\| : X \to \mathbb{R}$ is a continuous mapping ($\mathbb{R}$ with the usual metric).

(c) Show that on a normed space the vector sum and multiplication by scalars are continuous operations, i.e., the mappings $\mathcal{N} \times \mathcal{N} \to \mathcal{N}$, $(\eta, \xi) \mapsto \eta + \xi$, and $\mathbb{F} \times \mathcal{N} \to \mathcal{N}$, $(\alpha, \xi) \mapsto \alpha\xi$, are continuous.

EXERCISE 1.2. Let Ω be a compact subset of a Hausdorff topological space and $C(\Omega)$ the vector space of continuous functions $\psi : \Omega \to \mathbb{F}$. Show that

$$\|\psi\|_\infty = \sup_{t \in \Omega} |\psi(t)| = \max_{t \in \Omega} |\psi(t)|$$

defines a norm on $C(\Omega)$, and that the normed space $(C(\Omega), \|\cdot\|_\infty)$ is complete with the induced metric. This norm is also named as *norm of uniform convergence*. When no norm is specified on $C(\Omega)$, it is understood that $\|\cdot\|_\infty$ is being considered.

Recall that a metric space is complete if every Cauchy sequence converges to an element of this space.

Definition 1.2. A *Banach space* is a normed space which is complete with respect to the metric induced by its norm. $\mathcal{B}, \mathcal{B}_1, \mathcal{B}_2, \cdots$, will *always* denote such spaces.

$(C(\Omega), \|\cdot\|_\infty)$ is an example of Banach space. In the next exercises, other well-known examples of Banach spaces will appear.

EXERCISE 1.3. Denote by $\xi = (\xi_1, \cdots, \xi_n)$ the elements of $\mathbb{F}^n$.

(a) Show that $\mathbb{F}^n$ is a Banach space with each of the following norms $\|\xi\|_p = \left(\sum_{j=1}^n |\xi_j|^p\right)^{1/p}$, $1 \leq p < \infty$, and $\|\xi\|_\infty = \max_{1 \leq j \leq n} |\xi_j|$. Why for $p < 1$ it does not define a norm?

(b) Check that for every $\xi \in \mathbb{F}^n$, $\|\xi\|_\infty \leq \|\xi\|_p \leq n^{1/p}\|\xi\|_\infty$, for all $p \geq 1$.

EXERCISE 1.4. Show that $\|\psi\|_1 = \int_a^b |\psi(t)|\, dt$ is a norm on $C[a, b]$ (continuous functions $\psi : [a, b] \to \mathbb{F}$), and verify that $\|\psi\|_1 \leq (b - a)\|\psi\|_\infty$ for every function $\psi \in C[a, b]$. Show also that $(C[a, b], \|\cdot\|_1)$ is not complete. Does exist $A > 0$ so that $\|\psi\|_\infty \leq A\|\psi\|_1$ for all $\psi \in C[a, b]$?

Example 1.3. Denote by $\xi = (\xi_1, \xi_2, \cdots)$ an element of $\mathbb{F}^{\mathbb{N}}$, that is, the sequences in $\mathbb{F}$ indexed by $\mathbb{N}$. For $1 \leq p < \infty$, set

$$l^p(\mathbb{N}) = \left\{\xi \in \mathbb{F}^{\mathbb{N}} : \|\xi\|_p := \left(\sum_{j=1}^{\infty} |\xi_j|^p\right)^{1/p} < \infty\right\}$$

and $l^{\infty}(\mathbb{N}) = \{\xi \in \mathbb{F}^{\mathbb{N}} : \|\xi\|_{\infty} := \sup_{1 \leq j < \infty} |\xi_j| < \infty\}$. It is a standard approach to verify that $l^p(\mathbb{N})$, for $1 \leq p \leq \infty$, are Banach spaces (the cases $p = 1, 2, \infty$ are easier to check). In a similar way, the Banach spaces $l^p(\mathbb{Z})$ are defined.

If J is a set, $l^{\infty}(J)$ denotes all functions $\psi : J \to \mathbb{F}$ with $\|\psi\|_{\infty} := \sup_{t \in J} |\psi(t)| < \infty$, and for $1 \leq p < \infty$, $l^p(J)$ is the set of functions $\psi : J \to \mathbb{F}$ which vanish except on a countable subset of J, and so that $\|\psi\|_p := (\sum_{t \in J} |\psi(t)|^p)^{1/p} < \infty$. Those spaces are also Banach. Note that $l^{\infty}(J) \subset C(J)$, with J taken with the discrete topology.
●

Example 1.4. If $(\Omega, \mathcal{A}, \mu)$ is a (positive) measure space, with $\mathcal{A}$ being a σ-algebra on Ω, then it is a well-known result in the integration theory that, for $1 \leq p \leq \infty$, the set $L^p_{\mu}(\Omega)$ of (equivalent classes, by identifying functions that coincide μ-a.e.) measurable functions $\psi : \Omega \to \mathbb{F}$ with

$$\|\psi\|_p := \left(\int_{\Omega} |\psi(t)|^p \mathrm{d}\mu(t)\right)^{1/p} < \infty, \quad 1 \leq p < \infty,$$

and $\|\psi\|_{\infty} := \text{ess-sup}_{t \in \Omega} |\psi(t)| < \infty$, are Banach spaces. Again, the cases $p = 1, 2, \infty$ are easier to check. In the Additional Exercises section of this chapter, a way to prove such results is outlined. If the measure is Lebesgue measure of subsets of $\mathbb{R}^n$ the notation will simply be $L^p(\Omega)$; if Ω is the interval $[a, b]$ and the measure the Lebesgue one, then it will be used the notation $L^p[a, b]$. ●

EXERCISE **1.5.** Verify that $l^p(J)$, for $p = 1, 2, \infty$ and $L^{\infty}_{\mu}(\Omega)$ are Banach spaces, and that $L^p_{\mu}(\Omega)$ are normed spaces for $p = 1, 2$.

The *vector space generated* by a subset A of a vector space X is the set of finite linear combinations of elements of A, and it will be denoted by $\text{Lin}(A)$; note that $\text{Lin}(A)$ is the smaller vector subspace that contains A. Recall, also, that a *Hamel basis*, or simply basis, in a vector space X is a maximal linearly independent set A, i.e., $\text{Lin}(A) = X$. If there exists a finite basis of X with n elements, then the *algebraic dimension* of X, denoted by $\dim X$, is finite and equals to n (and all bases have n elements); otherwise, X is said to be of infinite dimension. A subset A in a normed space $\mathcal{N}$ is *total in* $\mathcal{N}$ if $\text{Lin}(A)$ is dense in $\mathcal{N}$.

Definition 1.5. Two norms $\|\cdot\|_1$ and $\|\cdot\|_2$ on a vector space X are *equivalent* if there exist $A, B > 0$ so that

$$A\|\xi\|_1 \leq \|\xi\|_2 \leq B\|\xi\|_1, \quad \forall \xi \in X.$$

EXERCISE **1.6.** (a) Verify that equivalent norms on a vector space generate the same (metric) topology and have the same Cauchy sequences; hence, if one of such metric spaces is complete, then the other one will also be complete.

(b) Show that if two norms generate the same topology, then they are equivalent.

Example 1.6. The norms $\| \cdot \|_\infty$ and $\| \cdot \|_1$ on the vector space of polynomials on $[0, 1]$ are not equivalent, since if $p_n(t) = t^{n-1}$, then $\|p_n\|_\infty = 1$, for all $n \geq 1$, while $\|p_n\|_1 = 1/n$ converges to zero as $n \to \infty$. ●

EXERCISE 1.7. (a) Show that on a finite-dimensional normed space a set is compact if and only if it is closed and bounded (recall that on a metric space every compact set is closed and bounded).

(b) Let $\xi_n = (\delta_{n,j})_{j=1}^\infty$, with $\delta_{k,i} = 0$ if $i \neq k$ and $\delta_{k,k} = 1$, the "Kronecker's delta." Use this sequence to show that in $l^p(\mathbb{N})$, $1 \leq p \leq \infty$, there are closed and bounded subsets that are not compact.

As a motivation, at this point, it would be interesting to comment on some particular properties of infinite-dimensional normed spaces, and compare them with the case of finite dimension. These properties will be discussed later on.

1. All norms on a finite-dimensional vector space are equivalent, and these spaces are Banach, but these properties do not generalize to the infinite-dimensional case.

2. The closed ball $\overline{B}(0; 1)$ is compact if and only if the normed space is of finite dimension.

3. All linear mappings from a finite-dimensional normed space into itself are continuous, which does not hold in infinite-dimensional spaces.

4. In normed spaces of infinite dimension, there are dense proper vector subspaces (with the possibility of two of these spaces with just the null vector in common), and also linear dense-defined mappings that have no linear extensions to the whole space.

Theorem 1.7. *Let X be a vector space of finite dimension. Then all norms on X are equivalent.*

Proof. Let $\{e_1, \cdots, e_n\}$ denote a basis of X, so that any $\xi \in X$ can be written in the form $\xi = \sum_{j=1}^n \alpha_j e_j$. It is enough to show that any norm $\| \cdot \|$ on X is equivalent to $\|\|\xi\|\| = \sum_{j=1}^n |\alpha_j|$. An inequality follows straightly from

$$\|\xi\| = \left\| \sum_{j=1}^n \alpha_j e_j \right\| \leq \sum_{j=1}^n |\alpha_j| \|e_j\| \leq \left(\max_{1 \leq j \leq n} \|e_j\| \right) \|\|\xi\|\| = B \|\|\xi\|\|,$$

with $B = \max_{1 \leq j \leq n} \|e_j\|$. In order to get the other inequality, suppose that there is no $A > 0$ with $A \|\|\xi\|\| \leq \|\xi\|$, for all $\xi \in X$. Thus, for each $N \in \mathbb{N}$ there exists $\xi_N \in X$ with $\|\|\xi_N\|\| = 1$ and $1 = \|\|\xi_N\|\| > N \|\xi_N\|$. Since $S(0; 1)$ is compact (finite dimension), there is a subsequence (ξ_{N_j}) of (ξ_N) which converges to ξ_0 in $(X, \|\| \cdot \|\|)$; the continuity of the norm yields $\|\|\xi_0\|\| = 1$. Hence, by using the above inequality,

$$\|\xi_0\| \leq \|\xi_0 - \xi_{N_j}\| + \|\xi_{N_j}\| \leq B \|\|\xi_0 - \xi_{N_j}\|\| + \frac{1}{N_j}$$

which converges to zero as $j \to \infty$, i.e., $\|\xi_0\| = 0$ and so $\xi_0 = 0$. The contradiction with $\|\|\xi_0\|\| = 1$ finishes the proof. □

Corollary 1.8. *All normed spaces of finite-dimension are Banach (thus a subspace of finite dimension in a normed space is closed).*

Proof. The notations introduced in the proof of Theorem 1.7 will be used. Let X be a finite-dimensional normed space. Since all norms are equivalent, it is possible to consider $\|\| \cdot \|\|$ (see Exercise 1.6). Let $\xi_k = \sum_{j=1}^{n} \alpha_j^k e_j$ be a Cauchy sequence in $(X, \|\| \cdot \|\|)$. Since $\sum_{j=1}^{n} |\alpha_j^k - \alpha_j^m| = \|\|\xi_k - \xi_m\|\|$, it follows that, for all $1 \leq j \leq n$, the sequence $(\alpha_j^k)_{k=1}^{\infty}$ is Cauchy in $\mathbb{F}$ and converges to some $\alpha_j^0 \in \mathbb{F}$. Defining $\xi_0 = \sum_{j=1}^{n} \alpha_j^0 e_j$ in X, one has

$$\lim_{k \to \infty} \|\|\xi_k - \xi_0\|\| = \lim_{k \to \infty} \sum_{j=1}^{n} |\alpha_j^k - \alpha_j^0| = 0,$$

i.e., $\xi_k \to \xi_0$ and the space is complete. □

Example 1.9. By Stone-Weierstrass' Theorem [Simm], the vector space of polynomials is dense in $C[a,b]$, so the space $C[a,b]$ is of infinite dimension. This is an example of a proper dense vector subspace of $C[a,b]$, since every polynomial is differentiable whereas there are nondifferentiable continuous functions on $[a,b]$; for instance $\psi(t) = |t|$ on $[-1, 1]$. •

In the particular class of the Banach spaces $L_\mu^p(\Omega)$, $1 \leq p \leq \infty$, the triangle inequality $\|\psi_1 + \psi_2\|_p \leq \|\psi_1\|_p + \|\psi_2\|_p$ is usually referred to as *Minkowski inequality*. p and q are *conjugate exponents* if $1/p + 1/q = 1$ (by convention, 1 and ∞ are conjugate exponents), and in this case the *Hölder inequality*

$$\left| \int \psi\varphi \, d\mu \right| \leq \int |\psi\varphi| \, d\mu \leq \|\psi\|_p \|\varphi\|_q$$

holds. See Exercise 1.20 for a guide to the proofs of these two standard inequalities.

Notes

The Functional Analysis arose from several problems in differential and integral equations, which required the use of vector spaces of infinite dimension. The systematic study of such spaces began in the first decade of the XXth century, mainly with works by S. Banach, M. R. Fréchet, E. Helly, D. Hilbert, F. Riesz, E. Schmidt, and others.

Presently it can appear to be a simple step, but it was outstanding to notice that the abstract definition of metric introduces precise notions of limit, continuity, compactness, etc., in spaces distinct from $\mathbb{F}^n$ (the concept of metric space was introduced by Fréchet in his thesis in 1906, although the expression "metric space" is due to Hausdorff). This, and the notion of norm on vector spaces, makes it possible to transport much of the Euclidean Geometry to infinite-dimensional systems; in one of his contributions, E. Schmidt in 1908 introduced the notation $\| \cdot \|$ for the norm on the particular space l^2, and then this notation was adopted by the mathematical community to denote norms in general. One of the first to introduce the abstract definition of norm was the Austrian mathematician Helly (for a while he was a war prisoner in Russia) around 1920, but with a distinct notation from that of Schmidt.

In 1932 the Polish mathematician S. Banach published a book with the then main known results on normed spaces, including many of his own theorems. Several notations used in that book were adopted and, after him, the term *Banach space* was introduced. It is curious to notice that Banach did not use complex numbers in his book!

Note that the spaces $l^p(J)$ are particular cases of $L^p_\mu(J)$, so it is possible to use results of integration theory to show that $l^p(J)$ are Banach. Indeed, choose the σ-algebra $\mathcal{A}$ as all subsets of J and, for $A \in \mathcal{A}$, define $\mu(A)$ as the cardinality of A if it is finite, otherwise $\mu(A) = \infty$; then, $l^p(J) = L^p_\mu(J)$.

Additional Exercises

EXERCISE **1.8.** Discuss the convergence of the sequences $(\psi_n(t) = t^n - t^{2n})$, $(\phi_n(t) = t^n - t^{n+1})$ and $(\varphi_n(t) = t^n/n - t^{n+1}/(n+1))$ in $L^p[0,1]$, for $p = 1$ and ∞.

EXERCISE **1.9.** Denote by $C^1[a,b]$ the set of continuously differentiable functions $\psi : [a,b] \to \mathbb{R}$ with the norm (the $'$ indicates differentiation)

$$\|\psi\|_{C^1} := \sup_{t \in [a,b]} |\psi(t)| + \sup_{t \in [a,b]} |\psi'(t)|.$$

Verify that $\|\cdot\|_{C^1}$ is a norm and $C^1[a,b]$ is Banach; generalize for functions of class C^r. Discuss the convergence of the sequence $(\varphi_n(t) = t^n/n - t^{n+1}/(n+1))$ in $C[0,1]$ and $C^1[0,1]$.

EXERCISE **1.10.** Which is the dimension of the vector space of matrixes $n \times m$? Consider different norms on this space.

EXERCISE **1.11.** Check that all Hamel bases on a given vector space have the same cardinality.

EXERCISE **1.12.** Let $c = c(\mathbb{N})$ and $c_0 = c_0(\mathbb{N})$ denote the set of sequences (α_n) in $\mathbb{F}$ whose limits $\lim_{n \to \infty} \alpha_n$ exist and $\lim_{n \to \infty} \alpha_n = 0$, respectively. Show that these sets are closed subspaces of $l^\infty(\mathbb{N})$, and therefore, are Banach spaces with the norm $\|\cdot\|_\infty$. Show that given an element $\xi \in c$ there are $\eta \in c_0$ and $\alpha \in \mathbb{F}$ such that $\xi = \zeta + \eta$, with $\zeta = (\alpha, \alpha, \alpha, \cdots) \in c$.

EXERCISE **1.13.** Show that in $l^\infty(\mathbb{N})$ the expression $\||\xi\|| = \limsup_{n \to \infty} |\xi_n|$ (with $\xi = (\xi_1, \xi_2, \xi_3, \cdots)$) is a seminorm and $\{\xi \in l^\infty(\mathbb{N}) : \||\xi\|| = 0\}$ coincides with c_0 (see Exercise 1.12).

EXERCISE **1.14.** Draw the unit sphere $S(0;1)$ in the normed space $(\mathbb{R}^2, \|\cdot\|_p)$ for $p = 1, 2, 5, \infty$. What about $0 < p < 1$? Consider $\xi = (1,0), \eta = (0,1)$ and show that $\|\xi + \eta\|_p > \|\xi\|_p + \|\eta\|_p$ if $0 < p < 1$, that is, the triangle inequality does not hold for such values of p.

EXERCISE **1.15.** Let $\xi_n = (1/n, 1/n, \cdots, 1/n, 0, 0, \cdots)$, with the n first entries equal to $1/n$ (resp. $\psi_n(t) = \chi_{[0,n]}(t)/n$, with χ_A being the characteristic function of the set A), a sequence in $l^1(\mathbb{N})$ $(L^1(\mathbb{R}))$. Show that $\xi_n \to 0$ ($\psi_n \to 0$) uniformly, but it does not converge in $l^1(\mathbb{N})$ (resp. $L^1(\mathbb{R})$). Adapt for $l^p(\mathbb{N})$ $(L^p(\mathbb{R}))$, $1 < p < \infty$.

EXERCISE **1.16.** Show that a metric d on a vector space X is induced by a norm if and only if $d(\xi + \zeta, \eta + \zeta) = d(\xi, \eta)$ and $d(\alpha\xi, \alpha\eta) = |\alpha| d(\xi, \eta)$, for all $\xi, \eta, \zeta \in X, \alpha \in \mathbb{F}$.

EXERCISE **1.17.** It is possible that two metrics generate the same topology, but just one of these spaces be complete (this does not occur in the case of two metrics induced by norms). Check this for the metrics $d(t,s) = |t - s|$ and $D(t,s) = |f(t) - f(s)|$ in $\mathbb{R}$, with $f(t) = t/(1 + |t|)$.

EXERCISE **1.18.** Let Ω be a nonempty open subset of $\mathbb{C}$. Show that the set of bounded holomorphic functions in Ω with the norm $\|\cdot\|_\infty$ is a Banach space.

EXERCISE **1.19.** Show that the set of absolutely continuous functions $\psi : [0,1] \to \mathbb{R}$, with $\psi(0) = 0$ and $\psi' \in L^2[0,1]$, is a Banach space with the norm $\||\psi\|| = \left(\int_0^1 |\psi'(t)|^2 \, dt \right)^{1/2}$. This result is not immediate, and it is related to the so-called Sobolev spaces.

EXERCISE **1.20.** Consider $1 < p < \infty$. Find the minimum of the function $\phi(t) = t^p/p - t$ and conclude that $t \le t^p/p + 1/q$, with $1/p + 1/q = 1$. By choosing $t = r/s^{q/p}$ conclude that $rs \le r^p/p + s^q/q$, for any $r, s > 0$. Use this to show Hölder inequality $|\int \psi\varphi \, d\mu| \le \int |\psi\varphi| \, d\mu \le \|\psi\|_p \|\varphi\|_q$ for $\psi \in L^p_\mu$ and $\varphi \in L^q_\mu$. Pick $\varphi = \psi^{p-1}/\|\psi\|_p^{p/q}$, and then show that $\|\psi\|_p = \sup_{\|\varphi\|_q = 1} \int |\psi\varphi| \, d\mu$, verify Minkowski inequality $\|\psi_1 + \psi_2\|_p \le \|\psi_1\|_p + \|\psi_2\|_p$ and that $\|\cdot\|_p$ is a norm. L^p is complete by Riesz-Fischer Theorem. Adapt for l^p.

CHAPTER 2

Compactness and Completion of Normed Spaces

In this chapter, two fundamental issues related to normed spaces are examined. The first concerns the compactness of the closed unit ball in $\mathcal{N}$, which is shown to occur if and only if the dimension of $\mathcal{N}$ is finite (recall that $\mathcal{N}$ always denotes a normed space). This characterization is established as a fundamental property that distinguishes finite-dimensional spaces from their infinite-dimensional counterparts. Following this, it is shown that every normed space can be extended to a Banach space through a process referred to as completion. In this construction, a larger space is systematically formed, within which the given normed space is densely embedded, ensuring that limits exist for all Cauchy sequences. This process guarantees that every normed space can be associated with a complete space, in which all Cauchy sequences converge, thereby providing a rigorous framework for further analysis.

2.1 COMPACTNESS AND DIMENSION

The following lemma serves as a fundamental tool for constructing bounded sequences without convergent subsequences in infinite-dimensional normed spaces. Despite the absence of an explicit notion of orthogonality, a geometric perspective plays a role in making its proof intuitive.

Lemma 2.1 (Riesz Lemma). *For each proper closed vector subspace X of $(\mathcal{N}, \|\cdot\|)$, given $0 < \alpha < 1$, there is $\xi \in \mathcal{N} \setminus X$ with $\|\xi\| = 1$ and $\inf_{\eta \in X} \|\xi - \eta\| \geq \alpha$.*

Proof. Pick $\zeta \in \mathcal{N} \setminus X$ and let $c = \inf_{\eta \in X} \|\eta - \zeta\|$, and note that $c > 0$ since X is closed. Now, for each $d > c$, there exists $\tau \in X$ with $c \leq \|\zeta - \tau\| \leq d$. The vector $\xi = (\zeta - \tau)/\|\zeta - \tau\|$ belongs to $\mathcal{N} \setminus X$ and $\|\xi\| = 1$. Furthermore, for any $\eta \in X$, one has

$$\|\xi - \eta\| = \frac{1}{\|\zeta - \tau\|} \big\| \zeta - (\tau + \|\zeta - \tau\| \eta) \big\| \geq \frac{c}{\|\zeta - \tau\|} \geq \frac{c}{d}.$$

Finally, for $0 < \alpha < 1$, it is enough to choose $d = c/\alpha$ to conclude the result. □

DOI: 10.1201/9781003656166-2

A notable consequence of Riesz Lemma is

Theorem 2.2. *The closed ball $\overline{B}(0;1)$ in a normed vector space $\mathcal{N}$ is compact if and only if* $\dim \mathcal{N} < \infty$.

Proof. If $\dim \mathcal{N} < \infty$, it is known that $\overline{B}(0;1)$ is compact (see Exercise 1.7). If $\dim \mathcal{N}$ is not finite, then Riesz Lemma will be used to construct a sequence in $\overline{B}(0;1)$ with no convergent subsequence.

Let $\xi_1 \in \mathcal{N}$, $\|\xi_1\| = 1$. By Riesz Lemma there exists $\xi_2 \in \mathcal{N}$, with $\|\xi_2\| = 1$, and $\|\xi_1 - \xi_2\| \geq 1/2$ (by choosing $\alpha = 1/2$ in Riesz Lemma). The vector space $\mathrm{Lin}(\{\xi_1, \xi_2\})$ is closed, since its dimension is finite. Again by Riesz Lemma, there exists $\xi_3 \in \mathcal{N}$, with $\|\xi_3\| = 1$, $\|\xi_3 - \xi_1\| \geq 1/2$ and $\|\xi_3 - \xi_2\| \geq 1/2$. In this way, a sequence $(\xi_n)_{n=1}^{\infty}$, $\|\xi_n\| = 1$, for all n, and $\|\xi_j - \xi_k\| \geq 1/2$ for all $j \neq k$ is constructed. Since such sequence has no convergent subsequence, the closed ball $\overline{B}(0;1)$ is not compact. □

EXERCISE **2.1.** Show that in an infinite-dimensional normed space any subset containing a nonempty open set is not compact. Conclude that normed spaces of infinite dimension have no locally compact subsets.

EXERCISE **2.2.** Discuss what happens if $S(0;1)$ replaces $\overline{B}(0;1)$ in Theorem 2.2?

EXERCISE **2.3.** A sequence $(\xi_n)_{n=1}^{\infty}$ in a normed space $\mathcal{N}$ is *absolutely summable* if $\sum_{n=1}^{\infty} \|\xi_n\| < \infty$. Show that $\mathcal{N}$ is a Banach space if and only if any absolutely summable sequence is summable (i.e., $\sum_{n=1}^{\infty} \xi_n$ converges) in $\mathcal{N}$.

EXERCISE **2.4.** Give examples of nonclosed subspaces of the following spaces: $l^1(\mathbb{Z})$, $l^{\infty}(\mathbb{Z})$, and $\mathrm{L}^1(\mathbb{R})$.

2.2 COMPLETION OF NORMED SPACES

Definition 2.3. Two metric spaces (X, d) and (Y, D) are *isometric* if there is a bijective isometry $\kappa : X \to Y$ (recall that κ is an isometry if $D(\kappa(\xi), \kappa(\eta)) = d(\xi, \eta)$, for any $\xi, \eta \in X$, and every isometry is injective).

EXERCISE **2.5.** Show that a metric space isometric to a complete metric space is also complete.

Definition 2.4. Two normed spaces $\mathcal{N}_1$ and $\mathcal{N}_2$ are *isomorphic* if there is a *linear* bijective isometry $\kappa : \mathcal{N}_1 \to \mathcal{N}_2$ (i.e., $\kappa(a\xi + \eta) = a\kappa(\xi) + \kappa(\eta)$, for all $\xi, \eta \in \mathcal{N}_1$, and all $a \in \mathbb{F}$ and κ is onto). The mapping κ is said to be an *isomorphism* between these normed spaces.

Theorem 2.5. *If (X, d) is a metric space, then it is isometric to a dense subset of a complete metric space $(\tilde{X}, \tilde{d})$; such $\tilde{X}$ is called a completion of X. Furthermore, any two completions of X are isometric.*

Proof. The construction of the completion will be similar to Cantor construction of real numbers; each real number can be identified with Cauchy sequences, of real or only rational numbers, that converge to that number.

Let $\tilde{X}$ the set of equivalence classes of Cauchy sequences in X, in which two sequences (ξ_n) and (ξ'_n) are equivalent if $\lim_{n\to\infty} d(\xi_n, \xi'_n) = 0$. By using the triangle inequality it follows that this relation actually defines equivalence classes, and moreover, that if $\tilde{\xi}, \tilde{\eta} \in \tilde{X}$, then the limit

$$\tilde{d}(\tilde{\xi}, \tilde{\eta}) := \lim_{n\to\infty} d(\xi_n, \eta_n)$$

exists and does not depend on the representatives $(\xi_n), (\eta_n)$ of $\tilde{\xi}$ and $\tilde{\eta}$, respectively, and $\tilde{d}$ defines a metric in $\tilde{X}$. To conclude that the limit exists, consider $d(\xi_n, \eta_n) \leq d(\xi_n, \xi_m) + d(\xi_m, \eta_m) + d(\eta_m, \eta_n)$, which implies that

$$|d(\xi_n, \eta_n) - d(\xi_m, \eta_m)| \leq d(\xi_n, \xi_m) + d(\eta_n, \eta_m),$$

and since $(\xi_n), (\eta_n)$ are Cauchy sequences, it follows that $(d(\xi_n, \eta_n))_n$ is a Cauchy sequence in $\mathbb{R}$, hence convergent. The independence on the representatives is shown in a similar way.

Define $\kappa : X \to \kappa(X) \subset \tilde{X}$, so that $(\xi, \xi, \xi, \cdots)$ is a representative of $\kappa(\xi)$; then κ is an isometry whose image is dense in $(\tilde{X}, \tilde{d})$. Indeed, if (ξ_n) is a representative of $\tilde{\xi} \in \tilde{X}$, then given $\varepsilon > 0$, one has

$$\tilde{d}(\tilde{\xi}, \kappa(\xi_m)) = \lim_{n\to\infty} d(\xi_n, \xi_m) < \varepsilon$$

for m large enough, since (ξ_n) is Cauchy; the density then follows.

By using this isometry, the above density and again the triangle inequality, it follows that $(\tilde{X}, \tilde{d})$ is complete. Indeed, if $(\tilde{\xi}_n) \subset \tilde{X}$ is a Cauchy sequence, for each n pick $\eta_n \in X$ so that $\tilde{d}(\tilde{\xi}_n, \kappa(\eta_n)) < 1/n$; thus,

$$\begin{aligned} \tilde{d}(\kappa(\eta_n), \kappa(\eta_m)) &\leq \tilde{d}(\kappa(\eta_n), \tilde{\xi}_n) + \tilde{d}(\tilde{\xi}_n, \tilde{\xi}_m) + \tilde{d}(\tilde{\xi}_m, \kappa(\eta_m)) \\ &< \frac{1}{n} + d(\tilde{\xi}_n, \tilde{\xi}_m) + \frac{1}{m}, \end{aligned}$$

which shows that $(\kappa(\eta_n))$ is Cauchy, since κ is an isometry, (η_n) is Cauchy in X and, so, it represents some $\tilde{\eta} \in \tilde{X}$. By the triangle inequality it is found that $\tilde{d}(\tilde{\xi}_n, \tilde{\eta}) \leq \tilde{d}(\tilde{\xi}_n, \kappa(\eta_n)) + \tilde{d}(\kappa(\eta_n), \tilde{\eta}) < 1/n + \lim_{m\to\infty} d(\eta_n, \eta_m)$ showing that, for $n \to \infty$, $(\tilde{\xi}_n)$ converges to $\tilde{\eta}$; hence $\tilde{X}$ is complete.

If there exists an other isometry onto $\iota : (X, d) \to (W, D)$, with W dense in the complete metric space (Z, D), then the composition $\iota \circ \kappa^{-1} : \kappa(X) \to W$ is a bijective isometry, which has a unique extension to an isometry between the closure of these spaces, namely, between $(\tilde{X}, \tilde{d})$ and (Z, D); therefore any two completions of (X, d) are isometric. Often one identifies the metric space X with $\iota(X)$ or $\kappa(X)$. □

EXERCISE **2.6.** Fill out the missing details in the proof of Theorem 2.5.

EXERCISE **2.7.** Which is the completion of a metric space with the discrete metric?

The next result shows that the above construction to complete a metric space can be translated into normed spaces and retaining the linear structure.

Theorem 2.6. *If $(\mathcal{N}, \|\cdot\|)$ is a normed space, then it is isomorphic to a dense subspace of a Banach space $(\mathcal{B}, \|\|\cdot\|\|)$; such $\mathcal{B}$ is called a completion of $\mathcal{N}$. Furthermore, any two completions of $\mathcal{N}$ are isomorphic.*

Proof. Borrow the notation from Theorem 2.5 and its proof. It is enough to show that, in this case, $(\tilde{X}, \tilde{d})$ is a vector space with $\tilde{d}$ generated by a norm compatible with $\kappa(\mathcal{N})$. The linear structure in $\tilde{X}$ is naturally defined as the equivalence class of pointwise sum of the representatives of the terms involved, so defining $\tilde{\zeta} = \tilde{\xi} + \tilde{\eta}$, and the product by scalar $\alpha\tilde{\xi}$ by the equivalence class for which $(\alpha\xi_n)$ is a representative (with (ξ_n) being a representative of $\tilde{\xi}$). Now, the isometry κ induces a norm $[\cdot]$ in $\kappa(\mathcal{N})$ with $[\kappa(\xi)] = \|\xi\|$ and since it is an isometry, the corresponding metric is the restriction of $\tilde{d}$ to $\kappa(\mathcal{N})$. This norm is extended to $\tilde{X}$ by defining $\|\|\tilde{\xi}\|\| = \tilde{d}(\tilde{0}, \tilde{\xi})$, so that $(\tilde{X}, \tilde{d})$ is a normed space; since it is also complete, it is a Banach space. □

EXERCISE **2.8.** Fill out the missing details in the proof of Theorem 2.6; in particular, show that $\|\|\cdot\|\|$ is a norm.

REMARK **2.7.** There is also an analogous to such completion to spaces with inner product (see Theorem 17.15). Another proof of Theorem 2.6 will be proposed when the dual of normed spaces is discussed; see Exercise 12.6.

REMARK **2.8.** In general, different norms on a vector space give rise to different completions; for instance, if in the vector space of sequences in $\mathbb{F}$ with only a finite number of nonzero entries one considers the norm $\|\cdot\|_p, 1 \leq p \leq \infty$, then the completion is isomorphic to the corresponding $l^p(\mathbb{N})$.

Notes

The idea of completion of metric spaces is very important; several operations and concepts only get a satisfactory form after some kind of extension or completion. For instance, the integral of Lebesgue can be defined via an extension of the integral of Riemann to certain completion of the space of continuous functions; with the Fourier transform something similar occurs, being $L^2(\mathbb{R})$ its natural space. It is fundamental in Mathematical Analysis, particularly to the theory of differential operators and equations, that there are subspaces of infinitely differentiable real functions whose completions (with adequate topologies) result in $L^p(\mathbb{R})$.

Although the completion of normed spaces can be elegantly proven by using the dual spaces (see Exercise 12.6), the proof discussed here is transparent and intuitive; that is the reason why it was included.

Note another particular characteristic of normed spaces of infinite dimension: it is possible that the intersection of two dense vector subspaces results only in the null vector, so that the sum of operators defined on each of these subspaces is, in principle, defined only at zero! For example, take in $L^1(\mathbb{R})$ the Schwartz space (infinitely differentiable functions fast decaying at infinity) and the space of simple functions.

The important Riesz Lemma 2.1 was published by F. Riesz in 1918, in a work whose main subject was the compact operators on Banach spaces and their spectra; see also the Notes in Chapters 24 and 30.

Additional Exercises

EXERCISE **2.9.** Let $\psi : \mathcal{N}_1 \to \mathcal{N}_2$ be uniformly continuous and $\tilde{\mathcal{N}}_1$ and $\tilde{\mathcal{N}}_2$ the completions of $\mathcal{N}_1$ and $\mathcal{N}_2$, respectively. Show that the mapping ψ has a unique uniformly continuous extension $\tilde{\psi} : \tilde{\mathcal{N}}_1 \to \tilde{\mathcal{N}}_2$ (clearly each normed space is being identified with a dense subspace of its completion).

EXERCISE **2.10.** Construct a sequence of functions $\psi_n : [0,1] \to \mathbb{R}$ so that $\|\psi_n\|_\infty = 1$, ψ_n converges to zero in $\mathrm{L}^p[0,1]$ for all $1 \le p < \infty$, and for each $t \in [0,1]$ the sequence of scalars $(\psi_n(t))$ is not convergent.

EXERCISE **2.11.** Given a continuous mapping $\psi : \mathcal{N} \to \mathbb{R}$ on a normed space $\mathcal{N}$, is $\psi(S(0;1))$ a bounded set?

EXERCISE **2.12.** Describe the completion of $\mathrm{C}^1[-1,1]$ with the norm

$$\|\psi\| = \sup_{t \in [-1,1]} |\psi(t)| + \sup_{t \in [-1,0]} |\psi'(t)| + \sup_{t \in [0,1]} |\psi'(t)|, \qquad \psi \in \mathrm{C}[-1,1].$$

CHAPTER 3

Separable Spaces and Linear Operators

The concept of separability in normed spaces is presented, and bounded linear operators are introduced as fundamental objects of study. The concepts of Hamel and Schauder bases are analyzed, with particular attention given to their relationship with the separability of vector spaces. Additionally, the separability of the Banach spaces $l^p(\mathbb{N})$ is considered, highlighting key properties of these spaces. An explicit example of a Schauder basis that does not form a Hamel basis is provided to illustrate the distinction between these notions. Moving on to linear operators between normed spaces, a detailed discussion of the continuity of such operators is included. This analysis serves as motivation for the study of unbounded operators, which is addressed in subsequent chapters.

3.1 SEPARABLE SPACES

The notion of Hamel basis works in any vector space since it needs just the definition of linear independent set; if the algebraic dimension of the vector space is infinite, generally the Hamel bases are not countable (see Proposition 6.11), so, in general they are not useful in practical calculations. This fact, associated with the need of a notion of basis related to the topology, has led Banach to include in his own book the following definition

Definition 3.1. A *Schauder basis* in a normed space $\mathcal{N}$ is a sequence (ξ_n) in $\mathcal{N}$ such that, to each vector $\xi \in \mathcal{N}$, it is associated a unique sequence $(\alpha_n)_{n=1}^{\infty} \subset \mathbb{F}$ so that

$$\xi = \sum_{j=1}^{\infty} \alpha_j \xi_j := \lim_{n \to \infty} \sum_{j=1}^{n} \alpha_j \xi_j.$$

Note that the uniqueness of the sequence of scalars (α_n) for each ξ implies that the Schauder basis (ξ_n) is linearly independent. Also, in case $\mathcal{N}$ is of finite dimension N one assumes that $\alpha_n = 0$ for $n > N$.

Definition 3.2. A metric space is *separable* if there exists a countable subset dense in this space.

DOI: 10.1201/9781003656166-3

Example 3.3. A discrete metric space is separable if and only if it is countable. •

Proposition 3.4. *(a) Any normed space $\mathcal{N}$ which has a Schauder basis is separable.*
(b) A normed space $\mathcal{N}$ is separable if and only if there exists a total countable linearly independent set of $\mathcal{N}$.

Proof. (a) If (ξ_n) is a Schauder basis, then the set of vectors of the form $r_1\xi_1 + \cdots + r_n\xi_n$, with r_j a rational number for all j, is countable and dense in $\mathcal{N}$. Therefore $\mathcal{N}$ is separable.

(b) A similar argument used in the proof of item (a) shows that there exists a countable subset $\{\xi_n\}$ which is total in $\mathcal{N}$, then this normed space is separable, because the set $r_1\xi_1 + \cdots + r_n\xi_n$, with r_j rational numbers, is countable and dense in $\mathcal{N}$.

Suppose now that $\mathcal{N}$ is separable, let $(\xi_n)_{n=1}^{\infty}$ be a dense sequence in $\mathcal{N}$. Define a total sequence (η_n) by the following induction: η_1 is the first nonzero element in $(\xi_n)_{n=1}^{\infty}$; after choosing $(\eta_n)_{n=1}^{j}$, η_{j+1} is the first element in $(\xi_n)_{n=j+1}^{\infty}$ such that $(\eta_n)_{n=1}^{j+1}$ is linearly independent (if such element does not exist, then the sequence terminates). By construction, $(\xi_n)_{n=1}^{\infty}$ and the set $(\eta_n)_{n=1}^{\infty}$ generate the same vector space, and the latter is countable and linearly independent. □

Every normed space which has a Schauder basis is separable; however, by using an elaborate construction, P. Enflo presented an example of a separable Banach space that has no Schauder basis; see the Notes. Many applications and abstract results require the separability of the underlying normed space. Furthermore, the presence of a countable dense set can significantly simplify some proofs.

EXERCISE **3.1.** Show that a subset E of a normed space is separable if and only if $\text{Lin}(E)$ is separable.

Example 3.5. By the density of rational numbers in $\mathbb{R}$, it follows that also $\mathbb{F}^n$ is separable. Another way to see this is to observe that the canonical basis of $\mathbb{F}^n$ is a Schauder basis. •

Example 3.6. $l^p(\mathbb{N})$ is separable for $1 \leq p < \infty$, since its canonical (or usual) basis $e_n = (\delta_{n,j})_{j=1}^{\infty}$ (Kronecker's δ) is a Schauder basis. Here it is also possible to use the density of the rational numbers in $\mathbb{R}$, so that the family of the elements of $l^p(\mathbb{N})$ with rational entries is enumerable and dense in $l^p(\mathbb{N})$. •

Example 3.7. $l^{\infty}(\mathbb{N})$ is not separable. Indeed, given a sequence $\xi^n = (\xi_j^n)_{j=1}^{\infty}$ in $l^{\infty}(\mathbb{N})$, the vector $\xi = (\xi_j)_{j=1}^{\infty}$ whose entries are $\xi_j = 0$ if $|\xi_j^j| \geq 1$ and $\xi_j = \xi_j^j + 1$ if $|\xi_j^j| < 1$; thus, $\|\xi\|_{\infty} \leq 2$ and $\|\xi - \xi^n\|_{\infty} \geq 1$ for all n. Therefore, there is no dense sequence in $l^{\infty}(\mathbb{N})$. Why $\{e_n\}$, as above, is not a Schauder basis of $l^{\infty}(\mathbb{N})$? •

EXERCISE **3.2.** Show that $l^p(J), 1 \leq p < \infty$, is separable if and only if J is countable. When is $l^{\infty}(J)$ separable?

EXERCISE **3.3.** Here is an alternative proof that $l^{\infty}(\mathbb{N})$ is not separable; fill out the missing details. Let $X \subset l^{\infty}(\mathbb{N})$ be the subset of sequences whose entries are only 0 and 1. Then the distance between any two distinct elements in X is equal to one and X is not countable (recall the representation of real numbers in the binary basis).

Example 3.8. [Schauder Basis that is not Hamel's] Denote by $\mathcal{N}$ the subspace of $l^\infty(\mathbb{N})$ of sequences with just a finite number of nonzero entries. The canonical basis (Schauder's) $\{e_j\}_{j=1}^\infty$ is also a Hamel basis of $\mathcal{N}$, since $\mathcal{N} = \mathrm{Lin}(\{e_j\})$. Now, the sequence $\{\eta_k = e_k/k - e_{k+1}/(k+1)\}_{k=1}^\infty$ is not a Hamel basis of the incomplete space $\mathcal{N}$, since the unique representation of $e_1 = \sum_{k=1}^\infty \eta_k$ is an infinite sum; nevertheless, this is a Schauder basis of $\mathcal{N}$, since each element $\xi = (\xi_1, \cdots, \xi_n, 0, 0, \cdots)$ of $\mathcal{N}$ can be written in the form

$$\xi = \sum_{k=1}^n a_k \eta_k + a_n \sum_{k=n+1}^\infty \eta_k,$$

with $\{a_k\}_{k=1}^n$ being the solution to the linear system

$$\begin{aligned} \xi_1 &= a_1, \\ \xi_j &= (a_j - a_{j-1})/j, \qquad 2 \le j \le n. \end{aligned}$$

This example shows that even if the Hamel basis is enumerable, it is possible to have a Schauder basis which is not Hamel's (compare this result with Proposition 6.11).
●

Example 3.9. $C[a,b]$ is separable by Proposition 3.4*(b)*, since the sequence of polynomials $(t^n)_{n=0}^\infty$ is total in $C[a,b]$ by Stone-Weierstrass' Theorem; more generally, $C(\Omega)$ is separable if Ω is a metric compact space [Simm]. From this it follows that $L^p[a,b]$, $1 \le p < \infty$, is also separable because the continuous functions form a dense subset in $L^p[a,b]$. ●

EXERCISE **3.4.** Suppose that there exists a countable family of measurable sets A_n in $(\Omega, \mathcal{A}, \mu)$ in such a way that $\Omega = \bigcup_n A_n$; identify $L^p(A_n)$ with a linear subspace of $L^p(\Omega)$. Show that if $L^p(A_n)$ is separable for all n, then $L^p(\Omega)$ is separable. Use this fact to show that $L^p(\mathbb{R})$ is separable for $1 \le p < \infty$.

EXERCISE **3.5.** Suppose that there exists an infinite family of subsets of strictly positive measure μ in (Ω, μ). Find a nonconstant family in $L_\mu^\infty(\Omega)$ whose distances between any two distinct of their elements is greater or equal to 1. Conclude that that $L_\mu^\infty(\Omega)$ is not separable. Apply also to $L^\infty[a,b]$.

3.2 LINEAR OPERATORS

Definition 3.10. Let X and Y be vectors spaces. A *linear operator* between them is a mapping $T : \mathrm{dom}\, T \subset X \to Y$, whose domain $\mathrm{dom}\, T$ is a vector subspace and $T(\xi + \alpha\eta) = T(\xi) + \alpha T(\eta)$, for all $\xi, \eta \in \mathrm{dom}\, T$ and all scalar values $\alpha \in \mathbb{F}$.

Note that for any linear operator T one has $T(0) = 0$, and that the set of linear operators with the same domain and codomain is a vector space with the usual pointwise operations; often, $T(\xi)$ will also be abbreviated to $T\xi$. Typical examples of linear operators include the *identity operator* $\mathbf{1} : X \to X$, defined by $\mathbf{1}(\xi) = \xi$, and the *null (or zero) operator* $T\xi = 0$, for every ξ; many more examples will appear throughout this book.

Example 3.11. The derivative acting on the space $\mathrm{C}^1(\mathbb{R})$ is a linear operator; indeed, for ϕ, ψ in this space, $(\phi + \alpha\psi)' = \phi' + \alpha\psi'$, for all scalars α. •

Example 3.12. The operator acting on $l^p(\mathbb{N})$, $1 \leq p \leq \infty$, $T(\xi_1, \xi_2, \xi_3, \cdots) = (\xi_1/1^2, \xi_2/2^2, \xi_3/3^2, \cdots)$ is linear. •

Example 3.13. Pick $\phi \in \mathrm{L}^\infty_\mu(\Omega)$, with μ denoting a positive σ-finite measure. The so-called multiplication operator by ϕ is defined as $\mathcal{M}_\phi : \mathrm{L}^p_\mu(\Omega) \to \mathrm{L}^p_\mu(\Omega)$,

$$(\mathcal{M}_\phi\psi)(t) := \phi(t)\psi(t), \quad \psi \in \mathrm{L}^p_\mu(\Omega),$$

which is a linear for all $1 \leq p \leq \infty$. For all $\psi \in \mathrm{L}^p_\mu$, one has $(\mathcal{M}_\phi\psi) \in \mathrm{L}^p_\mu$ •

Example 3.14. For compact metric spaces X and Y, and $u : Y \to X$ is a continuous mapping, one has the linear operator $T_u : \mathrm{C}(X) \to \mathrm{C}(Y)$, $(T_u\psi)(y) = \psi(u(y))$. •

EXERCISE **3.6.** Given a linear operator $T : \mathrm{dom}\, T \subset X \to Y$, verify that:

(a) The range (or image) of T, $\mathrm{rng}\, T := T(\mathrm{dom}\, T) \subset Y$, and the kernel of T, $\mathrm{N}(T) := \{\xi \in \mathrm{dom}\, T : T\xi = 0\}$, are both vector spaces.

(b) If $\dim(\mathrm{dom}\, T) = n < \infty$, then $\dim(\mathrm{rng}\, T) \leq n$.

(c) The inverse operator of T, denoted by $T^{-1} : \mathrm{rng}\, T \to \mathrm{dom}\, T$, exists if and only if $T\xi = 0 \Rightarrow \xi = 0$, and if it exists, it is again a linear operator.

(d) In case T, S are invertible linear operators, then $(TS)^{-1} = S^{-1}T^{-1}$ in case such manipulations are well defined.

A comprehensive theory emerges when linear operators are combined with the topology induced by norms. The following result exemplifies this synergy; it demonstrates that if a linear operator is continuous at any single point in its domain, then it is, indeed, uniformly continuous throughout its entire domain.

Theorem 3.15. *For linear operators $T : \mathcal{N}_1 \to \mathcal{N}_2$, the following statements are equivalent:*

(i) $\sup_{\|\xi\| \leq 1} \|T\xi\| < \infty$; that is, the range under T of the unity ball is bounded.

(ii) There exists a positive constant $C > 0$ such that $\|T\xi\| \leq C\|\xi\|$, for every $\xi \in \mathcal{N}_1$.

(iii) T is a uniformly continuous mapping.

(iv) T is a continuous mapping.

(v) T is continuous at zero.

Proof. *(i)⇒(ii)* Let $C = \sup_{\|\xi\|\leq 1} \|T\xi\|$. If $0 \neq \xi \in \mathcal{N}_1$, then $\|T(\xi/\|\xi\|)\| \leq C$, i.e., $\|T\xi\| \leq C\|\xi\|$, for every $\xi \in \mathcal{N}_1$.

(ii)⇒(iii) For $\xi, \eta \in \mathcal{N}_1$, one has $\|T\xi - T\eta\| = \|T(\xi - \eta)\| \leq C\|\xi - \eta\|$.

(iii)⇒(iv) and *(iv)⇒(v)* are clear.

(v)⇒(i) Since T is a continuous mapping at zero, then there exists $\delta > 0$ with $\|T\xi - 0\| = \|T\xi\| \leq 1$ if $\|\xi\| \leq \delta$. Hence, if $\|\xi\| \leq 1$, one finds that $\|\delta\xi\| \leq \delta$ and $\|T(\delta\xi)\| \leq 1$; therefore, $\|T\xi\| \leq 1/\delta$, and *(i)* holds true. □

Definition 3.16. A continuous linear operator is also known as a *bounded* operator, and the set of all bounded linear operators from $\mathcal{N}_1$ to $\mathcal{N}_2$ will be indicated by $\mathrm{B}(\mathcal{N}_1, \mathcal{N}_2)$. A usual short form of $\mathrm{B}(\mathcal{N}, \mathcal{N})$ is $\mathrm{B}(\mathcal{N})$.

Note the deliberate distinction in the terminology: Whereas a *bounded linear operator* refers to an operator whose norm is finite, the phrase *bounded application* in a broader context usually means a mapping with a bounded range. In that broader sense, any (nonzero) linear operator fails to be bounded.

Example 3.17. Let T_u be the operator introduces in Example 3.14; , since for every $\psi \in \mathrm{C}(X)$ one has $\|T_u\psi\|_\infty = \sup_{t\in Y} |\psi(u(t))| \leq \sup_{t\in X} |\psi(t)| = \|\psi\|_\infty$, it follows that T_u is continuous, by Theorem 3.15*(ii)*. ●

EXERCISE **3.7.** Show that for an operator $S : \mathcal{N}_1 \to \mathcal{N}_2$ to be continuous, it is sufficient to satisfy $S(\xi + \eta) = S(\xi) + S(\eta)$, for all $\xi, \eta \in \mathcal{N}_1$, and also be continuous at some point of $\mathcal{N}_1$.

Proposition 3.18. *Let $T : \mathcal{N}_1 \to \mathcal{N}_2$ be a linear operator. If $\dim \mathcal{N}_1 < \infty$, then T is bounded.*

Proof. The idea is to pick, in $\mathcal{N}_1$, the norm $\|\|\xi\|\| = \|\xi\|_1 + \|T\xi\|_2$; since all norms are equivalent (Theorem 1.7), there is $C > 0$ with $\|\|\xi\|\| \leq C\|\xi\|_1$, which implies that $\|T\xi\|_2 \leq \|\|\xi\|\| \leq C\|\xi\|_1$ and so T is bounded. □

Example 3.19. Let $T : \mathrm{dom}\, T \to l^p(\mathbb{N})$, with $1 \leq p < \infty$,

$$\mathrm{dom}\, T = \left\{ (\xi_n) \in l^p(\mathbb{N}) : \sum_n |n^2 \xi_n|^p < \infty \right\},$$

$T(\xi_n) = (n^2 \xi_n)$; this operator is linear, but it is not continuous; indeed, if $\{e_n\}_{n=1}^\infty$ denotes the canonical basis of $l^p(\mathbb{N})$, then $e_n/n \to 0$, while Te_n does not converge to zero. Another argument: T is not bounded since $\|e_n\|_p = 1$ and $\|Te_n\|_p = n^2$, for all $n \in \mathbb{N}$. ●

EXERCISE **3.8.** Let P_N be the normed space of polynomials $p : [-1, 1] \to \mathbb{R}$ of degree less than or equal to N, with the uniform convergence norm $\|\cdot\|_\infty$. Let $D : P_N \to P_N$ be the derivative operator $(Dp)(t) = p'(t)$. Show that D is bounded. Choose a basis of P_N and find the matrix that represents D.

Example 3.20. [Shifts] The *right (left) shift* operator acting on $l^p(\mathbb{Z})$, $1 \leq p \leq \infty$, is the linear mapping $S_r : l^p(\mathbb{Z}) \to l^p(\mathbb{Z})$ (S_l), whose action is given by $\eta = S_r\xi$ $(\eta = S_l\xi)$, with $\eta_j = \xi_{j-1}$ $(\eta_j = \xi_{j+1})$, $j \in \mathbb{Z}$. The shift operator is a bijective isometry, so it is bounded. There are versions of the shift operators acting on $l^p(\mathbb{N})$ in an analogous way, but in the case of $\eta = S_r\xi$, one defines $\eta_1 = 0$ (to guarantee that $S_r 0 = 0$); note that S_r in $l^p(\mathbb{N})$ is an isometry that is not surjective. ●

EXERCISE **3.9.** Use the concept of Hamel basis to show that if all linear operators $T : \mathcal{N} \to \mathcal{N}$ are continuous, then $\dim \mathcal{N} < \infty$.

Notes

Every vector space has a Hamel basis (see Proposition 10.9); this is shown by using the Zorn Lemma, which will be discussed in chapters concerning the Hahn-Banach Theorem. In 1973 Per Enflo [Enflo] presented an example of a separable Banach space which has no Schauder basis; see also the works by A. M. Davie [Davie1, Davie2], for technical simplifications, and P. R. Halmos [Halmos], where the problem was emphasized. See the Notes in Chapter 25 for other details of the Enflo's construction.

The conditions for $L_\mu^p(\Omega)$ to be separable depend on properties of Ω and of the measure μ as well; for example, it is enough that μ be σ-finite and has a countable basis.

Linear operators have applications to many areas of science; to mention a few cases: oscillations, diffusion, Ergodic Theory, Quantum Mechanics [de Oliv], Electromagnetism, studies of bifurcations, etc. A very popular technique in Mathematical Analysis and applications is the approximation of general operators by linear ones.

A linear operator is a generalization of matrices to infinite-dimensional spaces. This idea has appeared when Volterra, based on the construction of the integral, used discretization to find approximated solutions to an integral equation, which resulted in a finite linear system. The solution to the original problem is obtained in the limit of an infinite number of variables.

The Linear Algebra was developed in the XIX century (i.e., via linear equations, determinants, matrices and, finally, finite-dimensional vector spaces) in the reverse logical order that it is currently presented in text-books; this is natural, since typically one starts up from a particular problem to the future abstract formulation. However, the lessons from Linear Algebra did not speed up the development process of Functional Analysis; it is possible to understand this by noting that in the first years of the XX century the vector spaces were always associated with some fixed basis, consequently, the linear operators were represented through matrices in such bases. Summing up, the idea of a vector as a column of n numbers was abandoned relatively late in the development of Functional Analysis.

Additional Exercises

EXERCISE **3.10.** Let $T : X \to Y$ be a linear operator between finite-dimensional vector spaces X and Y. Select bases in X and Y and check that T can be represented by a matrix; also see how the matrix that represents T changes if other bases are selected. For more general vector spaces, what is the role of a Schauder basis in this matter?

EXERCISE **3.11.** Let A be a linearly independent subset of the vector space X. Show that A is a Hamel basis of X if and only if for all vector $\xi \in X$, there exists $n \in \mathbb{N}$ such that ξ can be written in the form $\xi = \sum_{j=1}^{n} \alpha_j \xi_j$, with $\xi_j \in A, \alpha_j \in \mathbb{F}$, $1 \leq j \leq n$ (justifying the term "maximal" in the definition of Hamel basis).

EXERCISE **3.12.** Show that a metric space is separable if and only if it has a countable topological basis.

EXERCISE **3.13.** Adapt Exercise 3.4 to show that $L^p(\mathbb{R}^n)$ is separable, for $1 \le p < \infty$.

EXERCISE **3.14.** Show that the subspaces c and c_0 of $l^\infty(\mathbb{N})$, defined in Exercise 1.12, are separable.

EXERCISE **3.15.** Show that a subset of a separable set (in a metric space) is also separable, and that the closure of a separable subset is separable. Use these facts, and Exercise 3.1 to show that a subset A of a normed space is separable if and only if the closure of the subspace generated by A, $\overline{\text{Lin}}(A)$, is separable.

EXERCISE **3.16.** Find a Schauder basis of $C[a, b]$.

EXERCISE **3.17.** Let X and Y be finite-dimensional vector spaces. If $\xi \in X$ is such that $T\xi = 0$ for all linear operator $T : X \to Y$, is it possible to conclude that $\xi = 0$? Note that this question is equivalent to: given $\xi, \eta \in X$, $\xi \ne \eta$, is there a linear operator $T : X \to Y$ with $T\xi \ne T\eta$?

EXERCISE **3.18.** Let X and Y be compact metric spaces, $u : Y \to X$ a continuous mapping, and consider the linear and bounded operator $T_u : C(X) \to C(Y)$, $(T_u\psi)(y) = \psi(u(y))$ (see Example 3.17). Show that:

(a) T_u is an isometry if and only if u is onto.

(b) T_u is onto if and only if u is injective.

CHAPTER 4

Bounded Operators and Dual Spaces

A natural norm is introduced in $B(\mathcal{N}_1, \mathcal{N}_2)$, many examples, in different contexts, are presented to illustrate the concept; for example, in Banach spaces, the exponential of a bounded operator is proposed, as well as the inverse of $(\mathbf{1} - T)$, for T with norm less than one, which will be used many times in this book. Then, the important idea of dual space of a normed space is defined, which will also be crucial sometimes. The reflexive spaces will be mentioned, although the precise definition will be postponed until Chapter 12.

It is important to note that $B(\mathcal{N}_1, \mathcal{N}_2)$ forms a vector space under pointwise operations, and

$$\|T\| := \sup_{\substack{\xi \in \mathcal{N}_1 \\ \|\xi\| \leq 1}} \|T\xi\|$$

is a norm on $B(\mathcal{N}_1, \mathcal{N}_2)$. Indeed, if $T \in B(\mathcal{N}_1, \mathcal{N}_2)$, $\|T\| = 0 \iff T\xi = 0$, for all $\xi \in \mathcal{N}_1$, that is, $T = 0$; $\|\alpha T\| = |\alpha| \|T\|$ is clear; if $S \in B(\mathcal{N}_1, \mathcal{N}_2)$, then

$$\|T + S\| = \sup_{\|\xi\| \leq 1} \|T\xi + S\xi\| \leq \sup_{\|\xi\| \leq 1} (\|T\xi\| + \|S\xi\|) \leq \|T\| + \|S\|.$$

In case a topology is not mentioned on $B(\mathcal{N}_1, \mathcal{N}_2)$, it is assumed that the topology in question is the one generated by this norm.

EXERCISE 4.1. (a) For each $T \in B(\mathcal{N}_1, \mathcal{N}_2)$, it follows that

$$\|T\| = \inf_{\xi \in \mathcal{N}_1} \{C > 0 : \|T\xi\| \leq C\|\xi\|\} = \sup_{\|\xi\| = 1} \|T\xi\| = \sup_{\xi \neq 0} \frac{\|T\xi\|}{\|\xi\|}.$$

(b) For bounded linear operators T, S with well-defined composition (or product of operators) TS, show that TS is bounded and for the norms $\|TS\| \leq \|T\| \|S\|$. Hence, if T^n denotes the n-th iterate of T, then $\|T^n\| \leq \|T\|^n$.

Example 4.1. It is clear that the identity operator has norm one and only the zero operator has a norm equals to zero. •

DOI: 10.1201/9781003656166-4

Example 4.2. The linear operator $\mathcal{M}_\phi$, with $\phi \in L^\infty_\mu(\Omega)$ (see Example 3.13) is bounded in $L^p_\mu(\Omega), 1 \leq p \leq \infty$, and $\|\mathcal{M}_\phi\| = \|\phi\|_\infty \, (= \text{ess-sup}|\phi|)$. •

Proof. It is enough to discuss the nontrivial case $\|\phi\|_\infty \neq 0$ and also suppose that $1 \leq p < \infty$. The possibilities $p = \infty$ and $\|\phi\|_\infty = 0$ are left as exercises. If $\|\varphi\|_p = 1$, since

$$\|\mathcal{M}_\phi\varphi\|^p_p = \int_\Omega |\phi(t)|^p |\varphi(t)|^p \, \mathrm{d}\mu(t) \leq \|\phi\|^p_\infty \|\varphi\|^p_p,$$

it follows that $\mathcal{M}_\phi$ is a bounded operator and $\|\mathcal{M}_\phi\| \leq \|\phi\|_\infty$.

Let $0 < \theta < \|\phi\|_\infty$; then there exists a measurable set A, with $0 < \mu(A) < \infty$ (recall that μ is σ-finite) so that $\|\phi\|_\infty \geq |\phi(t)| > \theta$, for all $t \in A$. If χ_A denotes the characteristic function of A, then χ_A is an element of $L^p_\mu(\Omega)$ and

$$\|\mathcal{M}_\phi\chi_A\|^p_p = \int_A |\phi(t)|^p |\chi_A(t)|^p \, \mathrm{d}\mu(t) \geq \theta^p \|\chi_A\|^p_p;$$

thus $\|\mathcal{M}_\phi\| \geq \theta$ and so $\|\mathcal{M}_\phi\| = \|\phi\|_\infty$. □

EXERCISE **4.2.** Let $T : X \to Y$ be a linear operator between finite-dimensional vector spaces X and Y. Select bases of X and Y and check that T can be represented by a matrix; also see how the matrix that represents T changes if other bases are selected. For more general vector spaces, what is the role of a Schauder basis in this matter?

Example 4.3. Assume that the measurable function $K : (\Omega, \mathcal{A}, \mu) \times (\Omega, \mathcal{A}, \mu) \to \mathbb{F}$, in a σ-finite space, satisfies

$$\int_\Omega |K(x, y)| \, \mathrm{d}\mu(x) \leq C, \quad \text{for} \quad y \, \mu - \text{a.e.},$$

for a constant $C > 0$. Then, $T_K : L^1_\mu(\Omega) \hookleftarrow$ with action

$$(T_K\varphi)(x) = \int_\Omega K(x, y)\varphi(y) \, \mathrm{d}\mu(y), \qquad \varphi \in L^1_\mu(\Omega),$$

is a bounded operator with norm bounded by C, that is, $\|T_K\| \leq C$. •

Proof. For $\varphi \in L^1_\mu(\Omega)$, one has

$$|(T_K\varphi)(x)| \leq \int_\Omega |K(x, y)\varphi(y)| \, \mathrm{d}\mu(y);$$

thus, $\|T_K\varphi\|_1 = \int_\Omega |(T_K\varphi)(x))| \, \mathrm{d}\mu(x) \leq \iint |K(x,y)| \, |\varphi(y)| \, \mathrm{d}\mu(y) \, \mathrm{d}\mu(x)$. An application of Fubini Theorem gives

$$\|T_K\varphi\|_1 \leq \iint_{\Omega\times\Omega} |K(x, y)| \, \mathrm{d}\mu(x) \, |\varphi(y)| \, \mathrm{d}\mu(y) \leq C\|\varphi\|_1.$$

Hence, $\|T_K\| \leq C$. □

Example 4.4. Let X denote the vector space of polynomials in C$[0,1]$ and $D : X \hookleftarrow$ the derivative operator $(Dp)(t) = p'(t)$, $p \in X$. Although D is a linear operator, it is unbounded; indeed, if $p_n(t) = t^n$, $n \geq 1$, then $(Dp_n)(t) = nt^{n-1}$, $\|p_n\|_\infty = 1$, whereas $\|Dp_n\|_\infty = n$. •

EXERCISE **4.3.** Denote by C$^1(0,1)$ the collection of continuously differentiable real functions on $(0,1)$, considered as a subspace of L$^2(0,1)$ (that is, use the norm of L^2). By using the derivative operator $(D\psi)(t) = \psi'(t)$, $D : C^1(0,1) \to L^2(0,1)$, on the functions $\psi_n(t) = \sin(n\pi t)$, conclude that D cannot be a bounded operator.

EXERCISE **4.4.** Check that the derivative operator $D : C^\infty[a,b] \hookleftarrow$ cannot be bounded for any chosen norm on C$^\infty[a,b]$.

EXERCISE **4.5.** Let $(e_n)_{n=1}^\infty$ be the usual basis of $l^2(\mathbb{N})$ and $(t_n)_{n=1}^\infty$ a sequence in $\mathbb{F}$. Show that there exists a bounded linear operator $T : l^2(\mathbb{N}) \hookleftarrow$ with $Te_n = t_n e_n$ if and only if $(t_n)_{n=1}^\infty$ is a bounded sequence. Verify that, in this case, $\|T\| = \sup_n |t_n|$.

Now it is presented conditions that guarantee that B$(\mathcal{N}_1, \mathcal{N}_2)$ is a Banach space (see also Proposition 12.6).

Theorem 4.5. B$(\mathcal{N}, \mathcal{B})$ *is a Banach space.*

Proof. Pick a Cauchy sequence $(T_n)_{n=1}^\infty$ in B$(\mathcal{N}, \mathcal{B})$. Since for each $\xi \in \mathcal{N}$ one has $\|T_n\xi - T_k\xi\| \leq \|T_n - T_k\| \|\xi\|$, it follows that $(T_n\xi)$ is a Cauchy sequence in $\mathcal{B}$ and so converges to some $\eta \in \mathcal{B}$. Consider the linear operator $T : \mathcal{N} \to \mathcal{B}$ given by $T\xi = \eta$. The goal is to show that T is bounded and one has the convergence $T_n \to T$ in B$(\mathcal{N}, \mathcal{B})$.

For each $\varepsilon > 0$, there exists $N(\varepsilon)$ such that, if $n, k \geq N(\varepsilon)$, then $\|T_n - T_k\| < \varepsilon$. Since the norm is a continuous mapping, one has

$$\|T_n\xi - T\xi\| = \lim_{k \to \infty} \|T_n\xi - T_k\xi\| \leq \varepsilon\|\xi\|, \qquad n \geq N(\varepsilon),$$

and $(T_n - T) \in$ B$(\mathcal{N}, \mathcal{B})$ with $\|T_n - T\| \leq \varepsilon$. By taking into account that B$(\mathcal{N}, \mathcal{B})$ is a vector space, and $T = T_n + (T - T_n)$, one finds $T \in$ B$(\mathcal{N}, \mathcal{B})$. The inequality $\|T_n - T\| \leq \varepsilon$, valid for all $n \geq N(\varepsilon)$, implies that $T_n \to T$ in B$(\mathcal{N}, \mathcal{B})$ and so B$(\mathcal{N}, \mathcal{B})$ is a complete normed space. □

EXERCISE **4.6.** Suppose that $T_n \to T$ in B$(\mathcal{N})$ and $\xi_n \to \xi$ in $\mathcal{N}$. Check that for the combined convergence $T_n\xi_n \to T\xi$.

Uniformly continuous functions on metric spaces can be extended uniformly to the closure of their domains. Likewise, there exists an analogous result for linear operators, which follows from the uniform continuity of bounded operators (see Theorem 3.15).

Definition 4.6. Let $f : X \to Z$ and $g : Y \to Z$. f is an *extension* of g, or g is a *restriction* of f, if $Y \subset X$ and for all $t \in Y$ one has $f(t) = g(t)$, which is denoted by $f|_Y = g$.

Theorem 4.7. *Let* $T : \operatorname{dom} T \subset \mathcal{N} \to \mathcal{B}$, *with* $\operatorname{dom} T$ *dense in* $\mathcal{N}$, *a bounded linear operator. Then* T *has a unique extension* $\overline{T} \in \mathrm{B}(\mathcal{N}, \mathcal{B})$. *Furthermore,* $\|\overline{T}\| = \|T\|$.

Proof. Let $\xi \in \mathcal{N}$ and $\xi_n \to \xi$, with $(\xi_n) \subset \operatorname{dom} T$. Since $\|T\xi_n - T\xi_m\| \leq \|T\| \|\xi_n - \xi_m\|$ then $(T\xi_n)$ is a Cauchy sequence in $\mathcal{B}$, so convergent to some $\eta \in \mathcal{B}$. Define

$$\overline{T}\xi = \eta = \lim_{n \to \infty} T\xi_n.$$

It is needed to show that $\overline{T}$ is well defined and that $\|\overline{T}\| = \|T\|$. If $\xi'_n \to \xi$, $(\xi'_n) \subset \operatorname{dom} T$, then the sequence $\xi_1, \xi'_1, \xi_2, \xi'_2, \cdots \to \xi$ and, by the same argument used above, $T\xi_1, T\xi'_1, T\xi_2, T\xi'_2, \cdots$, converges to η'. Since $(T\xi_n)$ is a subsequence of the latter, it is found that $\eta = \eta'$, and $\overline{T} : \mathcal{N} \to \mathcal{B}$ is well defined. $\overline{T}$ is evidently linear and an extension of T, since if $\xi \in \operatorname{dom} T$ consider the constant sequence $\xi, \xi, \xi, \cdots$, and $\overline{T}\xi = \lim T\xi = T\xi$.

Now, for $\xi \in \mathcal{N}$, by using the continuity of the norm,

$$\|\overline{T}\xi\| = \lim_{n \to \infty} \|T\xi_n\| \leq \lim_{n \to \infty} \|T\| \|\xi_n\| = \|T\| \|\xi\|,$$

so that $\|\overline{T}\| \leq \|T\|$. On the other hand,

$$\|T\| = \sup_{\substack{\xi \in \operatorname{dom} T \\ \|\xi\| = 1}} \|T\xi\| \leq \sup_{\substack{\xi \in \mathcal{N} \\ \|\xi\| = 1}} \|\overline{T}\xi\| = \|\overline{T}\|.$$

Therefore, $\|\overline{T}\| = \|T\|$. Suppose that $S \in \mathrm{B}(\mathcal{N}, \mathcal{B})$ is an extension of T and put $\xi \in \mathcal{N}$; then for all sequences $(\xi_n) \subset \operatorname{dom} T$, $\xi_n \to \xi$, one has $\overline{T}\xi_n = S\xi_n$, and by continuity, $\overline{T}\xi = S\xi$. Therefore $S = \overline{T}$ and the extension is unique. $\qquad \square$

EXERCISE 4.7. Let $T \in \mathrm{B}(\mathcal{N}, \mathcal{B})$. Show that T has a unique extension $\tilde{T} \in B(\tilde{\mathcal{N}}, \mathcal{B})$, with $\tilde{\mathcal{N}}$ being the completion of $\mathcal{N}$, and $\|\tilde{T}\| = \|T\|$ (see also Exercise 2.9).

EXERCISE 4.8. Show that if $T \in \mathrm{B}(\mathcal{N}_1, \mathcal{N}_2)$, then its kernel $\mathrm{N}(T)$ is a closed vector subspace. Confirm that $S : l^1(\mathbb{N}) \hookleftarrow$, $(S\xi)_n = \xi_n/n$ is bounded but rng S is not closed, and that its inverse operator $S^{-1} : \operatorname{rng} S \to l^1(\mathbb{N})$ exists and is not bounded.

Definition 4.8. For each normed space $\mathcal{N}$, the Banach space $\mathrm{B}(\mathcal{N}, \mathbb{F})$ is designated by $\mathcal{N}^*$ and called the *dual space* (or *conjugate space*) of $\mathcal{N}$. Every element of $\mathcal{N}^*$ is said to be a continuous *linear functional* on $\mathcal{N}$.

Example 4.9. The integral on $\mathrm{C}[a, b]$ is an element of the dual of $\mathrm{C}[a, b]$, since $\psi \mapsto \int_a^b \psi(t)\,\mathrm{d}t$ is linear and continuous. Indeed, every finite Borelian (complex) measure μ in $[a, b]$ defines an element of the dual of $\mathrm{C}[a, b]$ through the integral $\psi \mapsto \int_a^b \psi(t)\,\mathrm{d}\mu(t)$: $\left| \int_a^b \psi(t)\,\mathrm{d}\mu(t) \right| \leq \|\psi\|_\infty |\mu|([a, b])$. $\quad \bullet$

Example 4.10. This example is interesting due to its simplicity. Since the codomain of the norm is $\mathbb{R}$, it is a functional, however it is not linear. $\quad \bullet$

Example 4.11. The functional

$$f : C[-1, 1] \subset L^1[-1, 1] \to \mathbb{F}, \qquad f(\psi) = \psi(0)$$

is an example of an unbounded linear operator. To see this, choose a function $\psi \in$ C$[-1, 1]$ with $\psi(-1) = \psi(1) = 0$ and $\psi(0) \neq 0$. For integer numbers $n \geq 2$, put $\psi_n(t) = \psi(nt)$ if $|t| \leq 1/n$, and zero otherwise. Since

$$\|\psi_n\|_1 = \int_{-1}^{1} |\psi_n(t)| \, dt = \frac{\|\psi\|_1}{n},$$

it converges to zero as $n \to \infty$. On the other hand, $f(\psi_n) = \psi(0) \neq 0$ for every n, and so f is unbounded. •

EXERCISE **4.9.** Discuss if the kernel of the functional $f : (l^1(\mathbb{N}), \|\cdot\|_\infty) \to \mathbb{F}$, defined as $f(\xi_1, \xi_2, \cdots) = \sum_{j=1}^{\infty} \xi_j$, is closed.

Example 4.12. Let $1 < p < \infty$ and $1/p + 1/q = 1$. Every element $\phi \in L_\mu^q(\Omega)$ defines an element of the dual of $L_\mu^p(\Omega)$, since by Hölder inequality (see page 5) the product $\phi\psi \in L_\mu^1(\Omega)$, for all $\psi \in L_\mu^p(\Omega)$, and

$$\psi \mapsto \int_\Omega \phi\psi \, d\mu$$

is linear and bounded with $\|\phi\|_q$ as an upper bound for its norm (again by Hölder). Thus, $L_\mu^q(\Omega) \subset L_\mu^p(\Omega)^*$. The proof that $L_\mu^p(\Omega)^* = L_\mu^q(\Omega)$, for $1 < p < \infty$ appears in books on Integration Theory and if the measure μ is σ-finite, it also follows that $L_\mu^1(\Omega)^* = L_\mu^\infty(\Omega)$. •

REMARK **4.13.** The dual of the spaces l^p will be discussed in Chapter 13. It will be concluded that $(l^p)^* = l^q$, for $1 < p < \infty$ and $1/p + 1/q = 1$, and $(l^1)^* = l^\infty$.

Note that, for $1 < p < \infty$, the *second dual* $(L_\mu^p(\Omega)^*)^* = L_\mu^q(\Omega)^* = L_\mu^p(\Omega)$ (σ-finite measure); a normed space for which "$\mathcal{N}^{**} = \mathcal{N}$" is called a *reflexive space*. The precise definition of what is understood by this equality will be presented in Chapter 12. All reflexive spaces are Banach spaces (see the Notes ahead).

Example 4.14. Naively one could imagine, based on Example 4.12, that the dual of L^∞ would be L^1 (see the Notes). The following construction shows that this does not always hold. Let $f : C[-1, 1] \to \mathbb{F}$ be the continuous linear functional $f(\psi) = \psi(0)$, for $\psi \in C[-1, 1]$ (verify that $\|f\| = 1$), and $F : L^\infty[-1, 1] \to \mathbb{F}$ a continuous linear extension of f (such extension does exist; see Chapter 10, in particular Corollary 10.15). Suppose that there exists a function $\phi \in L^1[-1, 1]$ so that $F(\psi) = \int_{-1}^{1} \phi\psi \, dt$, for all $\psi \in L^\infty[-1, 1]$; the sequence of continuous functions $\psi_n(t) = e^{-nt^2}$ satisfy $\|\psi_n\|_\infty = 1$, $F(\psi_n) = f(\psi_n) = 1$, but $F(\psi_n) = \int_{-1}^{1} \phi\psi_n \, dt$ converges to zero (by dominated convergence). Therefore, there exists no function $\phi \in L^1[-1, 1]$ which represents $F \in L^\infty[-1, 1]^*$. Hence L^1 is not reflexive in general. •

Example 4.15. This example is important and references for its proof can be found in the Notes.

THEOREM OF RIESZ-MARKOV Let X be a Hausdorff compact topological space and $M(X)$ the set of complex Borelian measures on X with the norm $\|\mu\| = |\mu|(X)$, $\mu \in M(X)$. Then, $C(X)^* = M(X)$; more precisely, the mapping $M(X) \to C(X)^*$, $\mu \mapsto G_\mu$ with

$$G_\mu(\psi) = \int_X \psi \, \mathrm{d}\mu, \qquad \forall \psi \in C(X),$$

is a linear isometry and onto. Note that for each positive element $f \in C(X)^*$ (positive means that if $\psi \geq 0$, then $f(\psi) \geq 0$) is associated a unique positive (finite) Borelian measure μ on X. ●

EXERCISE 4.10. Show that if $f \in C[a,b]^*$ and $f(t^n) = 0$ for all integer $n \geq 0$, then $f = 0$. What can be said if a finite Borelian measure μ on $[a,b]$ satisfies $\int_a^b t^n \, \mathrm{d}\mu(t) = 0$, for all $n \geq 0$?

The concept of duality has become an important tool in the theory of Partial Differential Equations, Ergodic Theory and many branches of Mathematical Physics, and some of its details will be explored in the chapters ahead.

Notes

Duality in vector spaces appeared only around 1900, for finite-dimensional spaces, always associated with n-tuples of numbers. There is also the concept of *algebraic dual* of a vector space X, which is the set of linear functionals on X (without topology). However, in infinite-dimensional normed spaces it is possible to construct examples of (nonzero) linear functionals that vanish on dense subspaces; the exigency of continuity excludes such pathologies. Both duals coincide if and only if the dimension of the normed space is finite (see Exercise 4.24).

The equality of normed spaces must be interpreted with identification via an isomorphism (a linear isometric mapping), for instance, in $L^p[a,b]^* = L^q[a,b]$ (for $p, q \neq 1$); the latter relation was proven by F. Riesz around 1910, as a generalization of his own results for $l^2(\mathbb{N})$ and $C[a,b]$; it was the first example of reflexive spaces distinct from their duals. However, Riesz did not explicitly used the concept of duality, which was formally introduced by Hahn, around 1925. The fact $L^1[a,b]^* = L^\infty[a,b]$ was proven by H. Steinhaus in 1919; for the characterization of $L^1_\mu(\Omega)^*$ in general, see the original work [Schw].

The characterization of $L^\infty_\mu(\Omega)^*$, with elements represented by integrals with respect to finitely additive measures, can be found in §20 of [HewStr]. Another representation of such spaces can be obtained from Proposition 15.11, in case it is possible to find the space $C(X)$ that appears there. Since L^∞ is a commutative C^*-algebra with identity, it is possible to find $C(X)$ via the Gelfand transform [Rudin1] and the prominent Theorem of Gelfand-Naimark, and then an application of Riesz-Markov. For a characterization of the second dual of $C(X)$, i.e., $M(X)^*$, see [Conway], page 79.

Every Hilbert space is reflexive, as shown in Chapter 19.

There are some variants of the Theorem of Riesz-Markov; for example, it is possible to adapt it to locally compact topological spaces X. For the proof of this theorem, see [Royden]. In the particular case of $C[a,b]^*$ (as already mentioned, the case originally treated by F. Riesz), there are specific proofs, for instance in [Kreysz].

Additional Exercises

EXERCISE **4.11.** Let $T \in \mathrm{B}(\mathcal{B})$. Show that, for every scalar t, the operator e^{tT} given by the series

$$e^{tT} := \sum_{k=0}^{\infty} \frac{(tT)^k}{k!}$$

is an element of $\mathrm{B}(\mathcal{B})$ and $\|e^{tT}\| \leq e^{|t|\|T\|}$.

EXERCISE **4.12.** Let f be a linear functional on $\mathcal{N}$.

(a) Show that $f \in \mathcal{N}^*$ if and only if there exists $C > 0$ with $|f(\xi)| \leq C$ for all ξ in some open ball $B(\eta; r)$. Generalize for linear operators on normed spaces.

(b) Show that $f \in \mathcal{N}^*$ if and only if $\mathrm{N}(f)$ is closed (although it may seem trivial at first sight, it is not!). This may not be true for linear operators; see Exercise 9.11.

EXERCISE **4.13.** Let $T \in \mathrm{B}(\mathcal{B})$, with $\|T\| < 1$. Verify that the operator given by the series $S = \sum_{j=0}^{\infty} T^j$ is an element of $\mathrm{B}(\mathcal{B})$ and $S = (\mathbf{1} - T)^{-1}$.

EXERCISE **4.14.** If $S, T \in \mathrm{B}(\mathcal{B})$, with T invertible in $\mathrm{B}(\mathcal{B})$, and $\|T - S\| < 1/\|T^{-1}\|$, adapt Exercise 4.13 to show that S is invertible. Conclude that the set of invertible operators in $\mathrm{B}(\mathcal{B})$ is an open set.

EXERCISE **4.15.** If $T \in B(\mathcal{B})$, use series to define the operators $\sin T$ and $\cos T$ in $B(\mathcal{B})$.

EXERCISE **4.16.** Complete the proof in Example 4.2 for the case $p = \infty$.

EXERCISE **4.17.** Show that the operator I defined as $(I\psi)(t) = \int_a^t \psi(s)\,\mathrm{d}s$, $\psi \in \mathrm{C}[a,b]$, belongs to $\mathrm{B}(\mathrm{C}[a,b])$. Show also that this operator has no eigenvalues, i.e., that there is no $\lambda \in \mathbb{F}$ and $0 \neq \psi \in \mathrm{C}[a,b]$ so that $I\psi = \lambda\psi$.

EXERCISE **4.18.** Let $T : \mathcal{N}_1 \to \mathcal{N}_2$ be a linear operator (bounded or not). Show that if its inverse operator T^{-1} exists and belongs to $\mathrm{B}(\mathcal{N}_2, \mathcal{N}_1)$, then there exists $C > 0$ so that $\|\xi\| \leq C\|T\xi\|$ for all $\xi \in \mathcal{N}_1$.

EXERCISE **4.19.** For each $a \in \mathbb{R}$ consider the functional on $\mathrm{C}[-1,1]$ given by $f_a(\psi) = \int_{-1}^{1} \psi(t)\,\mathrm{d}t + a\psi(0)$. Show that f_a is an element of the dual of $\mathrm{C}[-1,1]$ and that $\|f_a\| = 2 + |a|$.

EXERCISE **4.20.** Let $\mathcal{N} = \{\psi \in \mathrm{C}[0,1] : \psi(0) = 0\}$. Show that $\mathcal{N}$ is complete and discuss for which values of $r > 0$ the linear functional

$$f_r(\psi) = \int_0^1 \frac{\psi(t)}{t^r}\,\mathrm{d}t, \qquad \psi \in \mathcal{N},$$

belongs to $\mathcal{N}^*$ and compute its norm on these cases.

EXERCISE **4.21.** Let X be a proper closed vector subspace of $\mathcal{N}$. If for $T \in \mathrm{B}(\mathcal{N})$ one has $(\mathbf{1}-T)\mathcal{N} \subset X$, show that for each $0 < \alpha < 1$ there exists $\xi \in \mathcal{N}$ with $\|\xi\| = 1$ and $\inf_{\eta \in X} \|T\xi - T\eta\| \geq \alpha$.

EXERCISE **4.22.** If $\phi : [-1,1] \to \mathbb{C}$ is Borelian, show that $f : \mathrm{L}^2[-1,1] \to \mathbb{C}$,

$$f(\psi) = \int_{-1}^{1} \phi(t)\psi(t)\,\mathrm{d}t, \qquad \psi \in \mathrm{L}^2[-1,1],$$

is continuous if and only if $\phi \in \mathrm{L}^2[-1,1]$.

EXERCISE **4.23.** Let $T, S \in \mathrm{B}(\mathcal{B})$. If $TS - ST = 1$, show that (S^n is the n-th iterate of S) $TS^n - S^n T = nS^{n-1}$, for all $n \in \mathbb{N}$, and hence

$$\|S^{n-1}\| \leq \frac{2}{n} \|T\| \|S\| \|S^{n-1}\|,$$

so that $S^N = 0$ for N large enough. Show, then, that $0 = S^N = S^{N-1} = \cdots = S^0 = 1$. From this contradiction, conclude that there are no continuous operators T, S satisfying $TS - ST = 1$ (this result is important in Quantum Mechanics).

EXERCISE **4.24.** Show that if $\dim \mathcal{N} < \infty$, then $\mathcal{N}^*$ coincides with the algebraic dual of $\mathcal{N}$ (see the Notes for the definition). If $\dim \mathcal{N} = \infty$, construct an unbounded linear functional in $\mathcal{N}$.

EXERCISE **4.25.** Let $\psi : \mathcal{N} \to \mathbb{R}$ be a continuous functional such that

$$\psi(\xi + \eta) = \psi(\xi) + \psi(\eta),$$

for all $\xi, \eta \in \mathcal{N}$. Show that $\psi(\alpha \xi) = \alpha \psi(\xi)$ for all $\alpha \in \mathbb{R}$.

CHAPTER 5

Banach Fixed Point

The Banach Fixed Point Theorem is recognized as a fundamental result in the theory of metric spaces, and it is associated with many important applications. In particular, it is widely used to prove the existence of solutions to both integral and differential equations. Several applications of the theorem are encountered in Banach spaces, where the scope is not restricted to linear transformations. It is likely that the reader has encountered this result and some of its applications previously; however, due to its simplicity, broad applicability, and fundamental importance, it is considered worthwhile to be revisited and further explored in detail in this chapter.

Definition 5.1. Let M be a set. A *fixed point* of a mapping $A : M \hookleftarrow$ is an element $\xi \in M$ so that $A(\xi) = \xi$.

Many problems in Mathematics are reduced to finding fixed points of mappings. Thus, it becomes important to give conditions for the existence of fixed points. A simple case is that every continuous mapping $\psi : [0, 1] \hookleftarrow$ has at least one fixed point; this follows by the Theorem of Intermediate Value or, in a more sophisticated way, because the intervals are the connected sets of $\mathbb{R}$. In order to check this result, consider the continuous function $\phi(t) = \psi(t) - t$; if $\psi(0) = 0$ or $\psi(1) = 1$, the result is immediate, then, suppose that such conditions do not occur. Hence, $\phi(0) > 0$ and $\phi(1) < 0$, and there exists $t_f \in (0, 1)$ with $\phi(t_f) = 0$, that is, $\psi(t_f) = t_f$ and it follows that ψ has a fixed point.

EXERCISE **5.1.** Find the fixed points of the mappings $\phi : \mathbb{R} \hookleftarrow$, $\phi(t) = t^3$ and $T : C^\infty(\mathbb{R}) \hookleftarrow$, $(T\psi)(t) = \psi'(t)$.

Example 5.2. The search for differentiable solutions ψ on $[t_0 - a, t_0 + a]$ $(a > 0)$ of the differential equation in $\mathbb{R}$, $\psi'(t) = F(t, \psi(t))$, with an initial condition $\psi(t_0) = \psi_0$, with F continuously differentiable in a neighborhood of (t_0, ψ_0), is reduced to find the fixed points of the operator $\Phi : C[t_0 - a, t_0 + a] \hookleftarrow$ given by

$$(\Phi\psi)(t) = \psi_0 + \int_{t_0}^{t} F(s, \psi(s)) \, \mathrm{d}s, \qquad t \in [t_0 - a, t_0 + a],$$

by the Fundamental Theorem of Calculus. ●

DOI: 10.1201/9781003656166-5

A natural idea to get uniqueness of fixed points for mappings $A : (X, d) \to (X, d)$ on metric spaces would be to require that A decreases distances, and a way to accomplish this appears in the following result by Edelstein:

Proposition 5.3. *Let (X, d) be a (nonempty) compact metric space and $A : X \to X$ a mapping that satisfies*

$$d(A(\xi), A(\eta)) < d(\xi, \eta), \quad \forall \xi, \eta \in X, \xi \neq \eta.$$

Then, A has a unique fixed point in X.

Proof. Define $f : X \to [0, \infty)$ by $f(\xi) = d(A(\xi), \xi)$, which is continuous and since X is compact f attains its minimum value at some $\eta \in X$. It will be shown that $f(\eta) = 0$, thus concluding that η is a fixed point of A.

If $f(\eta) > 0$, then, by the choice of η,

$$0 < f(\eta) \leq f(A(\eta)) = d(A(A(\eta)), A(\eta)) < d(A(\eta), \eta) = f(\eta),$$

and such contradiction implies that $f(\eta) = 0$.

If η_1, η_2 are two distinct fixed points of A, then the contradiction

$$0 < d(\eta_1, \eta_2) = d(A(\eta_1), A(\eta_2)) < d(\eta_1, \eta_2),$$

implies the uniqueness. □

Example 5.9 below shows that compactness is an essential ingredient in Edelstsein result, but for some applications one would like to remove such assumption. This is the central idea in what follows.

A possible way to find fixed points of $A : (X, d) \to (X, d)$ is to begin with a rather arbitrary point $\xi_0 \in X$ and apply A, successively, in order to get $\xi_1 = A(\xi_0)$, $\xi_2 = A(\xi_1)$, $\cdots$, $\xi_n = A(\xi_{n-1})$, and to take the limit $n \to \infty$. If such sequence ξ_n converges to some ζ and A is continuous, a fixed point is in fact obtained, since

$$\zeta = \lim_{n \to \infty} \xi_n = \lim_{n \to \infty} A(\xi_{n-1}) = A(\lim_{n \to \infty} \xi_{n-1}) = A(\zeta).$$

This idea is denominated method of *successive approximations*. In the case of complete metric spaces, a sufficient condition for the convergence of this iterative process is presented in the next.

Definition 5.4. Let (X, d) and (Y, D) be metric spaces. A mapping $A : X \to Y$ is a *contraction* if there exists a constant $0 \leq \alpha < 1$ such that, for all $\xi, \eta \in X$, $D(A(\xi), A(\eta)) \leq \alpha d(\xi, \eta)$.

EXERCISE **5.2.** Verify that every contraction on metric spaces is uniformly continuous.

EXERCISE **5.3.** Let $\psi : \mathbb{R} \to \mathbb{R}$ be a differentiable function so that, for all $t \in \mathbb{R}$, $|\psi'(t)| \leq \alpha < 1$. Show that ψ is a contraction. Generalize to mappings from $\mathbb{R}^n$ to $\mathbb{R}^m$.

Theorem 5.5 (Banach Fixed Point). *Let R be a closed subset in the complete metric space (X, d). If a mapping $A : R \to R$ is a contraction, then A has one, and only one, fixed point in R.*

Proof. If ζ_1 and ζ_2 are fixed points of A in R, then

$$d(\zeta_1, \zeta_2) = d(A(\zeta_1), A(\zeta_2)) \leq \alpha d(\zeta_1, \zeta_2),$$

that is, $(1 - \alpha)d(\zeta_1, \zeta_2) \leq 0$; since $\alpha < 1$ it follows that $d(\zeta_1, \zeta_2) = 0$ and the contraction A has at most one fixed point in R.

To show that there exists a fixed point, it is enough to show that, given $\xi \in R$, the sequence $(\xi_n = A^n(\xi))_{n=0}^{\infty}$ (with the notation $A^0(\xi) = \xi$) is Cauchy, since any contraction is continuous. By using induction one gets $d(\xi_1, \xi_2) \leq \alpha d(\xi, \xi_1)$, $d(\xi_2, \xi_3) \leq \alpha d(\xi_1, \xi_2) \leq \alpha^2 d(\xi, \xi_1)$ and $d(\xi_n, \xi_{n+1}) \leq \alpha^n d(\xi, \xi_1)$, $n \in \mathbb{N}$.

Thus, for all $n, m \in \mathbb{N}$,

$$\begin{aligned} d(\xi_n, \xi_{n+m}) &\leq d(\xi_n, \xi_{n+1}) + d(\xi_{n+1}, \xi_{n+2}) + \cdots + d(\xi_{n+m-1}, \xi_{n+m}) \\ &\leq \alpha^n \left(1 + \alpha + \alpha^2 + \cdots + \alpha^{m-1} \right) d(\xi, \xi_1) \\ &< \frac{\alpha^n}{1 - \alpha} d(\xi, \xi_1). \end{aligned}$$

Since $\alpha^n \to 0$ for $n \to \infty$, it is found that (ξ_n) is a Cauchy sequence in R, and A has a fixed point in the closed region R. □

The next result is a consequence of the proof of the Banach Fixed Point Theorem.

Corollary 5.6. *By using the notation of Theorem 5.5, with $A : R \to R$ a contraction, then for any $\xi \in R$ the sequence $(\xi_n = A^n(\xi))_{n=0}^{\infty}$ converges to the unique fixed point ζ of A and with error (and "speed of convergence") in the n-th iterate estimated by*

$$d(\xi_n, \zeta) \leq \frac{\alpha^n}{1 - \alpha} d(\xi, \xi_1).$$

Proof. It is enough to take $m \to \infty$ in the expression

$$d(\xi_n, \xi_{n+m}) < \frac{\alpha^n}{1 - \alpha} d(\xi, \xi_1)$$

and use the continuity of the metric. □

Example 5.7. The mapping $\psi : \mathbb{R} \to \mathbb{R}$, $\psi(t) = 2t$, is not a contraction in any open set of the real line, but it has a unique fixed point. •

Example 5.8. $\alpha = 1$ does not guarantee neither existence nor uniqueness of fixed points. Consider for instance, on the real line, the identity mapping (with infinitely many fixed points) and the translation $T\xi = \xi + 1$ (which has no fixed points). •

Example 5.9. The uniformity of the contraction, characterized by $0 \leq \alpha < 1$, is essential for the existence of a fixed point. This can be exemplified by $\psi, \phi : [1, \infty) \hookleftarrow$, $\psi(t) = t + 1/t$, $\phi(t) = t + e^{-t}$, which satisfies $|\psi(t) - \psi(s)| < |t - s|$, $|\phi(t) - \phi(s)| < |t - s|$, $\forall t, s \in [1, \infty)$, $t \neq s$, but they have no fixed points. •

Corollary 5.10. *Let R be a closed set in the complete metric space (X, d) and $A :$ $R \to R$. If there exists $m \in \mathbb{N}$ so that $A^m : R \to R$ is a contraction, then A has one, and only one, fixed point in R. Furthermore, for all $\xi \in R$, the sequence $(A^n(\xi))_{n=0}^{\infty}$ converges to this fixed point.*

Proof. Set $\Phi := A^m$, which has a unique fixed point $\zeta \in R$, since it is a contraction. Now, since every fixed point of A is also a fixed point of Φ, only ζ could be a fixed point of A in R. By using that A commutes with Φ, one has

$$\Phi(A(\zeta)) = A(\Phi(\zeta)) = A(\zeta),$$

i.e., $A(\zeta)$ is a fixed point of Φ; by uniqueness $A(\zeta) = \zeta$, and ζ is the unique fixed point of A in R.

Given $\xi \in R$, to show that $A^n(\xi)$ converges to ζ as $n \to \infty$, put $M = \max_{0 \leq k \leq m-1} d(A^k(\xi), \zeta)$ and $0 \leq \alpha < 1$ so that $d(\Phi(\omega), \Phi(\eta)) \leq \alpha d(\omega, \eta)$, for all $\omega, \eta \in R$. Each $n \in \mathbb{N}$ can be written in a unique way as $n = jm + k$, with $j \in \mathbb{N}$ and $0 \leq k < m$. Thus,

$$d(A^n(\xi), \zeta) = d(\Phi^j(A^k(\xi)), \Phi^j(\zeta)) \leq \alpha^j d(A^k(\xi), \zeta) \leq \alpha^j M.$$

For $n \to \infty$ one gets $j \to \infty$ and so $d(A^n(\xi), \zeta) \to 0$. $\square$

EXERCISE **5.4.** Let $A, B : M \hookleftarrow$ be given mappings. Discuss the relation between the fixed points of A and B if they are commuting, i.e., $AB = BA$.

Example 5.11. Let $\psi : [-1, 1] \hookleftarrow$, $\psi(t) = 1 - 2t^2$. The mapping $\phi = \psi \circ \psi$ has fixed points that do not occur for ψ. These new fixed points are "orbits of period two of ψ."
●

Now three standard applications of the Banach Fixed Point Theorem will be discussed: Fredholm integral equations, affine equations in Banach spaces, and ordinary differential equations on $\mathbb{R}$. Usually, the difficulty for applying this theorem is to find spaces or norms so that the operator of interest is a contraction.

Example 5.12. Let $K : Q \to \mathbb{R}$ be a continuous function in the region $Q = [a, b] \times [a, b] \times \mathbb{R}$ with the Lipschitz condition

$$|K(t, s, u) - K(t, s, v)| \leq L|u - v|, \qquad (t, s, u), (t, s, v) \in Q,$$

with $L > 0$. If $\varphi \in C[a, b]$, the nonlinear integral equation of Fredholm

$$\psi(t) = \int_a^b K(t, s, \psi(s)) \, ds + \varphi(t), \qquad t \in [a, b],$$

has a unique solution $\psi \in C[a, b]$ if $L(b - a) < 1$. ●

Proof. It is enough to verify that under such conditions the operator

$$S : C[a, b] \hookleftarrow, \qquad (S\psi)(t) = \int_a^b K(t, s, \psi(s)) \, ds + \varphi(t),$$

is a contraction on $C[a,b]$. If $\psi, \phi \in C[a,b]$, then for all $a \le t \le b$

$$|(S\psi)(t) - (S\phi)(t)| \le \int_a^b |K(t,s,\psi(s)) - K(t,s,\phi(s))|\,ds$$

$$\le L\int_a^b |\psi(s) - \phi(s)|\,ds$$

$$\le L(b-a)\|\psi - \phi\|_\infty,$$

so that $\|S\psi - S\phi\|_\infty \le L(b-a)\|\psi - \phi\|_\infty$, which shows that S is a contraction if $L(b-a) < 1$. $\quad\square$

If in the Fredholm integral equation $K(t,s,u) = \lambda\kappa(t,s)u$, $\lambda \in \mathbb{R}$, it is obtained the linear Fredholm integral equation

$$\psi(t) = \lambda\int_a^b \kappa(t,s)\psi(s)\,ds + \varphi(t), \qquad t \in [a,b].$$

EXERCISE **5.5.** Show that if $[(b-a)|\lambda|\sup_{t,s\in[a,b]} |\kappa(t,s)|] < 1$, then the linear Fredholm integral equation has a unique solution in $C[a,b]$.

Example 5.13. Let $T \in B(\mathcal{B})$. If $\|T\| < 1$, the equation $\xi - T\xi = \eta$, for given $\eta \in \mathcal{B}$, has a unique solution. Indeed, define the operator $S : \mathcal{B} \hookleftarrow$, $S\xi = T\xi + \eta$ so that

$$\|S\xi - S\omega\| \le \|T\|\,\|\xi - \omega\|,$$

for all $\xi, \omega \in \mathcal{B}$, which is a contraction whose unique fixed point ζ is the searched solution. $\quad\bullet$

EXERCISE **5.6.** Show that in Example 5.13 one has $\zeta = \sum_{j=0}^\infty T^j\eta$.

Example 5.14. [Theorem of Picard] Given a continuous function $F : U \to \mathbb{R}$ on $U = [t_0 - b, t_0 + b] \times \overline{B}(\psi_0; r) \subset \mathbb{R}^2$, $r, b > 0$, and satisfying the Lipschitz condition

$$|F(t,\xi) - F(t,\eta)| \le L|\xi - \eta|, \qquad L > 0,$$

for $(t,\xi), (t,\eta) \in U$, denote $M = \max_{(t,\xi)\in U} |F(t,\xi)|$. The Cauchy problem

$$\frac{d\psi}{dt}(t) = F(t,\psi(t)), \qquad \psi(t_0) = \psi_0 \in \mathbb{R},$$

has a unique differentiable solution $\psi : I \to \mathbb{R}$, with $I = [t_0 - a, t_0 + a]$ and $0 < a < \min(r/M, 1/L, b)$. $\quad\bullet$

Proof. Let $\mathcal{B} = \{\psi \in C(I) : |\psi(t) - \psi_0| \le r, \forall t \in I\}$ with the norm $\|\psi\| = \max_{t\in I} |\psi(t)|$. $\mathcal{B}$ is complete since it is a closed subset of $C(I)$. As already noted (Example 5.2), by the Fundamental Theorem of Calculus, a way to show that this Cauchy problem has a solution, and that is unique, is to show that the operator Φ given by

$$(\Phi\psi)(t) = \psi_0 + \int_{t_0}^t F(s,\psi(s))\,ds, \qquad t \in I,$$

is a contraction. It is left to the reader to check that since $a \leq r/M$, then one has $\Phi : \mathcal{B} \hookleftarrow$.

For $\psi, \phi \in \mathcal{B}$ one has

$$
\begin{aligned}
|(\Phi\psi)(t) - (\Phi\phi)(t)| &\leq \left| \int_{t_0}^{t} |F(s, \psi(s)) - F(s, \phi(s))| \, \mathrm{d}s \right| \\
&\leq |t - t_0| \sup_{s \in I} |F(s, \psi(s)) - F(s, \phi(s))| \\
&\leq aL \sup_{s \in I} |\psi(s) - \phi(s)|,
\end{aligned}
$$

so that $\|(\Phi\psi) - (\Phi\phi)\| \leq aL\|\psi - \phi\|$, and since $a < 1/L$, it follows that Φ is a contraction. □

EXERCISE **5.7.** Show that, if F is continuous and its partial derivative function $\partial F(t, \xi)/\partial \xi$ is continuous on U of Example 5.14, then F satisfies the Lipschitz condition.

EXERCISE **5.8.** Show that the real function $t \mapsto \sqrt{|t|}$, although continuous, does not satisfies the Lipschitz condition in any neighborhood of the origin.

EXERCISE **5.9.** Is the function $t \mapsto t^k$ Lipschitz on $\mathbb{R}$ for all $k \in \mathbb{N}$?

Notes

The proof of the Banach Fixed Point is constructive, so it gives approximations for the fixed point and an upper bound for the error at each iteration. Apparently, E. Picard and G. Peano (around 1890) were the first to consider in a systematic way this method of successive approximations, but in the particular case of differential equations. In his thesis, around 1920, Banach presented the abstract formulation of this method and the theorem discussed here; related results were also obtained by R. Caccioppoli.

The proof of the Theorem of Picard presented in this chapter also applies to differential equations in Banach spaces, in particular to equations in $\mathbb{R}^n$, the so-called systems of differential equations of first order. Just continuity of F assures the existence of solutions to the Cauchy problem, but not the uniqueness.

Edelstein result was originally published in J. London Math. Soc. **37** (1962) 74–79. By using Algebraic Topology, founded by Poincaré around 1900, L. Brouwer showed in 1910 that every continuous mapping on a set homeomorphic to the unit ball in $\mathbb{R}^n$ into itself, has a fixed point, and without mentioning the concept of contraction. Note that this is a generalization of the fact that every continuous mapping on $[0, 1]$ to itself has a fixed point. There are adaptations to Banach spaces of infinite dimensions, for example the theorems of J. Schauder of 1930 and Tychonov of 1935. There is an abundance of literature related to fixed points and some questions are active topics of research.

Additional Exercises

EXERCISE **5.10.** Let $\mathcal{B} = C[0, 1]$ (real), with the norm $\|\cdot\|_\infty$. Give sufficient conditions on $\varphi \in \mathcal{B}$ so that the operator $S : \mathcal{B} \to \mathcal{B}$,

$$
(S\psi)(t) = \psi(t) \int_0^1 \psi(s) \, \mathrm{d}s + \varphi(t), \qquad \psi \in \mathcal{B},
$$

has fixed points and find them in these cases.

EXERCISE **5.11.** Let $K : Q \to \mathbb{R}$ be a continuous function on $Q = \{t \in [a, b], s \in [a, t], u \in \mathbb{R}\}$, which satisfies the Lipschitz condition

$$|K(t, s, u) - K(t, s, v)| \le L|u - v|, \qquad (t, s, u), (t, s, v) \in Q,$$

with $L > 0$. If $\varphi \in C[a, b]$, consider the nonlinear integral equation of Volterra $S\psi = \psi$, where

$$(S\psi)(t) = \int_a^t K(t, s, \psi(s)) \, ds + \varphi(t), \qquad t \in [a, b].$$

Use induction to show that for all n

$$|(S^n \psi)(t) - (S^n \phi)(t)| \le \frac{L^n (t - a)^n}{n!} \|\psi - \phi\|_\infty,$$

and that for n large enough S^n is a contraction. Conclude that this nonlinear integral equation of Volterra has a unique solution in $C[a, b]$.

EXERCISE **5.12.** Let $T \in B(\mathcal{B})$. If there is n so that $\|T^n\| < 1$, show that the equation $\xi - T\xi = \eta$, for given $\eta \in \mathcal{B}$, has a unique solution.

EXERCISE **5.13.** Show that the Cauchy problem $dy/dt = 3y^{2/3}$, $y(0) = 0$, has infinitely many differentiable solutions in any neighborhood of the origin.

EXERCISE **5.14.** Show that $\psi(t) = e^{-t}$, $t \in \mathbb{R}$, is not a contraction, but $\varphi = \psi \circ \psi$ is a contraction. Use these facts to conclude that the equation $t + \ln t = 0$ has a unique real solution.

EXERCISE **5.15.** Let X be a complete metric space and $f : \mathbb{R} \times X \to X$ continuous; for each $a \in \mathbb{R}$ denote by $f_a : X \to X$ the mapping $f_a(\xi) = f(a, \xi)$. Assume that there exists $0 < \alpha < 1$ so that f_a is a contraction with constant α (as discussed in this chapter) for all $a \in \mathbb{R}$, and denote by ξ_a its unique fixed point. Show that the mapping $a \mapsto \xi_a$ is continuous, i.e., that the fixed point depends continuously on the parameter a.

EXERCISE **5.16.** Check that, for $a > 0$, the mapping $\psi(t) = (t + a/t)/2$, $0 \ne t \in \mathbb{R}$, has $\pm\sqrt{a}$ as fixed points. In which regions this mapping is a contraction? Compare with Example 5.9. Investigate $\varphi(t) = (t + a/t^s)/2$, $s \ge 0$, as well.

EXERCISE **5.17.** Let $\psi : [a, b] \hookleftarrow$ be of class C^2. If $r \in [a, b]$ is a simple root of ψ, show that the sequence (t_n) obtained by the iterative method of Newton

$$t_{n+1} = t_n - \psi(t_n)/\psi'(t_n), \qquad n = 0, 1, 2, 3, \cdots,$$

converges to r if the initial condition t_0 is sufficiently close to such root.

EXERCISE **5.18.** Set $0 < b < 1$. Verify that $T : C[0, b] \hookleftarrow$, $(T\psi)(t) = t(1 + \psi(t))$ has a unique fixed point. Find this fixed point straightly from the resultant equation $T\psi = \psi$ and also, by the method of successive approximations.

EXERCISE **5.19.** Verify that $T_a : C[0, 1] \hookleftarrow$,

$$(T_a\psi)(t) = \frac{3a\pi}{4} \int_0^t \psi(s) \cos(\pi s/2) \, ds$$

is not a contraction if $2/3 \le a < 4/(3\sqrt{2})$, but T_a^2 is. Conclude that T_a has a unique fixed point for a in this interval.

CHAPTER 6

Baire Theorem

The Baire Theorem and several of its significant consequences in Functional Analysis, such as the Uniform Boundedness Principle and the Closed Graph Theorem, are examined in this chapter and the following one. These results play an essential role in various areas of analysis and are basic tools for understanding the behavior of operators in Banach spaces. Additionally, the concept of robustness through the use of (total) Borelian measures, specifically Lebesgue measures in $\mathbb{F}^n$, is introduced. However, it is noted that there are certain cases where the two notions, namely total Lebesgue measure and the idea of being Baire generic, do not coincide.

There are different ways to quantify a set. The first one is simply the cardinality, and some property $\mathcal{P}$ would be *robust* (i.e., not negligible, taking into account this intuitive discussion) if it holds in a set of large cardinality. But this characterization is too simple for important applications in Functional Analysis; this becomes clear after considering, for instance, that both the set of prime numbers and the integers $\mathbb{Z}$ have the same cardinality, but there are sensible differences if a property of integer numbers holds (only) in each of such sets. An advantage of the classification through cardinality is that it applies to any set.

Restricting to topological concepts, an open set, or a set with nonempty interior, could be considered *robust*, as well as a dense set. The fusion of these concepts, i.e., an open and dense set, should indicate something really *robust* and could exclude *pathologies* of certain property.

Baire's Theorem states that in complete metric spaces the density is maintained after enumerable intersections of dense open sets, and a property holding in such intersection is called *generic*. Therefore, if $\mathcal{P}_1$ and $\mathcal{P}_2$ are generic properties, then the set for which both properties simultaneously hold is dense. Note that this generalizes to a countable number of generic properties! Recall that, in general, the intersection of two dense sets can be empty.

The Baire Theorem and some of its consequences in Functional Analysis (Uniform Bounded Principle, Closed Graph Theorem, etc.) are the subjects of this and next chapters. (There is also the idea of robustness via (total) Borelian measures, Lebesgue measures in $\mathbb{F}^n$ in particular; however, there are cases in which these two notions, i.e., total Lebesgue measure and to be generic, do not coincide; see the Notes.)

DOI: 10.1201/9781003656166-6

Definition 6.1. A subset of a topological space X is

(i) nowhere dense (or *rare*) in X if its closure has empty interior;

(ii) meager (or of the *first category*) in X if it is the union of countably many nowhere dense sets in X;

(iii) nonmeager (or of the *second category*) in X if it is not meager in X.

Example 6.2. The set of rational numbers $\mathbb{Q}$ is meager in $\mathbb{R}$, because each of its points forms a nowhere dense subset in $\mathbb{R}$ (with the usual topology). The ternary Cantor set is an uncountable nowhere dense set in $\mathbb{R}$. •

EXERCISE **6.1.** Which are the meager subsets in a discrete metric space? Find a meager and dense subset in $\mathbb{C}$.

EXERCISE **6.2.** Examine the following assertion: "$C[a, b]$ is meager in $L^\infty[a, b]$."

Note that subsets of a rare set are rare, and the union of finitely many rare sets are also rare. However, a meager set need not be rare; see the equivalences that follow, and Definition 6.4.

Proposition 6.3. *If X is a topological space, then the following assertions are equivalent:*

(a) The union of a countable number of rare closed sets in X is a set with empty interior, i.e., the enumerable union of closed sets with empty interior results in a set with empty interior as well.

(b) The intersection of denumerably many open dense sets in X is a dense set in X.

(c) Every meager set of X has empty interior.

(d) The complement of every meager subset in X is dense in X.

(e) Every nonempty open subset in X is nonmeager in X.

Proof. It is enough to consider the complements of the respective sets to conclude that (a) and (b) are equivalent. (c) is clearly equivalent to (d) and (e). Now, assertions (a) and (d) are equivalent by the definition of meager set. □

Definition 6.4. A topological space in which the assertions in Proposition 6.3 are satisfied is called a *Baire space*. A subset in a topological space is a G_δ if it is the enumerable intersection of open sets; it is an F_σ if it is a enumerable union of closed sets. A set is *generic* or *residual* if it is a dense G_δ.

So, in a Baire space X each G_δ obtained by the intersection of denumerably many dense open sets is also dense, and if X is the enumerable union of sets H_n, then the closure of at least one of such H_n has nonempty interior. Only from its statement, it is difficult to appreciate the power of Baire theorem; see, for instance, the applications presented in Propositions 6.9 and 6.11, and the Notes as well.

Theorem 6.5 (Baire Theorem). *Every complete metric space is a Baire space.*

Proof. It will be considered the condition (b) in Proposition 6.3. Let X be a complete metric space and $A = \bigcap_{n=1}^{\infty} A_n$, with A_n dense open subsets in X for all n. It will be shown that A is dense in X, i.e., if B_1 represents a nonempty open ball in X, then $A \cap B_1 \neq \emptyset$.

Clearly $A_1 \cap B_1 \neq \emptyset$ and it is open, so such intersection contains the closure $\overline{B}_2$ of an open ball B_2 with radius less than $1/2$. Again, there is an open ball B_3 with radius less than $1/3$ such that $\overline{B}_3$ is contained in $A_2 \cap B_2$. In this way it is obtained a sequence of open balls B_n, with radius less than $1/n$, satisfying

$$(A_1 \cap B_1) \supset \overline{B}_2 \supset \overline{B}_3 \supset \cdots,$$

and with $\overline{B}_{n+1} \subset A_n \cap B_n$. The sequence of the centers of these balls is Cauchy and since X is complete, there exists a unique $\xi \in X$ such that $\{\xi\} = \bigcap_{n=2}^{\infty} \overline{B}_n$; it then follows that $\xi \in A \cap B_1 \neq \emptyset$. $\qquad \square$

EXERCISE **6.3.** Adapt the proof Theorem 6.5 to verify that a Hausdorff and locally compact topological space is a Baire space (recall that such spaces are regular).

Proposition 6.6. *If X is a Baire space, then all nonempty open sets in X and all generic sets of X are Baire spaces.*

Proof. Let $U \subset X$ be an open set; if U is not a Baire space, there exists a sequence of open sets $U_n \subset U$, each one dense in U, whose intersection is not dense in U. So, if $\overline{U}$ denotes the closure of U, then the sets $V_n = U_n \cup (X \backslash \overline{U})$ are open and dense in X, but the intersection of them is not dense in X. The contradiction with the fact of X being a Baire space shows that U is also a Baire space.

Consider now a sequence A_n of open and dense sets in X whose intersection $G = \bigcap_n A_n$ is dense in X, i.e., G is generic in X. If $\{B_j\}$ is a sequence of open dense sets in G, there exist C_j open and dense in X such that $B_j = C_j \cap G$. Thus, $\bigcap_j B_j = \bigcap_j (C_j \cap G) = \bigcap_{j,n} C_j \cap A_n$ is dense in X, and therefore, dense in G. This shows that G is a Baire space. $\qquad \square$

Proposition 6.7. *If H is a meager subset of a Baire space X, then its complement $X \backslash H$ is a Baire space.*

Proof. Since H is meager in X, the set $X \backslash H$ contains a generic subset G in X. If (A_n) is a sequence of open dense subsets in $X \backslash H$, then $A_n \cap G$ is open and dense in G for all n. Since G is dense in X, the intersection $\bigcap_n (A_n \cap G) = (\bigcap_n A_n) \cap G$ is dense in G and in $X \backslash H$. This shows that $X \backslash H$ is a Baire space. $\qquad \square$

REMARK **6.8.** As indicated in Exercise 6.3 and Proposition 6.6, Baire Theorem is a topological fact (and not just a metric one). For example, $\mathbb{R}$ is a (complete metric) Baire space, homeomorphic to the interval $(-1, 1)$, which is also Baire, although the latter is not a complete metric space.

EXERCISE **6.4.** Show that a topological space homeomorphic to a Baire space is also a Baire space.

EXERCISE **6.5.** Show that the set $\mathbb{R} \backslash \mathbb{Q}$ is generic in $\mathbb{R}$.

EXERCISE **6.6.** Show that an enumerable subset E of a metric space is meager if and only if E does not contain any isolated point.

In other chapters, there are applications of Baire Theorem to Analysis, with emphasis toward the theory of linear operators. The next result is a standard example, which answers the question: Are there continuous real functions that are not derivable at all points of their domains? And how large is such set?

Proposition 6.9. *Set $\mathcal{B} = \mathrm{C}[0, 1]$ real. Then the set of functions in $\mathcal{B}$ which do not have (finite) derivative at any point of $[0, 1]$ is generic in $\mathcal{B}$.*

Proof. Set $I = [0, 1]$, $u_\psi^n(t, h) = |\psi(t + h) - \psi(t)| - n|h|$ (for $\psi \in \mathcal{B}$ and $(t + h) \in I$) and

$$A_n = \left\{ \psi \in \mathcal{B} : \forall t \in I \text{ there exists } h \text{ so that } u_\psi^n(t, h) > 0 \right\}.$$

If $\psi \in \bigcap_{n=1}^\infty A_n$, then ψ is not derivable at any point of $[0, 1]$. Since $\mathcal{B}$ is a complete metric space, by Baire Theorem, it is enough to show that each A_n is open and dense in $\mathcal{B}$.

If $\psi \in A_n$, there is $\varepsilon > 0$ such that for any $t \in I$ there exists $\bar{h} = \bar{h}(t)$ with $u_\psi^n(t, \bar{h}) > \varepsilon$. Indeed, if such $\varepsilon > 0$ does not exist, then for any $j \in \mathbb{N}$ one can find t_j with $u_\psi^n(t_j, h) \leq 1/j$ for all admissible h. Since I is compact, the sequence (t_j) has an accumulation point t_0 and $u_\psi^n(t_0, h) \leq 0$ for all h, contradicting $\psi \in A_n$. Thus, the assertion is verified. Now, if $\phi \in \mathcal{B}$,

$$\begin{aligned} \varepsilon + n|\bar{h}| \quad &< \quad |\psi(t + \bar{h}) - \psi(t)| \\ &\leq \quad |\psi(t + \bar{h}) - \phi(t + \bar{h})| + |\phi(t + \bar{h}) - \phi(t)| + |\phi(t) - \psi(t)|, \end{aligned}$$

so that if $\|\psi - \phi\|_\infty < \varepsilon/2$, it follows that $u_\phi(t, \bar{h}) > 0$ for all $t \in [0, 1]$, i.e., $\phi \in A_n$. Therefore A_n is open.

To show that A_n is dense in $\mathcal{B}$, two facts will be used. Note that any continuous function on $[0, 1]$, formed by a finite number of linear parts, whose absolute value of each slope is greater than n, belongs to A_n. The other fact is that every $\psi \in \mathrm{C}[0, 1]$ is uniformly continuous; thus, given $\varepsilon > 0$, there exists $\delta(\varepsilon) > 0$ such that if $t, s \in [0, 1]$ and $|t - s| < \delta(\varepsilon)$, then $|\psi(t) - \psi(s)| < \varepsilon$.

Consider a finite partition of $[0, 1]$ by intervals with length less than $\delta(\varepsilon)$; the plot of $\psi \in \mathcal{B}$ in each of these intervals is contained in a rectangle of basis less than $\delta(\varepsilon)$ and height less then ε; in each such intervals take a continuous function whose range is inside the corresponding rectangle, linear by parts with absolute values of their slopes greater than n, and such that at the points of the partition (i.e., the ends of such intervals) this function coincides with ψ. This procedure defines a continuous function φ on $[0, 1]$, belonging to A_n and with $\|\psi - \varphi\|_\infty < \varepsilon$ (plot ψ and φ). Therefore A_n is dense in $\mathcal{B}$. $\qquad\square$

Corollary 6.10. *The set of continuous real functions that are derivable at some point is meager in $\mathrm{C}[a, b]$ (real).*

In case the normed space is not complete, Example 3.8 shows that it is possible to have a enumerable Hamel basis even in infinite dimension. As another application of Baire Theorem:

Proposition 6.11. *If $\mathcal{B}$ is complete and $\dim \mathcal{B} = \infty$, then any Hamel basis of $\mathcal{B}$ is uncountable.*

Proof. Suppose that the sequence $(e_j)_{j \in \mathbb{N}}$ is a Hamel basis of $\mathcal{B}$. It is possible to assume that $\|e_j\| = 1$, for all j. Let

$$E_{n,m} = \left\{ \left(\sum_{j=1}^{n} a_j e_j \right) \in \mathcal{B} : \sum_{j=1}^{n} |a_j| \leq m \right\},$$

which is a closed set and $\mathcal{B} = \cup_{n,m} E_{n,m}$. By Baire's Theorem, some $E_{n,m}$ contains an open ball $B(\eta; r)$, $r > 0$; but this is impossible because the vector $\eta + r e_{n+1}/2$ belongs to this ball but not to $E_{n,m}$. This contradiction shows that a Hamel basis of $\mathcal{B}$ cannot be countable. □

Notes

The work of Baire about the theorem discussed here was published in 1899, and only $\mathbb{R}^n$ was considered (notice that Osgood had presented a version of that result for $\mathbb{R}$ in 1897); Baire's original proof was adapted to complete metric spaces by Kuratowski and Banach in 1930 (these independent works appeared in the same issue of the journal Fund. Math.). This result became well known particularly after its use by Banach and Steinhaus in a proof of the Uniform Bounded Principle. The nomenclature "first and second categories" was due to Baire himself, but currently it is not widely used; sometimes Baire's Theorem is referred to as Category Theorem.

Baire's Theorem has applications in different contexts: it is a basic ingredient in Dynamical System Theory in order to guarantee that certain properties are generic; it can be used to show that Cantor ternary set is not enumerable; to show that given a point in $\mathbb{R}$ there are many continuous functions whose Fourier series diverge at this point (see Corollary 7.6); that the pointwise convergence of real functions is not metric, etc.

Robust sets from the topological and Lebesgue measure points of view may not coincide. There are nonmeager sets of $\mathbb{R}$ with null Lebesgue measure (see Exercise 6.13).

The first published example of a nondifferentiable real continuous function (at every point) was due to Weierstrass in 1872 (although B. Bolzano seemed to be aware of an example about 40 years before that publication). The existence of real functions which are continuous only on the irrational numbers (Exercise 6.7) was discovered by K. J. Thomae in 1875; given an arbitrary F_σ subset of $\mathbb{R}$, in 1903 W. H. Young showed how to construct a function discontinuous exactly on this set.

Additional Exercises

EXERCISE **6.7.** Show that $\mathbb{Q}$ is not a G_δ in $\mathbb{R}$ and that, for each function $f : \mathbb{R} \to \mathbb{R}$, the set of its points of continuity is also a G_δ. Conclude that there is no continuous function only in $\mathbb{Q}$. On the other hand: if $\mathbb{Q} = \{p/q\}$ (irreducible ratios, with $q \in \mathbb{N}$ and $p \in \mathbb{Z}$), verify that $g : \mathbb{R} \to \mathbb{R}$, $g(0) = 1$, $g(p/q) = 1/q$, and zero in $\mathbb{R} \setminus \mathbb{Q}$, is continuous only on the set of irrational numbers.

EXERCISE **6.8.** Show that the border of an open (or closed) set of a metric space is a nowhere dense set.

EXERCISE **6.9.** If X is a complete metric space and there exists a sequence of closed sets (F_n) whose union is X, show that $\cup_n \mathrm{int} F_n$ is open and dense in X ($\mathrm{int} F_n$ indicates the interior of F_n).

EXERCISE **6.10.** Show that the subset of functions in C$[a,b]$ that have left (right) derivative at some point of (a,b) (of $[a,b)$) is meager in C$[a,b]$.

EXERCISE **6.11.** Let $\psi : \mathbb{R} \to (0,\infty)$. Show that there exists a nonempty interval (a,b) and $n_0 \in \mathbb{N}$ such that $\{t \in (a,b) : \psi(t) > 1/n_0\}$ is dense in (a,b).

EXERCISE **6.12.** A homeomorphism $h : X \hookleftarrow$, on the complete and separable metric space X, is *transitive* if there exists $\xi \in X$ whose *orbit*

$$\mathcal{O}(\xi) := \{h^m(\xi) : m \in \mathbb{Z}\}$$

is dense in X. Show that if h is transitive, the set of points whose orbits are not dense in X is a meager F_σ.

EXERCISE **6.13.** Let (q_n) be an enumeration of the rational numbers in $\mathbb{R}$. Show that the set

$$A = \bigcap_{j=1}^{\infty} \bigcup_{n=1}^{\infty} \left(q_n - \frac{1}{2^{j+n}}, q_n + \frac{1}{2^{j+n}} \right)$$

is nonmeager in $\mathbb{R}$, although it is mensurable and has zero Lebesgue measure.

CHAPTER 7

Uniform Boundedness Principle

Some of the main consequences of the Baire Theorem, including the Banach-Steinhaus Theorem and the Uniform Boundedness Principle (with the term "Principle" being of historical origin), are presented in this chapter. These important results will be applied, in particular, to the concept of weak convergence in other chapters. The Uniform Boundedness Principle provides conditions under which a finite upper bound for the set of norms of a family of linear operators can be guaranteed from pointwise boundedness, even without any restrictions on the cardinality of the family. These results play a role in understanding the behavior of operators in Banach spaces.

Theorem 7.1 (Uniform Boundedness Principle). *If a collection of linear operators* $\{T_j\}_{j \in J}$ *in* $\mathrm{B}(\mathcal{B}, \mathcal{N})$ *is pointwise bounded, that is, for every* $\xi \in \mathcal{B}$ *one has*

$$\sup_{j \in J} \|T_j \xi\| < \infty,$$

it follows that it is uniformly bounded, that is, $\sup_{j \in J} \|T_j\| < \infty$.

Proof. Consider the closed set $F_k = \{\xi \in \mathcal{B} : \|T_j \xi\| \leq k, \ \forall j \in J\}$; it is closed because T_j is continuous, and F_k is the intersection of the closed sets $T_j^{-1} \overline{B_{\mathcal{N}}}(0; k)$ for all $j \in J$. One has $\mathcal{B} = \bigcup_{k=1}^{\infty} F_k$, and so, by Baire's Theorem, there is an F_m with nonempty interior. Pick an open ball $B_{\mathcal{B}}(\xi_0; r)$ $(r > 0)$ contained in F_m; then, for any $j \in J$, it follows that $\|T_j \xi\| \leq m$ for all $\xi \in B_{\mathcal{B}}(\xi_0; r)$.

If $\xi \in \mathcal{B}, \|\xi\| = 1$, then $\eta = \xi_0 + r\xi/2$ is an element of $B_{\mathcal{B}}(\xi_0; r)$ and

$$\|T_j \xi\| = \frac{2}{r} \|T_j \eta - T_j \xi_0\| \leq \frac{2}{r} \left(\|T_j \eta\| + \|T_j \xi_0\| \right) \leq \frac{4m}{r};$$

thus $\|T_j \xi\| \leq 4m/r$ for all $j \in J$ and $\|\xi\| = 1$; hence $\sup_j \|T_j\| \leq 4m/r < \infty$. $\square$

Corollary 7.2. *A subset* $S \subset \mathcal{B}^* = \mathrm{B}(\mathcal{B}, \mathbb{F})$ *is bounded if and only if for all* $\xi \in \mathcal{B}$, $\sup_{f \in S} |f(\xi)| < \infty$.

DOI: 10.1201/9781003656166-7

Proof. If S is bounded, then $M = \sup_{f \in S} \|f\| < \infty$ and for every $\xi \in \mathcal{B}$ one finds $\sup_{f \in S} |f(\xi)| \leq M\|\xi\| < \infty$. In order to conclude the reverse assertion, by using the notation of the Uniform Boundedness Principle, it is enough to consider S as the collection $\{T_j\}$ in the Banach space $\mathcal{B}^*$. □

Corollary 7.3 (Banach-Steinhaus Theorem). *Suppose that for the sequence* $(T_n)_{n=1}^{\infty}$ *in* $\mathrm{B}(\mathcal{B}, \mathcal{N})$, *for every* $\xi \in \mathcal{B}$, *there exists the limit*

$$T\xi := \lim_{n \to \infty} T_n\xi.$$

Then $\sup_n \|T_n\| < \infty$ *and* T *is also an operator in* $\mathrm{B}(\mathcal{B}, \mathcal{N})$.

Proof. The linearity of T is immediate. Since for each $\xi \in \mathcal{B}$ there exists $\lim_{n \to \infty} T_n\xi$, then $\sup_n \|T_n\xi\| < \infty$, and by the Uniform Boundedness Principle, $\sup_n \|T_n\| < \infty$. The defined action of T implies that

$$\|T\xi\| = \lim_{n \to \infty} \|T_n\xi\| \leq \left(\sup_n \|T_n\| \right) \|\xi\|, \qquad \forall \xi \in \mathcal{B},$$

and so one concludes that T is bounded. □

EXERCISE **7.1.** In the Banach-Steinhaus Theorem, show that

$$\|T\| \leq \liminf_{n \to \infty} \|T_n\|.$$

Example 7.4. Denote by $\mathcal{N}$ the normed space of $\xi = (\xi_j) \in l^{\infty}(\mathbb{N})$ with $\xi_j \neq 0$ only for j in a finite set of indices. Consider $T_n : \mathcal{N} \to l^{\infty}$ with action $T_n\xi = (n\xi_n)_{j \in \mathbb{N}}$. Then $T_n \in \mathrm{B}(\mathcal{N}, l^{\infty})$ for every n, and for each $\xi \in \mathcal{N}$ there exists the limit $\lim_{n \to \infty} T_n\xi = 0$, however $\lim_{n \to \infty} \|T_n\| = \infty$. This exemplifies that the conclusions of the Banach-Steinhaus Theorem (and of the Uniform Boundedness Principle) may fail if the domain of the operators is not a Banach space. •

EXERCISE **7.2.** Let $S_l : l^2(\mathbb{N}) \hookleftarrow$ be the left shift

$$S_l(\xi_1, \xi_2, \xi_3, \cdots) = (\xi_2, \xi_3, \xi_4, \cdots)$$

and denote its iterates $T_n = S_l^n$. Find $\|T_n\xi\|$ and the limit operator as in the Banach-Steinhaus Theorem.

Proposition 7.5. *Pick a collection* $\{T_j\}_{j \in J}$ *in* $\mathrm{B}(\mathcal{B}, \mathcal{N})$ *with*

$$\sup_{j \in J} \|T_j\| = \infty.$$

Then the set $\mathcal{S} = \{\xi \in \mathcal{B} : \sup_j \|T_j\xi\| < \infty\}$ *is meager in* $\mathcal{B}$.

Proof. By employing the notation in the proof of the Uniform Boundedness Principle, one finds $\mathcal{S} = \cup_{k=1}^{\infty} F_k$, and by that proof, it follows that the interior of every F_k is empty, since if not one would get $\sup_{j \in J} \|T_j\| < \infty$. By using that F_k is closed, it follows that $\mathcal{S}$ is a meager set. □

Now, the above results will be applied to the convergence (or not!) of Fourier series of real periodic continuous functions; in the case of periodic continuously differentiable functions, that convergence is uniform (see Corollary 22.6 and Exercises 22.10 and 22.11). Set $C_p[0, 2\pi] = \{\psi \in C[0, 2\pi] : \psi(0) = \psi(2\pi)\}$, which is a closed subspace of $C[0, 2\pi]$, so it is Banach, and set

$$(\mathcal{F}\psi)_n = \frac{1}{2\pi} \int_0^{2\pi} e^{-int} \psi(t) \, dt, \quad \psi \in C_p[0, 2\pi].$$

Corollary 7.6. *The set of functions* $\psi \in C_p[0, 2\pi]$ *whose Fourier series* $\sum_{n \in \mathbb{Z}} (\mathcal{F}\psi)_n e^{int}$ *converges at* $t = 0$ *is meager.*

Proof. Trigonometric relations imply that, for each N, the partial sum $(S_N\psi)(t) = \sum_{|n| \leq N} (\mathcal{F}\psi)_n e^{int}$ can be rewritten as

$$(S_N\psi)(t) = \frac{1}{2\pi} \int_0^{2\pi} \frac{\sin[(2N+1)(t-s)/2]}{\sin[(t-s)/2]} \psi(s) \, ds.$$

Now, $f_N : C_p[0, 2\pi] \to \mathbb{C}$, $f_N(\psi) = (S_N\psi)(0)$, belongs to the dual space of $C_p[0, 2\pi]$; so, to conclude this proof it is enough to check that $\sup_N \|f_N\| = \infty$ and apply Proposition 7.5 with f_N in place of T_j.

Set $\phi_N(t) = \sin[(2N+1)t/2]$, which is an element of $C_p[0, 2\pi]$ with unit norm; hence

$$
\begin{aligned}
f_N(\phi_N) &= \frac{1}{2\pi} \int_0^{2\pi} \frac{\sin^2[(2N+1)s/2]}{\sin(s/2)} \, ds \\
&\geq \frac{1}{\pi} \int_0^{2\pi} \frac{\sin^2[(2N+1)s/2]}{s} \, ds = \frac{1}{\pi} \int_0^{(2N+1)\pi} \frac{\sin^2 u}{u} \, du \\
&\geq \frac{1}{\pi} \sum_{n=1}^{2N+1} \int_{(n-1)\pi}^{n\pi} \frac{\sin^2 u}{n\pi} \, du = \frac{1}{2\pi} \sum_{n=1}^{2N+1} \frac{1}{n}.
\end{aligned}
$$

By using that the harmonic series is divergent, it follows that $\lim_{N \to \infty} \|f_N\| = \infty$, and this concludes the proof. $\qquad\square$

EXERCISE **7.3.** Show the following version of the Uniform Boundedness Principle: Let $\{\psi_j\}_{j \in J}$ be a family of (real or complex) continuous functions defined on a complete metric space X, so that, for each $\xi \in X$, $\sup_{j \in J} |\psi_j \xi| < \infty$. Then there exists a nonempty open set $A \subset X$ with

$$\sup_{\substack{j \in J \\ \xi \in A}} |\psi_j(\xi)| < \infty.$$

EXERCISE **7.4.** Show that every complete metric space with no isolated points has an uncountable number of points. Note that this gives a proof that the set of real numbers is uncountable.

The next application refers to the continuity of bilinear mappings. However, note that there are two-variable mappings that are continuous in each variable but that are not continuous; see Example 7.9.

Corollary 7.7. *If $b : \mathcal{B}_1 \times \mathcal{B}_2 \to \mathbb{F}$ is a bilinear mapping continuous in each variable (i.e., $b(\cdot, \eta)$ and $b(\xi, \cdot)$ are linear and continuous for each $\eta \in \mathcal{B}_2$ and each $\xi \in \mathcal{B}_1$, respectively), then b is continuous, that is, if $\xi_n \to \xi$ and $\eta_n \to \eta$, then $b(\xi_n, \eta_n) \to b(\xi, \eta)$.*

Proof. By (bi)linearity it is enough to consider $\xi_n \to 0$ and $\eta_n \to 0$, and the goal is to show that $b(\xi_n, \eta_n) \to 0$. For each $\xi_n \in \mathcal{B}_1$ define $T_n : \mathcal{B}_2 \to \mathbb{F}$ by $T_n \eta = b(\xi_n, \eta)$, which is continuous and $T_n \eta$ converges to zero as $n \to \infty$. Thus, by Banach-Steinhaus, there is $C > 0$ so that $|T_n \eta| \leq C\|\eta\|$, for all $\eta \in \mathcal{B}_2$, in particular for η_n, and one gets $|b(\xi_n, \eta_n)| = |T_n \eta_n| \leq C\|\eta_n\|$ which converges to zero as $n \to \infty$. $\square$

In the sequel two examples are presented, one that is not linear, but is included here due to its simplicity (and it is a typical example in Differential Calculus), and the second one is a bilinear mapping defined on a space that is not complete.

Example 7.8. The function $\psi : \mathbb{R}^2 \to \mathbb{R}$ given by $\psi(t, s) = ts/(t^2 + s^2)$ if $(t, s) \neq (0, 0)$, and $\psi(0, 0) = 0$ is continuous in each variable, but is not continuous at the origin (consider $s(t) = mt$ and check that the value of the limit at the origin depends on m). •

Example 7.9. Let $\mathcal{N} = (C[0, \pi], \|\cdot\|_1)$, with $\|\psi\|_1 = \int_0^\pi |\psi(t)| \, dt$, for $\psi \in \mathcal{N}$. Denote by $b(\cdot, \cdot)$ the bilinear form $b : \mathcal{N} \times \mathcal{N} \to \mathbb{C}$, $b(\psi, \phi) = \int_0^\pi \psi(t)\phi(t) \, dt$, which is continuous in each variable. By defining

$$\psi_n(t) = \begin{cases} \sqrt{n} \, \sin(nt) & \text{if } 0 \leq t \leq \pi/n \\ 0 & \text{if } \pi/n < t \leq \pi \end{cases},$$

it follows that $\|\psi_n\|_1 = 2/\sqrt{n}$ which converges to zero as $n \to \infty$, while $b(\psi_n, \psi_n) = \pi/2$ for all n, and so b is not continuous. •

EXERCISE **7.5.** Check that the bilinear form b in Example 7.9 is continuous in each variable.

The Uniform Boundedness Principle, as well as the Banach-Steinhaus Theorem, is important in the study of weak convergence in normed spaces, which will be discussed in other chapters.

Notes

There is no general agreement in the literature about what is named Uniform Boundedness or Banach-Steinhaus, with the two denominations usually considered synonymous, and also both referring to the set of the main results presented in this chapter. From the historical point of view, the first version of such results was due to Hahn in 1922, but only for linear continuous functionals, and he used an argument of contradiction, which was later adapted to continuous linear operators between Banach spaces in Banach's thesis. The proof by virtue of Baire's Theorem is due to Banach and Steinhaus in 1927. For details on the history of such results, see [Swartz] and [Hochst].

Additional Exercises

EXERCISE **7.6.** Use the results discussed in this chapter to show that the space $\mathcal{N} = (C[a,b], \|\cdot\|_1)$, with $\|\psi\|_1 = \int_a^b |\psi(t)| \, dt$, is not complete.

EXERCISE **7.7.** Let $T : \mathcal{N}_1 \to \mathcal{N}_2$ be linear. Show that T is bounded if and only if $T^{-1}\overline{B}_{\mathcal{N}_2}(0;1)$ has nonempty interior.

EXERCISE **7.8.** Check that $C_p[0, 2\pi]$ is a Banach space, and also the expression for partial sums of the Fourier series employed in the proof of Corollary 7.6.

EXERCISE **7.9.** Let X be a compact Hausdorff space and $\psi_n, \psi \in C(X)$.

(a) By using the Theorem of Riesz-Markov 4.15 and verifying that for each $t \in X$, $\delta_t : C(X) \to \mathbb{F}$, $\delta_t \phi = \phi(t)$, $\forall \phi \in C(X)$, is an element of $C(X)^*$, show that if $\int \psi_n d\mu \to \int \psi d\mu$, $\forall \mu \in M(X)$, then $\sup_n \|\psi_n\| < \infty$ and that the pointwise convergence $\psi_n \to \psi$ takes place.

(b) By the Theorem of Dominated Convergence, show that the converse of the result in item (a) holds, namely, if $\sup_n \|\psi_n\| < \infty$ and the pointwise convergence $\psi_n \to \psi$ takes place, then $\int \psi_n d\mu \to \int \psi d\mu$, $\forall \mu \in M(X)$.

EXERCISE **7.10.** Can Corollary 7.7 be generalized to

(a) multilinear mappings $b : \mathcal{B}_1 \times \cdots \times \mathcal{B}_k \to \mathbb{F}$ that are continuous in each variable?

(b) bilinear mappings that are continuous in each variable $b : \mathcal{N}_1 \times \mathcal{N}_2 \to \mathcal{N}_3$, with $\mathcal{N}_1$ and/or $\mathcal{N}_2$ complete?

EXERCISE **7.11.** Let $(T_n)_{n=1}^\infty$ be a bounded sequence in $B(\mathcal{N}, \mathcal{B})$ (i.e., $\sup_n \|T_n\| < \infty$) so that, for any ξ in a dense set in the sphere $S_{\mathcal{N}}(0;1)$, there exists the limit $\lim_{n\to\infty} T_n \xi$. Show that for any $\xi \in \mathcal{N}$ there exists

$$T\xi := \lim_{n\to\infty} T_n \xi$$

and that $T \in B(\mathcal{N}, \mathcal{B})$.

EXERCISE **7.12.** Show that there exists a generic subset A in $C_p[0, 2\pi]$ and a set B dense in $[0, 2\pi]$, so that the Fourier series of any element of A diverge in every point of B.

CHAPTER 8

Open Mapping Theorem

An important technical result in Functional Analysis, known as the Open Mapping Theorem, will now be discussed. This theorem, attributed to Banach, provides a insight into the behavior of continuous linear mappings between Banach spaces. One of its key consequences is that sufficient conditions are given for a bijective continuous linear mapping to be continuously invertible; in other words, it ensures that the mapping is a linear homeomorphism. This result is of great importance, as it will be utilized in the next chapter to prove the Closed Graph Theorem, which is another fundamental result in the theory of Banach spaces.

Recall that a mapping between topological spaces is open if the image under this mapping of every open subset is also open. There are invertible continuous mappings that are not open, as the following examples show.

Example 8.1. The identity mapping from $\mathbb{R}$ equipped with the discrete topology to $\mathbb{R}$ with the usual topology is continuous and invertible; however, its inverse fails to be continuous; thus, this continuous and invertible mapping is not open. •

Example 8.2. Let $u : [0,1) \cup [2,\infty) \to [0,\infty)$ in $\mathbb{R}$, $u(t) = t$ if $0 \leq t < 1$ and $u(t) = t - 1$ if $t \geq 2$. Then u is a continuous bijection, but its inverse $u^{-1} : [0,\infty) \to [0,1) \cup [2,\infty)$, whose action is

$$u^{-1}(t) = \begin{cases} t & \text{if } 0 \leq t < 1 \\ t+1 & \text{if } t \geq 1 \end{cases} ,$$

is not continuous at 1, so u is not an open mapping. •

EXERCISE **8.1.** Show that $\psi : [0,2\pi) \to S^1 = \{z \in \mathbb{C} : |z| = 1\}$, $\psi(t) = e^{it}$, is continuous, whereas its inverse (which exists) is discontinuous at $t = 0$.

EXERCISE **8.2.** Show that the linear operator $T : l^1(\mathbb{N}) \hookleftarrow$ with action

$$T(\xi_1, \xi_2, \xi_3, \cdots) = (\xi_1/1, \xi_2/2, \xi_3/3, \cdots)$$

continuous and invertible, however its (linear) inverse T^{-1}, defined on the range of T, is not continuous.

DOI: 10.1201/9781003656166-8

Theorem 8.3 (Open Mapping). *Let $T \in \mathrm{B}(\mathcal{B}_1, \mathcal{B}_2)$ with range given by the Banach space $\mathcal{B}_2$, then T is an open mapping.*

Proof. First, a sketch of the proof. It will be reduced to show that there exists an open ball $B(0; r)$ so that $TB(0; r)$ contains an open set (that is, this will imply that T is open); the linearity of T will imply that this open set can be considered a neighborhood of the origin and that one may assume that $r = 1$. Since T is surjective, by Baire Theorem and linearity, there exists $\delta > 0$ with $B(0; \delta) \subset \overline{T(B(0; 1))}$, and an interesting final argument will show that $B(0; \delta) \subset \overline{T(B(0; 1))} \subset TB(0; 2)$, which will complete the proof.

The proof will make use of the following properties, of which only the final one requires justification:

(a) For all $r, s > 0$, one has $TB(0; r) = \frac{r}{s} TB(0; s)$.

(b) For all $\xi \in \mathcal{B}_1$ and $r > 0$, one has $TB(\xi; r) = T\xi + TB(0; r)$, which is seen as a sum of sets.

(c) If $B(0; \varepsilon) \subset \overline{TB(0; r)}$, $\varepsilon > 0$, then $B(0; \alpha\varepsilon) \subset \overline{TB(0; \alpha r)}$, for all $\alpha > 0$. Thus, if there is $r > 0$ so that $\overline{TB(0; r)}$ contains a neighborhood of the origin, then $\overline{TB(0; s)}$ also contains a neighborhood of the origin for all $s > 0$; observe that such conclusions remain valid without the closure of the sets.

(d) If $B(\eta_0; \varepsilon) \subset \overline{TB(0; r)}$, then there exists $\delta > 0$ so that $B(0; \delta) \subset \overline{TB(0; r)}$; observe that such conclusions remain valid without the closure of the sets.

In order to verity property **(d)**, take $\xi_1 \in B(0; r)$ so that $\|\eta_1 - \eta_0\| < \varepsilon/2$, with $\eta_1 = T\xi_1$. Hence,

$$B(\eta_1; \varepsilon/2) \subset B(\eta_0; \varepsilon) \subset \overline{TB(0; r)},$$

and so

$$
\begin{aligned}
B(0; \varepsilon/2) &= B(\eta_1; \varepsilon/2) - \eta_1 \subset \{B(\eta_0; \varepsilon) - T\xi_1\} \\
&\subset \left\{\overline{TB(0; r)} - T\xi_1\right\} \subset \overline{T[B(0; r) - \xi_1]} \\
&\subset \overline{TB(0; 2r)}.
\end{aligned}
$$

It then follows that $B(0; \varepsilon/2) \subset \overline{TB(0; 2r)}$ and so $B(0; \delta) \subset \overline{TB(0; r)}$ with $\delta = \varepsilon/4$, so concluding **(d)**.

The next Lemma is a key step in the proof.

Lemma 8.4. *Pick $T \in \mathrm{B}(\mathcal{N}_1, \mathcal{N}_2)$ and suppose that there is $r > 0$ in such way that the interior of $TB(0; r)$ is nonempty; then the operator T is an open mapping.*

Proof. Use the fact that the interior of $TB(0; r)$ is not empty and the above properties to conclude that for all $s > 0$, $TB(0; s)$ contains an open ball centered at the origin. To establish that T is an open mapping, it suffices to show that for every $\xi \in \mathcal{N}_1$ and every $s > 0$, the image $TB(\xi; s)$ contains an open neighborhood of $T\xi$. Since $TB(\xi; s) = T\xi + TB(0; s)$, one may take $\xi = 0$ and thus it only remains to check that for every $s > 0$ the image $TB(0; s)$ contains a neighborhood of the origin, but this is exactly what was noted above. $\square$

By the preceding lemma, the Open Mapping Theorem follows once one shows that there is an $r > 0$ so that $TB(0; r)$ contains an open ball centered at the origin. Henceforth, only the completeness of of $\mathcal{B}_1, \mathcal{B}_2$, and that T is surjective are used and the Baire Category Theorem is called upon as essential.

Since T is surjective $\mathcal{B}_2 = \bigcup_{n=1}^{\infty} \overline{TB(0; n)}$, and so Baire's Theorem implies that there exists an m so that the interior of $\overline{TB(0; m)}$ is nonempty. Property **(c)** above implies that one may take $m = 1$.

By property **(d)** one may suppose that there is $\delta > 0$ so that $B(0; \delta) \subset \overline{TB(0; 1)}$. It remains to establish that the relation $\overline{TB(0; 1)} \subset TB(0; 2)$ holds, which, by Lemma 8.4, will complete the proof of the theorem.

Let $\eta \in \overline{TB(0; 1)}$. Pick $\xi_1 \in B(0; 1)$ with

$$(\eta - T\xi_1) \in B(0; \delta/2) \subset \overline{TB(0; 1/2)}.$$

Note that in the last step property **(c)** was used. Select ξ_2 in $B(0; 1/2)$ so that (property **(c)** again)

$$(\eta - T\xi_1 - T\xi_2) \in B(0; \delta/2^2) \subset \overline{TB(0; 1/2^2)}.$$

By using induction, choose $\xi_n \in B(0; 1/2^{n-1})$ satisfying

$$\left(\eta - \sum_{j=1}^{n} T\xi_j\right) \in B(0; \delta/2^n) \subset \overline{TB(0; 1/2^n)}.$$

$(\sum_{j=1}^{n} \xi_j)_n$ is a Cauchy sequence and since $\mathcal{B}_1$ is complete, there is $\xi = \sum_{j=1}^{\infty} \xi_j$; by using the continuity of T, one finds that $\eta = T\xi$. Since $\|\xi\| < 2$, it follows that $\overline{TB(0; 1)} \subset TB(0; 2)$. $\qquad\square$

By the Open Mapping Theorem, the following result is immediate and is referred to as the *Inverse Mapping Theorem*.

Corollary 8.5 (Open Mapping 2). *If the operator $T \in B(\mathcal{B}_1, \mathcal{B}_2)$ is a bijection between two Banach spaces $\mathcal{B}_1$ and $\mathcal{B}_2$, then its inverse T^{-1} is also a bounded linear operator. In summary: $T^{-1} \in B(\mathcal{B}_2, \mathcal{B}_1)$.*

EXERCISE **8.3.** Verify the properties **(a)**, **(b)**, and **(c)** at the beginning of the proof of Theorem 8.3.

EXERCISE **8.4.** Let $T : C[-1, 1] \to C[-1, 1]$ be the operator $(T\psi)(t) = \int_{-1}^{t} \psi(s)\,ds$. Check that T is bounded and invertible, but T^{-1} is not bounded. Discuss such result in terms of the Open Mapping Theorem.

Example 8.6. In this application, it will be shown that there exists no "test sequence" of complex numbers $b = (b_n)_{n \in \mathbb{N}}$, so that a sequence of complex numbers $a = (a_n)_n$ is absolutely summable if and only if $(a_n b_n)_n$ is bounded; in other symbols,

$$a \in l^1(\mathbb{N}) \iff (a_n b_n) \in l^{\infty}(\mathbb{N}).$$

Suppose that such sequence b exists; since $(1, 1/2, 1/3, \cdots)$ is not absolutely summable, then $\lim_{n\to\infty} |b_n| = \infty$, and it is possible to suppose that $b_n \neq 0$, for all n. Then the bounded operator $\tau : l^\infty(\mathbb{N}) \to l^1(\mathbb{N})$, $(\tau c)_n = c_n/b_n$, is well defined, bounded, invertible, and rng $\tau = l^1(\mathbb{N})$. By the Open Mapping Theorem 8.5, it follows that $\tau^{-1} : l^1(\mathbb{N}) \to l^\infty(\mathbb{N})$, $(\tau^{-1}d)_n = b_n d_n$ is bounded. If $(e^m)_{m=1}^\infty$ is the canonical basis of $l^1(\mathbb{N})$, then $\|e^m\|_1 = 1$, whereas

$$\|\tau^{-1}e^m\|_\infty = |b_m| \overset{m\to\infty}{\longrightarrow} \infty,$$

and τ^{-1} cannot be bounded. This contradiction shows that such test sequence b does not exist. •

EXERCISE **8.5.** Verify that, under the initial hypotheses in Example 8.6, $\lim_{n\to\infty} |b_n| = \infty$, and the operator τ would be bounded and with rng $\tau = l^1(\mathbb{N})$.

As another application of the Open Mapping Theorem, it will be presented an interesting result that indicates, in another way (see Corollary 6.10), that the set of differentiable elements in $C[a, b]$ is "small."

EXERCISE **8.6.** Let $\|\cdot\|_1$ and $\|\cdot\|_2$ be two norms in the vector space X so that both resulting normed spaces are Banach. Show that if there is $C > 0$ with

$$\|\xi\|_1 \leq C\|\xi\|_2, \qquad \forall \xi \in X,$$

then these two norms are in fact equivalent.

Proposition 8.7. *Every closed vector subspace of continuously differentiable functions in* $C[-1, 1]$ *is finite dimensional.*

Proof. Let E be a closed vector subspace of $C[-1, 1]$ whose elements are continuously differentiable functions. Note that E is also a closed vector subspace of $C^1[-1, 1]$ with the norm

$$\|\!|\psi|\!\| = \sup_{t\in[-1,1]} |\psi(t)| + \sup_{t\in[-1,1]} |\psi'(t)|;$$

indeed, if $\psi_n \to \psi$ and $\psi_n' \to \phi$, both with uniform convergence, and since

$$\psi_n(t) = \psi_n(-1) + \int_{-1}^t \psi_n'(s)\,\mathrm{d}s,$$

one gets $\psi(t) = \psi(-1) + \int_{-1}^t \phi(s)\,\mathrm{d}s$ and $\phi = \psi'$. Hence E is a complete subspace of $(C^1[-1,1], \|\!|\cdot|\!\|)$ (see Exercise 1.9). Denote these Banach spaces by $\mathcal{B}_1 = (E, \|\!|\cdot|\!\|)$ and $\mathcal{B}_2 = (E, \|\cdot\|_\infty)$.

Consider the identity mapping $\mathbf{1} : \mathcal{B}_1 \to \mathcal{B}_2$, which is linear and bounded; by the Inverse Mapping Theorem (Corollary 8.5) it is a homeomorphism. Denote by $\overline{B_1}(0;1)$ the closed ball in $\mathcal{B}_1$. The set $\mathbf{1}(\overline{B_1}(0;1))$ is clearly bounded in $\mathcal{B}_2$ and it is also equicontinuous, since for all $\psi \in \mathbf{1}(\overline{B_1}(0;1))$ one has $(s < t)$

$$|\psi(t) - \psi(s)| = \left| \int_s^t \psi'(u)\,\mathrm{d}u \right| \leq \int_s^t |\psi'(u)|\,\mathrm{d}u \leq |t - s|.$$

Hence, by Ascoli's Theorem, $\mathbf{1}(\overline{B_1}(0;1))$ has compact closure in $\mathcal{B}_2$.

Since $\mathbf{1}$ is a homeomorphism, $\overline{B_1}(0;1)$ is also compact in $\mathcal{B}_1$; by Theorem 2.2, the (closed) unit ball in a normed space is compact if and only if this space is of finite dimension. □

Additional Exercises

EXERCISE **8.7.** If $T \in \mathrm{B}(\mathcal{B}_1, \mathcal{B}_2)$ is a bijection between $\mathcal{B}_1$ and $\mathcal{B}_2$, show that there exist $C_1, C_2 > 0$ so that
$$C_1\|\xi\|_{\mathcal{B}_1} \leq \|T\xi\|_{\mathcal{B}_2} \leq C_2\|\xi\|_{\mathcal{B}_1}, \qquad \forall \xi \in \mathcal{B}_1.$$

EXERCISE **8.8.** With respect to Corollary 8.5, it is worth considering the following result from General Topology: if X and Y are topological spaces, with X compact and Y Hausdorff, and $f : X \to Y$ is a continuous bijection, show that f is a homeomorphism.

EXERCISE **8.9.** Analyze the statement: "Every linear functional in $\mathcal{N}^*$ is an open mapping."

EXERCISE **8.10.** From Exercise 8.2, deduce that the hypothesis that rng T is a Banach space is essential in the Open Mapping Theorem.

EXERCISE **8.11.** Let $T : \mathcal{B}_1 \to \mathcal{B}_2$ be linear, surjective and so that $T(B_{\mathcal{B}_1}(0;r))$ is contained in a compact set of $\mathcal{B}_2$ for each $r > 0$. Show that $\dim \mathcal{B}_2 < \infty$.

EXERCISE **8.12.** If $T \in \mathrm{B}(\mathcal{B}_1, \mathcal{B}_2)$ is surjective, show that there is $C > 0$ so that for each $\eta \in \mathcal{B}_2$ the equation $T\xi = \eta$ has a solution $\xi_\eta \in \mathcal{B}_1$ with $\|\xi_\eta\| \leq C\|\eta\|$.

EXERCISE **8.13.** Given $T \in \mathrm{B}(\mathcal{B})$ and $\eta \in \mathcal{B}$, consider the equation $T\xi = \eta$. Show that the following statements are equivalent:

(a) If $T\xi = 0$, then $\xi = 0$ and this equation has a solution for all $\eta \in \mathcal{B}$.

(b) For all $\eta \in \mathcal{B}$, this equation has the unique solution $\xi = \xi(\eta)$, which depends continuously on η.

EXERCISE **8.14.** Let $\mathcal{F} : \mathcal{B} = \mathrm{L}^1[-\pi, \pi] \to l^\infty(\mathbb{Z})$ be the operator that associates the Fourier series $((\mathcal{F}\psi)_n)_{n\in\mathbb{Z}}$, of $\psi \in \mathcal{B}$, with
$$(\mathcal{F}\psi)_n = \frac{1}{2\pi} \int_{-\pi}^{\pi} e^{-int}\psi(t)\,\mathrm{d}t.$$
The well-known Riemann-Lebesgue Lemma (see Exercise 21.7) states that $\mathcal{F}(\mathcal{B}) \subset c_0$. By using the known fact that $\mathcal{F}$ is injective, use the Open Mapping Theorem to show that there are sequences in c_0 that are not Fourier series of functions in $\mathcal{B}$.

EXERCISE **8.15.** Use the following guide to verify that the set of surjective elements in $\mathrm{B}(\mathcal{B}_1, \mathcal{B}_2)$ is an open subset in this space. Notation: the elements denoted by ξ_j and η_j belong to $\mathcal{B}_1$ and $\mathcal{B}_2$, respectively.

(a) If $T \in \mathrm{B}(\mathcal{B}_1, \mathcal{B}_2)$ is surjective, show that there exists $C > 0$ so that for each $\|\eta_1\| \leq 1$, there is ξ_1 with $T\xi_1 = \eta_1$ and $\|\xi_1\| \leq C\|\eta_1\|$.

(b) Let $S \in \mathrm{B}(\mathcal{B}_1, \mathcal{B}_2)$ with $\|T - S\| \leq 1/(2C)$. Show that if $\eta_2 = (T - S)\xi_1$, then $\|\eta_2\| \leq 1/2$ and there exists ξ_2 with $\|\xi_2\| \leq C/2$, so that $T\xi_2 = \eta_2$. By defining $\eta_3 = (T - S)\xi_2$, then $\|\eta_3\| \leq 1/2^2$.

(c) By induction, find $\|\xi_j\| \leq C/2^{j-1}$ so that $T\xi_j = \eta_j$; define $\eta_{j+1} = (T - S)\xi_j$ and show that $\|\eta_{j+1}\| \leq 1/2^j$.

(d) Check that
$$\eta_1 = S(\xi_1 + \cdots + \xi_j) + \eta_{j+1},$$
the vector $\xi = \sum_{j=1}^{\infty} \xi_j$ is well defined and satisfies $\|\xi\| \leq 2C$, and $T\xi = \eta_1$, so that $S(\overline{B}_{\mathcal{B}_1}(0;2C)) \supset \overline{B}_{\mathcal{B}_2}(0;1)$. Conclude that S is an open mapping and so the proposed result.

EXERCISE **8.16.** This is a good opportunity to underline that all nonconstant holomorphic mappings $F : G \to \mathbb{C}$, defined on an open set $G \subset \mathbb{C}$, is open. Verify this.

CHAPTER 9

Closed Graph Theorem

There are many ways of constructing discontinuous operators. The most abrupt discontinuity of a linear operator T (between normed spaces) would be the existence of sequences $\xi_n \to \xi$ with $T\xi_n \to \eta$ and $\eta \neq T\xi$. The notion of closed operator avoids such kind of discontinuity by imposing that, if $\xi_n \to \xi$, then either $T\xi_n$ does not converge or (if it converges) the convergence is necessarily to $T\xi$. Thus, among the unbounded operators the closed ones are, in some sense, near the continuous ones. This excludes operators that rarely arise in applications.

It is known that the Cartesian product $\mathcal{N}_1 \times \mathcal{N}_2$ of two normed spaces has a natural structure of vector space given by $\alpha(\xi, \eta) = (\alpha\xi, \alpha\eta)$, $\alpha \in \mathbb{F}$, and $(\xi_1, \eta_1) + (\xi_2, \eta_2) = (\xi_1 + \xi_2, \eta_1 + \eta_2)$; additionally, this Cartesian product becomes a normed space in case the following norm $\|(\xi, \eta)\| = \|\xi\|_{\mathcal{N}_1} + \|\eta\|_{\mathcal{N}_2}$ is considered.

EXERCISE 9.1. Verify that the Cartesian product $\mathcal{B}_1 \times \mathcal{B}_2$, with the norm $\|(\xi, \eta)\|$ mentioned above, is a Banach space.

Definition 9.1. Let $T : \operatorname{dom} T \subset \mathcal{N}_1 \to \mathcal{N}_2$ be a linear operator. Its *graph* is the vector subspace $\mathcal{G}(T) := \{(\xi, T\xi) : \xi \in \operatorname{dom} T\}$ of $\mathcal{N}_1 \times \mathcal{N}_2$.

Definition 9.2. One says that a linear operator $T : \operatorname{dom} T \subset \mathcal{N}_1 \to \mathcal{N}_2$ is *closed* if for all convergent sequence $(\xi_n) \subset \operatorname{dom} T$, say $\xi_n \to \xi \in \mathcal{N}_1$, and also with $(T\xi_n) \subset \mathcal{N}_2$ convergent, say $T\xi_n \to \eta$, then $\xi \in \operatorname{dom} T$ and $\eta = T\xi$. In other words, the linear operator T is closed if its graph $\mathcal{G}(T)$ is a closed subspace of $\mathcal{N}_1 \times \mathcal{N}_2$.

EXERCISE 9.2. Verify that $\mathcal{G}(T)$ is a vector subspace of $\mathcal{N}_1 \times \mathcal{N}_2$ and check the equivalence stated in the definition above of a closed operator.

REMARK 9.3. The distinction between a continuous operator and a closed operator should be clear: a linear operator T is continuous if for $\xi_n \to \xi$ in $\operatorname{dom} T$, then necessarily $T\xi_n \to T\xi$, whereas for T be a closed operator, it is required that if both $(\xi_n) \subset \operatorname{dom} T$ and $(T\xi_n)$ are convergent, say $\xi_n \to \xi$, then necessarily ξ belongs to $\operatorname{dom} T$ and $T\xi_n \to T\xi$.

EXERCISE 9.3. For a linear operator $T : \operatorname{dom} T \subset \mathcal{N}_1 \to \mathcal{N}_2$, let $\pi_1 : \mathcal{G}(T) \to \operatorname{dom} T$ and $\pi_2 : \mathcal{G}(T) \to \operatorname{rng} T$ denote the natural projections $\pi_1(\xi, T\xi) = \xi$ and $\pi_2(\xi, T\xi) = T\xi$, for $\xi \in \operatorname{dom} T$. Prove that such projections are continuous linear operators.

DOI: 10.1201/9781003656166-9

It is important to establish conditions under which closed operators are continuous, as verifying that an operator is closed is often simpler. The Closed Graph Theorem, which will be presented later, shows that these two notions are equivalent for linear operators between Banach spaces. As a first step in this direction, consider the following result.

Proposition 9.4. *Every linear operator $T \in \mathrm{B}(\mathcal{B}_1, \mathcal{B}_2)$ is closed.*

Proof. Pick $\xi_n \to \xi$ with $T\xi_n \to \eta$. Since $\xi \in \mathrm{dom}\, T$ and T is continuous, then $T\xi_n \to T\xi = \eta$; hence T is a closed operator. $\qquad\square$

EXERCISE **9.4.** Check that any linear operator $T : \mathrm{dom}\, T \subset \mathcal{N}_1 \to \mathcal{N}_2$, whose domain $\dim \mathcal{N}_1 < \infty$, is closed.

Example 9.5. [Bounded and Nonclosed] Let $\mathbf{1} : \mathrm{dom}\, \mathbf{1} \to \mathcal{B}$ be the identity operator $\mathbf{1}(\xi) = \xi$, for $\xi \in \mathrm{dom}\, \mathbf{1}$, with $\mathrm{dom}\, \mathbf{1}$ a proper dense subspace of $\mathcal{B}$; such an operator is bounded. Now, pick $(\xi_n) \subset \mathrm{dom}\, \mathbf{1}$ with $\xi_n \to \xi \in \mathcal{B} \backslash \mathrm{dom}\, \mathbf{1}$. Since $\xi_n \to \xi$ and $\mathbf{1}(\xi_n) \to \xi$, but $\xi \notin \mathrm{dom}\, \mathbf{1}$, this operator is not closed. Although somewhat artificial, this example gives emphasis to the differences between bounded and closed linear operators. •

REMARK **9.6.** If $T \in \mathrm{B}(\mathcal{N}_1, \mathcal{B}_2)$ with $\mathcal{N}_1 \subset \mathcal{B}_1$, then its unique continuous linear extension $\overline{T} : \overline{\mathcal{N}} \to \mathcal{B}_2$ is a closed operator (Proposition 9.4). Thus, every continuous linear operator is essentially closed, making the artificial nature of Example 9.5 inevitable.

EXERCISE **9.5.** If $\mathcal{N} \subset \mathcal{B}$, show that $T \in \mathrm{B}(\mathcal{N}, \mathcal{B})$ is closed if and only if its domain $\mathcal{N}$ is a Banach space.

Example 9.7. [Unbounded and Closed] Let $\mathrm{C}^1[0, \pi] \subset \mathrm{C}[0, \pi]$ (both with the uniform convergence norm) be the subspace of continuously differentiable functions on $[0, \pi]$ and consider the linear operator $D : \mathrm{C}^1[0, \pi] \to \mathrm{C}[0, \pi]$, $(D\psi)(t) = \psi'(t)$. D is not continuous, since the sequence $\psi_n(t) = \sin(nt)/n \to 0$, whereas $(D\psi_n)(t) = \cos(nt)$ does not converge uniformly to zero. But note that this operator is closed. Indeed, if $\psi_n \to \psi$ and $D\psi_n = \psi'_n \to \varphi$, then, by taking into account that these limits are uniform,

$$\int_0^t \varphi(s)\,\mathrm{d}s = \int_0^t \lim_{n \to \infty} \psi'_n(s)\,\mathrm{d}s = \lim_{n \to \infty} \int_0^t \psi'_n(s)\,\mathrm{d}s = \psi(t) - \psi(0).$$

By the Fundamental Theorem of Calculus, $\psi \in \mathrm{dom}\, D = \mathrm{C}^1[0, \pi]$ and $(D\psi)(t) = \varphi(t)$, for every $t \in [0, \pi]$, and so D is closed. •

Example 9.8. [Unbounded and Nonclosed] Let $\mathrm{dom}\, T$ be the subspace of continuous functions in the Banach space $\mathrm{L}^1[-1, 1]$ and the operator action $(T\psi)(t) = \psi(0)$, for every t, as element of $\mathrm{L}^1[-1, 1]$. Since $\psi_n(t) = e^{-|t|n} \to 0$ in $\mathrm{L}^1[-1, 1]$, whereas $(T\psi_n)(t) = 1$, for every t and every n, consequently, the operator fails to be continuous and moreover, it is not closed. •

Finally:

Theorem 9.9 (Closed Graph). *Let $T : \mathcal{B}_1 \to \mathcal{B}_2$ be a linear operator (note that its domain is a Banach space). Then T is continuous if and only if T is closed.*

Proof. Proposition 9.4 presents one of the implications of the Closed Graph Theorem, so it remains to show that, under such hypotheses, if the linear operator T is closed, then it is bounded; the Open Mapping Theorem will be employed in the arguments.

Suppose that $\mathcal{G}(T)$ is closed in $\mathcal{B}_1 \times \mathcal{B}_2$, then $\mathcal{G}(T)$ is also a Banach space. The projection operators π_1 and π_2 (see Exercise 9.3) are linear and continuous. Furthermore, π_1 is a bijection between the Banach spaces $\mathcal{G}(T)$ and $\mathcal{B}_1$; thus, by the Open Mapping Theorem (Corollary 8.5), its inverse $\pi_1^{-1} : \mathcal{B}_1 \to \mathcal{G}(T)$ is continuous. Now, T is the composition

$$T = \pi_2 \circ \pi_1^{-1},$$

and so it follows that T is a continuous operator. □

Example 9.10. [Unbounded and Closed] For the Closed Graph Theorem to hold, it is required that the range of the operator be a complete space. Indeed, the operator $T^{-1} : \mathrm{rng}\, T \to l^1(\mathbb{N})$ in Exercise 8.2 has a closed graph but it is not continuous. ●

REMARK 9.11. It might seem that a linear operator fails to be closed simply because its domain is too small, and that extending it by considering the closure $\overline{\mathcal{G}(T)}$ in $\mathcal{N}_1 \times \mathcal{N}_2$ could yield a closed operator. However, this approach does not work in general, since $\overline{\mathcal{G}(T)}$ is not necessarily the graph of an operator; see Example 9.8 where the point $(0, 1)$ belongs to $\overline{\mathcal{G}(T)}$, but it is not of the form $(0, S0)$ for any linear operator S.

EXERCISE 9.6. Pick a subspace E of the Cartesian product $\mathcal{N}_1 \times \mathcal{N}_2$. Conclude that E is the graph of a linear operator if and only if E does not contain any element of the form $(0, \eta)$, with $\eta \neq 0$.

EXERCISE 9.7. Let $T : \mathrm{dom}\, T \subset \mathcal{B}_1 \to \mathcal{B}_2$ be linear and consider the *graph norm* of T, $\|\cdot\|_T$, given by

$$\|\xi\|_T := \|\xi\| + \|T\xi\|, \qquad \xi \in \mathrm{dom}\, T.$$

Show that if T is closed, then $(\mathrm{dom}\, T, \|\cdot\|_T)$ is a Banach space.

Definition 9.12. If $\overline{\mathcal{G}(T)}$ is the graph of a linear extension $\overline{T}$ of a linear operator T, then T is said to be a *closable operator* and $\overline{T}$ is its *closure*.

Definition 9.13. If T, S are closed linear operators on $\mathcal{N}$, then T is called S-bounded if $\mathrm{dom}\, S \subset \mathrm{dom}\, T$ and there exist $a, b \geq 0$ such that

$$\|T\xi\| \leq a\|S\xi\| + b\|\xi\|, \qquad \xi \in \mathrm{dom}\, S;$$

the infimum of the $a \geq 0$ such that this relation holds is the *S-bound of T* (in general b is a nondecreasing function of a).

EXERCISE **9.8.** (a) If $||| \cdot |||$ is a norm on $\mathcal{X} = \mathrm{B}(\mathcal{N}, \mathcal{B})$ for which $(\mathcal{X}, ||| \cdot |||)$ is Banach and $|||T_n||| \to 0$ implies strong convergence of $T_n \to 0$ (see Definition 14.5), show that $||| \cdot |||$ is equivalent to the usual norm on $\mathrm{B}(\mathcal{N}, \mathcal{B})$.

(b) Verify that $|||\xi|||_\infty := \max\{\|\xi\|, \|T\xi\|\}$ and, for each $1 < p < \infty$,

$$|||\xi|||_p := (\|\xi\|^p + \|T\xi\|^p)^{1/p},$$

are norms equivalent to the corresponding graph norm of $T : \mathrm{dom}\, T \subset \mathcal{N}_1 \to \mathcal{N}_2$ (see Exercise 9.7).

In some settings this concept allows one to treat S as a perturbation of T, even for closed and unbounded operators, as in the well-known Kato-Rellich Theorem [de Oliv]. Next, there is a condition guaranteeing that T is S-bounded on Banach spaces.

Proposition 9.14. *If T, S are closed linear operators on the Banach space $\mathcal{B}$, with $\mathrm{dom}\, S \subset \mathrm{dom}\, T$, and there exists $\lambda \in \mathbb{C}$ such that $S - \lambda \mathbf{1}$ is invertible in $\mathrm{B}(\mathcal{B})$, then T is S-bounded.*

Proof. Since $\mathrm{dom}\, S \subset \mathrm{dom}\, T$ and T is closed, $T(S - \lambda\mathbf{1})^{-1}$ is defined on all $\mathcal{B}$ and is closed (verify!). So, by the Closed Graph Theorem, this operator is bounded and there exists $a \geq 0$ such that

$$\|T(S - \lambda\mathbf{1})^{-1}\eta\| \leq a\|\eta\|$$

for all $\eta \in \mathcal{B}$.

If $\xi \in \mathrm{dom}\, S$, then $\xi = (S - \lambda\mathbf{1})^{-1}\eta$ for some $\eta \in \mathcal{B}$; hence, $\eta = (S - \lambda\mathbf{1})\xi$ and

$$\|T\xi\| \leq a\|\eta\| = a\|(S - \lambda\mathbf{1})\xi\| \leq a\|S\xi\| + a|\lambda|\|\xi\|,$$

for all $\xi \in \mathrm{dom}\, S$, and so T is S-bounded with $b = a|\lambda|$. $\qquad\square$

EXERCISE **9.9.** If S is a closed linear operator on $\mathcal{B}$ and $T \in \mathrm{B}(\mathcal{B})$, show that T is S-bounded and with null S-bound.

Notes

The notion of closed operator was introduced by J. von Neumann, around 1930. A proof of how the Open Mapping Theorem follows from the Closed Graph Theorem appears in Chapter 23.

The introduction of the Graph Norm (Exercise 9.7) allows one to study closed operators "as bounded ones" (an idea of Friedrichs), and this is another way of seeing that in the world of unbounded linear operators the closed ones are, in some sense, near to the set of continuous operators.

An operator T that is S-bounded is also said to be *relatively bounded* with respect to S. The Kato-Rellich Theorem is one of the main tools used to show that certain perturbations of self-adjoint operators in Hilbert spaces are also self-adjoint; in such context, the concept of a relatively bounded operator, with $a < 1$, plays a key role [de Oliv].

Additional Exercises

EXERCISE **9.10.** Let dom T and dom S be the sets of $\psi \in L^2[-1,1]$ satisfying, respectively, $\psi(t) = 0$ in a neighborhood of zero and $\psi(t)/t \in L^2[-1,1]$, with

$$(T\psi)(t) = (S\psi)(t) = \frac{\psi(t)}{t}, \qquad \forall\, t \in [-1,1],$$

for ψ in the appropriate domains. Show that T is not closed, while S is a closed operator.

EXERCISE **9.11.** Let $T : \mathcal{N}_1 \to \mathcal{N}_2$ be a linear closed operator. Show that:

(a) If T^{-1} exists, then it is also closed.

(b) If $S \in B(\mathcal{N}_1, \mathcal{N}_2)$, then $T + S$ is closed.

(c) If $\mathcal{N}_1 = \mathcal{N}_2 = \mathcal{N}$, then for all $\lambda \in \mathbb{F}$, the kernel $N(T - \lambda\mathbf{1})$ is a closed vector subspace of $\mathcal{N}$.

EXERCISE **9.12.** Show that the identity operator

$$\mathbf{1} : (C[0,1], \|\cdot\|_1) \to (C[0,1], \|\cdot\|_\infty), \qquad \psi \overset{\mathbf{1}}{\longmapsto} \psi,$$

is closed, however it is not continuous. Conclude that $(C[0,1], \|\cdot\|_1)$ is not complete (recall that $\|\psi\|_1 = \int_0^1 |\psi(t)|\, dt$ and $\|\cdot\|_\infty$ is the norm of uniform convergence).

EXERCISE **9.13.** Use the Closed Graph Theorem to prove Corollary 8.5.

EXERCISE **9.14.** Fix $s > 0$. Show that the linear operator $T : \text{dom}\, T \to C[0,\infty)$, with

$$\text{dom}\, T = \{\psi \in C[0,\infty) : \exists A > 0 \text{ with } |\psi(t)| \le A/(1+t^s), \forall t \ge 0\},$$

and action $(T\psi)(t) = t^s \psi(t)$ is closed and unbounded.

EXERCISE **9.15.** Use the Closed Graph Theorem to show that if $T : l^p(\mathbb{N}) \hookleftarrow, 1 \le p \le \infty$, is linear and commutes with the shift operator

$$S_r(\xi_1, \xi_2, \cdots) = (0, \xi_1, \xi_2, \cdots),$$

then T is bounded.

EXERCISE **9.16.** Let T, S be closed linear operators in $\mathcal{B}$ with dom $S \subset$ dom T; consider dom S with the graph norm (Exercise 9.7). Show that T is S-bounded if and only if $T : \text{dom}\, S \to \mathcal{B}$ is bounded.

EXERCISE **9.17.** If T, S are closed linear operators in $\mathcal{B}$ with dom $S \subset$ dom T, show that T is S-bounded if and only if there exist $c, d \ge 0$ such that

$$\|T\xi\|^2 \le c^2 \|S\xi\|^2 + d^2 \|\xi\|^2, \qquad \forall \xi \in \text{dom}\, S.$$

Verify also that the infimum of $c \ge 0$ for which this relation holds coincide with the S-bound of T.

EXERCISE **9.18.** If T, S are closed linear operators in $\mathcal{N}$ and T is S-bounded with S-bound h, show that if $h < 1$ then T is $(T+S)$-bounded with $(T+S)$-bound $h/(1-h)$. Conclude that if $h < 1/2$, then the $(T+S)$-bound of T is less than 1.

CHAPTER 10

Hahn-Banach Theorem

The Hahn-Banach Theorem is discussed. It is is an extension result, of functionals defined on subspaces, to the entire vector space; note that no topology is mentioned. Later on, this theorem will be used to give affirmative answers to the following questions: Given a normed space, are there nonzero continuous linear functionals defined on such space? Are there a sufficient number of continuous functionals to distinguish points of the underlining space? Is there a (nonzero) continuous functional that vanishes on a given proper closed subspace?

Important applications, such as the notion of adjoint operator and weak convergences will be dealt with in other chapters; these are among the most important sets of results in classical Functional Analysis. The proof of Hahn-Banach theorem will be the object of the next chapter. An important tool in this context, which deals with families of infinite sets, is Zorn's Lemma.

10.1   MAX ZORN'S LEMMA

Definition 10.1. A *partial ordering* in a set X is a binary relation $\prec$ in X which is reflexive ($\xi \prec \xi$, for all $\xi \in X$), transitive ($\xi \prec \eta$ and $\eta \prec \zeta \Rightarrow \xi \prec \zeta$) and antisymmetric ($\xi \prec \eta$ and $\eta \prec \xi \Rightarrow \xi = \eta$).

REMARK **10.2.** (a) The term "partial" appears due to the possibility of noncomparable elements with respect to the given ordering $\prec$.

(b) If $\prec$ is a partial ordering in X, then $(X, \prec)$ is called a *partially ordered set*.

Definition 10.3. A *totally ordered set* is a partially ordered set for which any two elements are comparable with respect to the given partial ordering.

Definition 10.4. Let $(X, \prec)$ be a partially ordered set. $\zeta \in X$ is a *maximal element* in X if for any $\xi \in X$ with $\zeta \prec \xi$, it follows that $\xi = \zeta$. An element $\eta \in X$ is a *superior limit* of $Y \subset X$ if $\xi \prec \eta$, for all $\xi \in Y$.

The next example is standard and nicely illustrates the above definitions.

Example 10.5. Denote by $\mathcal{P}(X)$ the collection of subsets of X. The inclusion of sets, that is, if $A, B \in \mathcal{P}(X)$, then $A \prec B$ if and only if $A \subset B$, defines a partial ordering in $\mathcal{P}(X)$. The unique maximal element of $\mathcal{P}(X)$ in this case is X.   •

DOI: 10.1201/9781003656166-10

Example 10.6. The set of real numbers $\mathbb{R}$ with the usual ordering $\leq$ is totally ordered; note that in this case $\mathbb{R}$ has no maximal element. •

Axiom 10.7 (Zorn's Lemma). *A nonempty partially ordered set, for which every totally ordered subset has a superior limit, has a maximal element.*

REMARK **10.8.** The term "Lemma" appears for historical reasons. Zorn's Lemma is equivalent to the Axiom of Choice, which for many people is intuitively acceptable, at least in comparison with Zorn's Lemma. The Cartesian product of an arbitrary family of sets $\{X_t\}_{t \in J}$, denoted by $\prod_{t \in J} X_t$, is the set of functions $\psi : J \to \bigcup_{t \in J} X_t$, with $\psi(t) \in X_t$ for all $t \in J$; each ψ is also called a *choice function*. The Axiom of Choice was proposed by E. Zermelo in the beginning of the XX century, and it states that "The Cartesian product of any family of nonempty sets is nonempty." In 1963, P. J. Cohen showed that this axiom is independent of the other axioms of Set Theory. The notion of Cartesian product of an arbitrary family of sets will appear in other occasions in this book.

The next result illustrates how usually Zorn's Lemma is applied.

Proposition 10.9. *Every nontrivial vector space (i.e., which contains at least one nonzero element) has a Hamel basis.*

Proof. Let $X \neq \{0\}$ be a vector space and V the collection of all linearly independent subsets of X. Clearly $V \neq \emptyset$ and the inclusion of sets defines a partial ordering in V. The union of all elements in a totally ordered subset of V is its superior limit (which is also linearly independent; verify!). By Zorn's Lemma, V has a maximal element M. It will be checked that M is a Hamel basis of X.

Set $W = \text{Lin}(M)$; then $W \subset X$. If $W \neq X$ there is $0 \neq \xi \in X \backslash W$ and so $M \cup \{\xi\}$ would be a linearly independent set that contains properly M; this contradicts its maximality. Therefore, $W = X$ and M is a Hamel basis of X. □

10.2 HAHN-BANACH

It will be presented two traditional versions of Hahn-Banach theorem, one for real and other for complex vector spaces; observe that some of their applications, on existence of extensions of continuous linear functionals in normed spaces, are sometimes also referred to as Hahn-Banach. The concept of sublinear functional will be central.

Definition 10.10. A *sublinear functional* on a vector space X is a mapping $p : X \to \mathbb{R}$ that satisfies, for all $\xi, \eta \in X$,

$$\begin{aligned} p(\xi + \eta) &\leq p(\xi) + p(\eta) \quad \text{(subadditivity)} \\ p(\alpha\xi) &= \alpha p(\xi), \quad \alpha \geq 0. \end{aligned}$$

Example 10.11. A norm, as well as a seminorm, on a vector space are sublinear functionals. •

Theorem 10.12 (Hahn-Banach (real)). *Let X be a real vector space and $p : X \to \mathbb{R}$ a sublinear functional. If $f : Z \to \mathbb{R}$ is a linear functional defined on the subspace $Z \subset X$ and dominated by p, that is,*

$$f(\zeta) \leq p(\zeta), \qquad \forall \zeta \in Z,$$

then f has a linear extension $F : X \to \mathbb{R}$ that is also dominated by p, that is,

$$F(\xi) \leq p(\xi), \qquad \forall \xi \in X.$$

F is said to be a Hahn-Banach extension of f.

Theorem 10.13 (Hahn-Banach (complex)). *Let X be a vector space (real or complex) and $p : X \to [0, \infty)$ satisfying*

$$
\begin{aligned}
p(\xi + \eta) &\leq p(\xi) + p(\eta), & \forall \xi, \eta \in X, \\
p(\alpha \xi) &= |\alpha| p(\xi), & \forall \xi \in X, \forall \alpha \in \mathbb{F}.
\end{aligned}
$$

If $f : Z \to \mathbb{F}$ is a linear functional defined on the subspace $Z \subset X$ (X, Z and f over the same field) with $|f(\zeta)| \leq p(\zeta)$, for all $\zeta \in Z$ (i.e., f is dominated by p), then f has a linear extension $F : X \to \mathbb{F}$ dominated by p, that is,

$$|F(\xi)| \leq p(\xi), \qquad \forall \xi \in X.$$

F is said to be a Hahn-Banach extension of f.

REMARK **10.14.** (a) Note that Theorem 10.13 is also applicable to the case of real functionals.

(b) In the real version of Hahn-Banach the functional $p(\xi)$ is not necessarily positive for all ξ; see Exercise 10.6.

(c) The main point in Hahn-Banach theorem is not the simple existence of an extension of a linear functional, but the existence of a linear extension dominated by p. For example, if $\{e_j\}$ is a Hamel basis of Z and $\{e_j\} \cup \{g_k\}$ a Hamel basis of X, then for any collection of scalars $\{\alpha_k\}$, the functional F defined by

$$F\Big(\sum_j a_j e_j + \sum_k b_k g_k\Big) = f\Big(\sum_j a_j e_j\Big) + \sum_k b_k \alpha_k$$

is a linear extension of f to all X.

An important consequence of Hahn-Banach Theorem is the existence of extensions of bounded functionals defined on subspaces of normed spaces. Often this result is also stated as "the Hahn-Banach Theorem."

Corollary 10.15. *Let M be a subspace of $\mathcal{N}$ (both over the same field). Then every f in M^*, the dual space of M, has an extension $F \in \mathcal{N}^*$ with $\|F\| = \|f\|$, that is, F is a norm-preserving extension.*

Proof. Consider the sublinear functional $p : \mathcal{N} \to \mathbb{R}$, $p(\xi) = \|f\|\|\xi\|$, which satisfies $|f(\eta)| \leq p(\eta)$, for all $\eta \in M$. By Hahn-Banach theorem, there exists an extension $F \in \mathcal{N}^*$ of f with $|F(\xi)| \leq p(\xi) = \|f\|\|\xi\|$, for all $\xi \in \mathcal{N}$; it then follows that $\|F\| \leq \|f\|$. Since F is an extension of f, one has $\|f\| \leq \|F\|$ and therefore, $\|f\| = \|F\|$. $\square$

Example 10.16. In $\mathbb{F}^2$ with the norm $\|(\xi_1, \xi_2)\|_\infty = \max\{|\xi_1|, |\xi_2|\}$, consider $X = \mathrm{Lin}(\{(1,1)\})$; then the functional $f : X \to \mathbb{F}$, $f(\xi_1, \xi_1) = \xi_1$ has the extensions $g(\xi_1, \xi_2) = \xi_1$ and $h(\xi_1, \xi_2) = \xi_2$ satisfying the conclusions of Corollary 10.15. Compare with Example 11.5. See the Notes in Chapter 11 for a remark on the uniqueness of the Hahn-Banach extension. •

Example 10.17. In the complex case, the conclusions of Corollary 10.15 may fail if the vector subspace M is real. It will be shown that in any complex infinite dimensional normed space $\mathcal{N}$, there is a real vector subspace M on which there is a real-valued complex and bounded linear functional $f : M \to \mathbb{R}$ which admits no complex bounded linear extension to the whole space $\mathcal{N}$.

Given an infinite-dimensional complex normed space $\mathcal{N}$, consider a normalized and linearly independent sequence $(\xi_1, \eta_2, \eta_3, \cdots)$ in this space. Now, for $j \geq 2$ define

$$\zeta_j = \frac{\eta_j - j\xi_1}{\|\eta_j - j\xi_1\|},$$

and note that $(\xi_1, \zeta_2, \zeta_3, \cdots)$ is also normalized and linearly independent. In what follows r_j will always denote real numbers, and ahead $(\theta_j)_{j=1}^\infty$ and complex $(\xi_j := e^{i\theta_j}\zeta_j)_{j=1}^\infty$ sequences will be constructed real, with $\theta_1 = 0$ and $\zeta_1 = \xi_1$, such that for all n the *key condition*

$$|r_1| \leq \|r_1\xi_1 + r_2\xi_2 + \cdots + r_n\xi_n\|, \qquad \forall r_1, \cdots, r_n \in \mathbb{R},$$

is satisfied.

Suppose that such sequences were constructed and let M be the vector subspace obtained by real linear combinations of $(\xi_j)_{j=1}^\infty$ and $f : M \to \mathbb{C}$ given by $f(r_1\xi_1 + r_2\xi_2 + \cdots + r_n\xi_n) = r_1$; then, f is linear, by the key condition one has $\|f\| \leq 1$ and since $f(\xi_1) = 1$, it follows that $\|f\| = 1$. However, if $F : \mathcal{N} \to \mathbb{C}$ is a complex linear extension of f, then

$$F(\eta_j) = F\left(j\xi_1 + \|\eta_j - j\xi_1\|e^{-i\theta_j}\xi_j\right) = j, \qquad \forall j \geq 2,$$

and F is not bounded.

To finish the example it is only remains to construct $(\theta_j)_{j=2}^\infty$ and verify the key condition, what will be done by induction. This condition is trivial for $n = 1$; suppose that $\theta_1, \cdots, \theta_{n-1}$ (and so $\xi_1, \cdots, \xi_{n-1}$) were constructed. Let Λ_n be the region in the complex plane given by $z \in \mathbb{C}$ so that there exist $r_2, \cdots, r_{n-1}$ with

$$\|\xi_1 + r_2\xi_2 + \cdots + r_{n-1}\xi_{n-1} + z\zeta_n\| < 1,$$

which is an open and convex set; since the origin does not belong to Λ_n, there is a line $\{z(r) = re^{i\theta_n} : r \in \mathbb{R}\}$ in $\mathbb{C}$, determined by some fixed θ_n, whose intersection with Λ_n

is empty (so, on this line the above inequality does not hold). By defining $\xi_n = e^{i\theta_n}\zeta_n$, one has $re^{i\theta_n}\zeta_n = r\xi_n$. Thus, if $r_1 = 0$ the key condition is straightforward, and if $r_1 \neq 0$, by denoting $s_j = r_j/r_1$ it is found, by the construction of θ_n, that for any $r_2, \cdots, r_n$ the following relation holds

$$\|r_1\xi_1 + r_2\xi_2 + \cdots + r_n\xi_n\| = |r_1| \, \|\xi_1 + s_2\xi_2 + \cdots + s_n\xi_n\| \geq |r_1|,$$

which is exactly the key condition for $r_1 \neq 0$. ●

EXERCISE **10.1.** Verify that the region Λ_n in Example 10.17 is convex.

Notes

Paul J. Cohen's proof that the Axiom of Choice is independent of others such as Zermelo-Fraenkel axioms of Set Theory, including the continuum hypothesis, was stated in [Cohen1, Cohen2]. There is a famous construction (which appears in most texts on Measure and Integration) of Lebesgue nonmeasurable sets, due to Vitali, which makes use of the Axiom of Choice; if this "axiom" is not supposed to be valid, R. Solovay has shown that there is a mathematical model in which every subset of $\mathbb{R}$ is Lebesgue measurable. An excellent discussion about the Axiom of Choice and its main equivalences can be found in the book [LinLin].

Based on the Hahn-Banach's Theorem, one could imagine that it is always possible to find (nonzero) continuous linear functionals on vector spaces, in case it is assumed that the sum of operators and scalar multiplication were continuous in the given topology, i.e., the so-called Topological Vector Spaces; as mentioned at the beginning of this chapter, and showed in Chapter 12, this holds in normed spaces, but can fail in general. In 1941, La Salle [LaSalle] showed that such functionals exist if and only if the space has an open set, containing the origin, which is convex and distinct from all space; one year before that publication, M. M. Day [Day] had already shown that the unique continuous linear functional on L^p, with $0 < p < 1$, is the null functional.

It is natural to wonder about the extensions of continuous linear operators previously defined on vector subspaces of $\mathcal{B}_1$ taking values in $\mathcal{B}_2$. If $\mathcal{B}_2 = \mathbb{F}$ it is the Hahn-Banach Theorem, but in this more general setting there is no far-reaching result, although it exists if $\dim \mathcal{B}_2 < \infty$ or if $\mathcal{B}_1$ is a Hilbert space; see page 14 of [Brezis].

The complex version of Hahn-Banach Theorem was published in 1938 by H. F. Bohnenblust and A. Sobczyk [BohSob]; the construction in Example 10.17 had also appeared there. Finally, a detailed description of Hahn-Banach Theorem and generalizations can be found in the nice article [Buskes] (I thank Prof. Buskes for sending me a copy of his work).

Additional Exercises

EXERCISE **10.2.** Show that the set of complex numbers $\mathbb{C}$ becomes partially ordered with the relation $\zeta \prec \eta$ defined by $\text{Re}\,\zeta \leq \text{Re}\,\eta$ and $\text{Im}\,\zeta \leq \text{Im}\,\eta$.

EXERCISE **10.3.** Show that a partially ordered set $(X, \prec)$ can have at most one *minimum element*, i.e., a $\eta \in X$ such that $\eta \prec \xi$, for all $\xi \in X$. Idem for *maximum element* (definition left to the reader).

EXERCISE **10.4.** Use Proposition 10.9 to show that all vector spaces X admit a norm.

EXERCISE **10.5.** This exercise illustrates how ordering can be related to the Axiom of Choice. A totally ordered set X is *well-ordered* if all $\emptyset \neq A \subset X$ contain a minimum element in A (see Exercise 10.3). Zorn's Lemma is also equivalent to Zermelo's Well-Ordering Theorem: "Any set admits an ordering for which it becomes well-ordered." Use the following construction to show that the Axiom of Choice follows from such fact: Given a family of nonempty sets $\{X_t\}_{t\in J}$, let $X = \bigcup_{t\in J} X_t$, which can be well-ordered with, say, $\prec$; show that $\psi(t)$ defined as the minimum element of X_t, with respect to this ordering $\prec$, is a choice function for $\{X_t\}_{t\in J}$.

EXERCISE **10.6.** Show that $p : l^\infty(\mathbb{N})(\text{real}) \to \mathbb{R}$ defined by

$$p(\xi) = \limsup_{n\to\infty} \xi_n,$$

with $\xi = (\xi_1, \xi_2, \xi_3, \cdots)$, is a sublinear functional. Note that this is an example of sublinear functional p which is not always positive.

EXERCISE **10.7.** Let p be as in the complex Hahn-Banach Theorem. Show that

$$|p(\xi) - p(\eta)| \le p(\xi - \eta), \qquad \forall \xi, \eta \in X.$$

EXERCISE **10.8.** Show that if a sublinear functional in a normed space $\mathcal{N}$ is continuous at $\xi = 0$, then it is continuous at every point of $\mathcal{N}$.

EXERCISE **10.9.** If the normed space $\mathcal{N} \ne \{0\}$, show that its dual $\mathcal{N}^* \ne \{0\}$.

EXERCISE **10.10.** Verify the following properties of a sublinear functional $p : X \to \mathbb{R}$: (a) $p(0) \ge 0$; (b) $\max\{p(\xi), p(-\xi)\} \ge 0$, for all $\xi \in X$; (c) $q(\xi) := \sup_{|\alpha|=1} p(\alpha\xi)$ is a seminorm; (d) if $\{p_j\}$ is a family of sublinear functionals such that $\{p_j(\xi)\}$ is bounded for all $\xi \in X$, then $p(\xi) := \sup_j p_j(\xi)$ is also a sublinear functional.

CHAPTER 11

Proof of Hahn-Banach Theorem

The main objective of this chapter is to establish a proof of the Hahn-Banach Theorem. In accordance with traditional approaches, the proof will first be developed for the real case, which will subsequently be utilized to derive the complex version. These proofs rely on the concept of sublinear functionals, which were introduced by Banach and play a role in the general argument. The connection between the real and complex cases will be established through a key lemma, which provides a relationship between real and complex functionals. This lemma is not only essential for the proof but is also of independent mathematical interest.

Proof. [**Hahn-Banach (real)**] Let $G = \{g_t\}$ be the collection of linear extensions $g_t : Z_t \to \mathbb{R}$ of f (with $Z \subset Z_t \subset X$, for all t) such that $g_t(\xi) \leq p(\xi)$ on their respective domains Z_t. Since $f \in G$, then $G \neq \emptyset$, and it is partially ordered with $g_t \prec g_s$ if $Z_t \subset Z_s$ and $g_t(\xi) = g_s(\xi)$ for $\xi \in Z_t$.

If $\{g_t\}_{t \in J}$ is a totally ordered subset of G, then $\Lambda = \bigcup_{t \in J} Z_t$ is also a vector subspace since it the union of an increasing family of vector subspaces, and the functional

$$g : \Lambda \to \mathbb{R}, \quad g(\xi) := g_t(\xi) \qquad \text{if} \qquad \xi \in Z_t,$$

is well defined and furthermore, $g(\xi) \leq p(\xi)$, for all $\xi \in \Lambda$. Since for all $t \in J$ one has $g_t \prec g$, it is concluded that a totally ordered family in G has an upper bound and, by Zorn Lemma, G has a maximal element F defined on a certain subspace $X_0 \subset X$ and satisfying $F(\xi) \leq p(\xi)$ if $\xi \in X_0$. Now it will be shown that $X_0 = X$, for otherwise a contradiction with the maximality of F would be found.

If $X_0 = X$ there is nothing to prove, so suppose that there exists $0 \neq \eta \in X \backslash X_0$. It is enough to show that in this case would exist a linear extension $\tilde{F}$ of F to the vector space $Y = \text{Lin}(X_0 \cup \{\eta\})$ and with $\tilde{F} \leq p$.

Denote by $\tilde{F}$ a general linear extension of F to Y (which always exists; it is enough to assign a value to $\tilde{F}(\eta)$; see Remark 10.14). Since each ξ in Y can be written as $\xi = \zeta + \alpha\eta$, for some $\alpha \in \mathbb{R}$ and $\zeta \in X_0$, then for each element $\xi \in Y$ one has

$$\tilde{F}(\xi) = \tilde{F}(\zeta + \alpha\eta) = F(\zeta) + \alpha\tilde{F}(\eta), \qquad \forall \zeta \in X_0, \alpha \in \mathbb{R};$$

DOI: 10.1201/9781003656166-11

the task now is to show that it is possible to choose $\tilde{F}(\eta)$ in an appropriate way so that $\tilde{F} \leq p$. If $\zeta_1, \zeta_2 \in X_0$, then

$$F(\zeta_1) + F(\zeta_2) = F(\zeta_1 + \zeta_2) \leq p(\zeta_1 + \zeta_2) \leq p(\zeta_1 - \eta) + p(\zeta_2 + \eta),$$

or

$$F(\zeta_1) - p(\zeta_1 - \eta) \leq p(\zeta_2 + \eta) - F(\zeta_2).$$

Therefore, there exists $\lambda \in \mathbb{R}$ so that

$$\sup_{\zeta \in X_0} \{F(\zeta) - p(\zeta - \eta)\} \leq \lambda \leq \inf_{\zeta \in X_0} \{p(\zeta + \eta) - F(\zeta)\}.$$

By using the above construction, define $\tilde{F} : Y \to \mathbb{R}$ as a linear extension of F with $\tilde{F}(\eta) = \lambda$. To verify that for all $\xi \in Y$ one has $\tilde{F}(\xi) \leq p(\xi)$, the representation $\xi = \zeta + \alpha\eta$ will be used again. If $\alpha > 0$, then

$$\tilde{F}(\xi) = F(\zeta) + \alpha\lambda \leq F(\zeta) + \alpha \left[p\left(\frac{\zeta}{\alpha} + \eta\right) - F\left(\frac{\zeta}{\alpha}\right) \right] = p(\zeta + \alpha\eta),$$

while for $\alpha < 0$ (note that the case $\alpha = 0$ is trivial),

$$
\begin{aligned}
\tilde{F}(\zeta + \alpha\eta) &= \tilde{F}(\zeta - |\alpha|\eta) = F(\zeta) - |\alpha|\lambda \\
&\leq F(\zeta) - |\alpha| \left[F(\zeta/|\alpha|) - p(\zeta/|\alpha| - \eta) \right] \\
&= |\alpha| \, p(\zeta/|\alpha| - \eta) = p(\zeta + \alpha\eta).
\end{aligned}
$$

Therefore, $\tilde{F}(\xi) \leq p(\xi)$ for each $\xi \in Y$, and a contradiction with the maximality of F (if $X_0 \neq X$) is obtained. Hereby Theorem 10.12 has been proven. □

Note that the above proof of the real case made use of the natural ordering in $\mathbb{R}$, a structure missing in $\mathbb{C}$.

Definition 11.1. A *real linear functional* on a complex vector space X is a functional $h : X \to \mathbb{R}$, so that $h(\xi + \eta) = h(\xi) + h(\eta)$ and $h(\alpha\xi) = \alpha h(\xi)$, for all $\xi, \eta \in X$ and all $\alpha \in \mathbb{R}$.

Lemma 11.2. *Let X be a complex vector space.*

(i) If $h : X \to \mathbb{R}$ is a real linear functional, then $f : X \to \mathbb{C}$, given by

$$f(\xi) = h(\xi) - ih(i\xi), \qquad \xi \in X,$$

is a complex linear functional.

(ii) If $f : X \to \mathbb{C}$ is a complex linear functional, then there exists a real linear functional $h : X \to \mathbb{R}$ so that

$$f(\xi) = h(\xi) - ih(i\xi), \qquad \xi \in X.$$

In both cases one has $h = \mathrm{Re}\, f$.

Proof. The motivation for the construction in this proof comes from the following remark. If $g : X \to \mathbb{C}$ is a linear functional, then $g(i\xi) = \text{Re } g(i\xi) + i\text{Im } g(i\xi) = ig(\xi) = i\text{Re } g(\xi) - \text{Im } g(\xi)$, that is, the relation $\text{Im } g(\xi) = -\text{Re } g(i\xi)$ is valid.

(i) Given a real linear functional h define $f : X \to \mathbb{C}$, $f(\xi) = h(\xi) - ih(i\xi)$. To show that f is complex linear it is sufficient to check that $f(i\xi) = if(\xi)$, since h is real linear. Explicitly:

$$f(i\xi) = h(i\xi) - ih(-\xi) = h(i\xi) + ih(\xi) = if(\xi).$$

(ii) Given a complex linear functional f, set $h = \text{Re } f$, which is clearly real linear; by the above motivation,

$$\text{Im } f(\xi) = -\text{Re } f(i\xi) = -h(i\xi),$$

and so $f(\xi) = h(\xi) - ih(i\xi)$. □

Proof. [**Hahn-Banach (complex)**] Put $h = \text{Re } f$, which is a real linear functional on Z. From the linearity of h it follows that for all $\zeta \in Z$ (with adequate choices of sign)

$$|h(\zeta)| = \pm h(\zeta) = h(\pm\zeta),$$

and since $p(-\zeta) = p(\zeta)$, it is found that $h(\zeta) \le p(\zeta)$ is equivalent to $|h(\zeta)| \le p(\zeta)$. Thus, since

$$h(\zeta) \le |f(\zeta)| \le p(\zeta), \qquad \zeta \in Z,$$

by the real Hahn-Banach (if X is complex, consider X and Z restricted to the multiplication by real scalars, so that they become real vector spaces) there exists a real linear extension $H : X \to \mathbb{R}$ of h with $H(\xi) \le p(\xi)$, for all $\xi \in X$. If f is real the proof has finished. Suppose then that f is complex; by Lemma 11.2, $f(\zeta) = h(\zeta) - ih(i\zeta)$.

Define $F : X \to \mathbb{C}$ by $F(\xi) = H(\xi) - iH(i\xi)$, that clearly extends f and, by Lemma 11.2, is complex linear. To complete the proof it is enough to show that $|F(\xi)| \le p(\xi)$, for all $\xi \in X$. If $F(\xi) = 0$ such property is clear, since $p(\xi) \ge 0$. If $F(\xi) \ne 0$, then there exists $0 \le \theta < 2\pi$ so that $F(\xi) = e^{i\theta}|F(\xi)|$; the linearity of F implies that

$$|F(\xi)| = F(e^{-i\theta}\xi) = \text{Re } (F(e^{-i\theta}\xi)) = H(e^{-i\theta}\xi) \le p(e^{-i\theta}\xi) = p(\xi),$$

and Theorem 10.13 is proven. □

Now a short discussion about the uniqueness of the Hahn-Banach extension is presented.

Example 11.3. Let Z be a subspace of the vector space X, p and $f : Z \to \mathbb{C}$, as in the Hahn-Banach Theorem. If there exist two distinct Hahn-Banach extensions $F_0, F_1 : X \to \mathbb{C}$ of f, then for any $s \in [0, 1]$ consider the functional F_s given by

$$F_s(\xi) := sF_1(\xi) + (1 - s)F_0(\xi), \qquad \xi \in X.$$

Note that F_s is linear and for all $\zeta \in Z$ one has

$$F_s(\zeta) = sF_1(\zeta) + (1-s)F_0(\zeta) = sf(\zeta) + (1-s)f(\zeta) = f(\zeta),$$

and so F_s is an extension of f. Since $|F_0(\xi)| \le p(\xi)$ and $|F_1(\xi)| \le p(\xi)$, it follows that

$$|F_s(\xi)| \le s|F_1(\xi)| + (1-s)|F_0(\xi)| \le p(\xi),$$

and F_s, for each $s \in [0,1]$, is in fact a Hahn-Banach extension of f. Hence, there are infinitely many of such extensions. ●

A normed space is *strictly convex* if $\|(\xi + \eta)/2\| < 1$ for all $\xi \ne \eta$ with $\|\xi\| = \|\eta\| = 1$; $l^p(\mathbb{N})$ is strictly convex if and only if $1 < p < \infty$ (see Exercise 11.8). The Hilbert spaces, discussed in Chapter 17, are distinguished examples of strictly convex spaces. From the geometric point of view, $\mathcal{N}$ is strictly convex if the midpoint, of any line segment joining two distinct points on the unit sphere $S(0;1)$, is in the interior of the unit ball $\overline{B}(0;1)$; indeed, the unit radius is immaterial, and there are similar interpretations for any sphere $S(0;r)$.

Proposition 11.4. *If $\mathcal{N}^*$ is strictly convex, then every norm-preserving linear Hahn-Banach extension (as in Corollary 10.15) is unique.*

Proof. If M is a subspace of $\mathcal{N}$, $f \in M^*$, and let $F_1, F_2 \in \mathcal{N}^*$ be two norm-preserving extensions of f. Since $(F_1 + F_2)/2 \in \mathcal{N}^*$ is another linear extension of f, one has $\|f\| \le \|(F_1 + F_2)/2\|$. Now,

$$\left\|\frac{1}{2}(F_1+F_2)\right\| \le \frac{1}{2}(\|F_1\| + \|F_2\|) = \|f\|,$$

and so $\|(F_1+F_2)/2\| = \|f\|$, that its, $(F_1+F_2)/2$ is a norm-preserving extension of f as well. Since $\mathcal{N}^*$ is strictly convex, it follows that $F_1 = F_2$. □

Example 11.5. Let $\mathcal{N} = \mathbb{R}^2$ with the Euclidean norm, $a \in \mathbb{R}$, $Z_a := \{(\xi_1, \xi_2) \in \mathbb{R}^2 : \xi_2 = a\xi_1\}$, and the bounded linear functional $f : Z_a \to \mathbb{R}$ given by

$$f(\xi_1, \xi_2) = \xi_1, \qquad \forall(\xi_1, \xi_2) \in Z_a.$$

Note that $|f(\xi_1, \xi_2)|^2 = |\xi_1|^2 = \frac{1}{1+a^2}\|(\xi_1, \xi_2)\|^2$, so that $\|f\| = 1/\sqrt{1+a^2}$. Let $F : \mathbb{R}^2 \to \mathbb{R}$ be a norm-preserving Hahn-Banach extension of f as in Corollary 10.15. It will be checked that such extension is unique and F will be explicitly found; compare with Example 10.16.

Write $c = F(1,0), d = F(0,1)$, and note that $F(\xi_1, \xi_2) = c\xi_1 + d\xi_2$, for all $(\xi_1, \xi_2) \in \mathbb{R}^2$. Each linear extension of f is described by a suitable pair c, d of real numbers. Since F is an extension of f, one has

$$c\xi_1 + ad\xi_1 = F(\xi_1, a\xi_1) = f(\xi_1, a\xi_1) = \xi_1, \qquad \forall \xi_1 \in \mathbb{R},$$

and a first equation $c + ad = 1$ results. Since $\|F\|^2 = c^2 + d^2$ equals $\|f\|^2 = 1/(1+a^2)$, a second equation $c^2 + d^2 = 1/(1 + a^2)$ is obtained. Now, these equations together have just one solution, that is,

$$c = \frac{1}{1 + a^2}, \qquad d = \frac{a}{1 + a^2},$$

and so

$$F(\xi_1, \xi_2) = \frac{\xi_1}{1 + a^2} + \frac{a\,\xi_2}{1 + a^2},$$

which is the unique Hahn-Banach extension of f. In case $a = 0$, the unique Hahn-Banach extension is simply $F(\xi_1, \xi_2) = \xi_1$. •

Example 11.6. Let $\mathcal{N} = C[-1, 1]$, M be the subspace of $\mathcal{N}$ of constant functions, and $f : M \to \mathbb{C}$ the continuous functional $f(\psi) = \psi(0)$, for all $\psi \in \mathcal{N}$. Then, for any fixed $t_0 \in [-1, 1]$, the functional $F_{t_0} : \mathcal{N} \to \mathbb{C}$, $F_{t_0}(\psi) = \psi(t_0)$, is a linear extension of f with $\|F_{t_0}\| = \|f\| = 1$. Hence, in this case, f has infinitely many norm-preserving Hahn-Banach extensions (parametrized by t_0), as described in Corollary 10.15. •

Notes

The complex version of the Hahn-Banach Theorem appeared approximately ten years after the real version. The real case was due to Hahn in 1927 and Banach wrote a more general form in 1929 (apparently Banach did not know Hahn's work on the subject); however, a preliminary version has appeared in 1922 in a work of Helly [Hochst]. In his work, Banach made use of the sublinear functional p, an important contribution for the development of the theory of Locally Convex Spaces.

In general there are many extensions compatible with the Hahn-Banach Theorem, as discussed in [Phelps]. In [Foguel], it is concluded that in a normed space $\mathcal{N}$ every norm-preserving linear Hahn-Banach extension (see Proposition 11.4) is unique if and only if its dual $\mathcal{N}^*$ is strictly convex.

Additional Exercises

EXERCISE **11.1.** Prove Corollary 10.15 in case $\mathcal{N}$ is separable, but without appealing to Zorn Lemma.

EXERCISE **11.2.** If X is a real vector space and $p : X \to \mathbb{R}$ a sublinear functional, show that for each $\eta \in X$ there exists a linear functional $f : X \to \mathbb{R}$ so that $f(\eta) = p(\eta)$ and $f(\xi) \le p(\xi)$, for all $\xi \in X$. State and prove a version of this result for complex X.

EXERCISE **11.3.** Check that any seminorm in a normed space is a sublinear functional. Does the converse hold?

EXERCISE **11.4.** If $p : \mathcal{N} \to \mathbb{R}$ is subadditive (i.e., $p(\xi + \eta) \le p(\xi) + p(\eta)$), show that if $p(\xi) \ge 0$ for all $\xi \in \mathcal{N}$ with $\|\xi\| \ge r > 0$, then $p(\xi) \ge 0$ for all $\xi \in \mathcal{N}$.

EXERCISE **11.5.** Let $\mathcal{N}$ be a complex normed space and $h : \mathcal{N} \to \mathbb{R}$ a bounded real linear functional. Show that $f : \mathcal{N} \to \mathbb{C}$, given by $f(\xi) = h(\xi) - ih(i\xi)$, belongs to the dual of $\mathcal{N}$ and $\|f\| = \|h\|$.

EXERCISE **11.6.** [A Version of the Real Hahn-Banach] Let X be a real vector space, Y a subspace of X and $\mathcal{C}$ a *positive cone* in X (i.e., $(\xi + \eta)$ and $t\xi$ belong to $\mathcal{C}$ if $\xi, \eta \in \mathcal{C}$ and $t \ge 0$), so that for each $\xi \in X$ there exists $\eta \in Y$ with $(\eta - \xi) \in \mathcal{C}$. Suppose that $f : Y \to \mathbb{R}$ is a linear functional with $f(\eta) \ge 0$ if $\eta \in (\mathcal{C} \cap Y)$. Define $p : X \to \mathbb{R}$ by $p(\xi) = \inf\{f(\eta) : \eta \in Y, (\eta - \xi) \in \mathcal{C}\}$. Verify that p is a sublinear functional and show that there exists a linear functional $F : X \to \mathbb{R}$ that extends f and $F(\xi) \ge 0$ if $\xi \in \mathcal{C}$.

EXERCISE **11.7.** Let $\mathcal{N}_1, \mathcal{N}_2$ be nontrivial normed spaces. Use the Hahn-Banach Theorem to show that if any nonzero linear and bounded operator $T : \mathcal{N}_1 \to \mathcal{N}_2$ is surjective, then $\dim \mathcal{N}_2 = 1$.

EXERCISE **11.8.** Use Minkowski inequality to show that $l^p(\mathbb{N})$ is strictly convex if and only if $1 < p < \infty$.

EXERCISE **11.9.** Show that if there are two distinct Hahn-Banach extensions of $f \in M^*$, as in Corollary 10.15, then there are infinitely many such norm-preserving extensions.

EXERCISE **11.10.** Let $M = \{\xi \in l^1(\mathbb{N}) : \xi_{2j} = 0, \ \forall j \in \mathbb{N}\}$, and $0 \neq f \in M^*$. Use that $l^1(\mathbb{N})^* = l^\infty(\mathbb{N})$ (see Proposition 13.4) to explicitly find infinitely many linear norm-preserving extensions of f (as in Corollary 10.15).

CHAPTER 12

Applications of Hahn-Banach Theorem

In this chapter, as well as in the subsequent ones, several significant applications of the Hahn-Banach Theorem are presented. Initially, the theorem was introduced as an extension result, but its applications are further extended to fundamental concepts such as dual spaces and separability. Through these tools, various important relationships between normed spaces and their corresponding duals, equipped with appropriate topologies, are explored. The implications of these results are examined, underscoring their broad relevance in Functional Analysis—for instance, a characterization of the denseness of a subspace in terms of the dual space. The essential role of the Hahn-Banach Theorem in these developments is highlighted throughout the discussion and illustrated with examples.

The next result will be used in many occasions in the text.

Theorem 12.1. *Let $\mathcal{N}$ be a nontrivial normed space and $\mathcal{N}^*$ its dual.*

(i) If $0 \neq \xi \in \mathcal{N}$, then there exists $f \in \mathcal{N}^*$ with $f(\xi) = \|\xi\|$ and $\|f\| = 1$.

(ii) (separation of points) If $\eta \neq \xi$ are elements of $\mathcal{N}$, then there exists $f \in \mathcal{N}^*$ so that $f(\xi) \neq f(\eta)$.

(iii) If $\xi \in \mathcal{N}$ satisfies $f(\xi) = 0$, for all $f \in \mathcal{N}^*$, then $\xi = 0$.

(iv) If $\xi \in \mathcal{N}$, then

$$\|\xi\| = \sup_{0 \neq f \in \mathcal{N}^*} \frac{|f(\xi)|}{\|f\|} = \max_{0 \neq f \in \mathcal{N}^*} \frac{|f(\xi)|}{\|f\|}.$$

Proof. *(i)* It is enough to apply Hahn-Banach under the following conditions. The sublinear functional $p : \mathcal{N} \to \mathbb{R}$ is given by $p(\eta) = \|\eta\|$; the subspace $Z = \text{Lin}(\{\xi\})$, and the linear functional $g : Z \to \mathbb{F}$ is defined by $g(\alpha\xi) = \alpha\|\xi\|$, for all $\alpha \in \mathbb{F}$.

(ii) If $\xi \neq \eta$, then $\xi - \eta \neq 0$; by item *(i)* above, there exists $f \in \mathcal{N}^*$ such that $0 \neq f(\xi - \eta) = f(\xi) - f(\eta)$.

(iii) It is immediate from *(ii)*.

(iv) If $\xi = 0$, the result is clear. Assume that $\xi \neq 0$; by item *(i)* there exists $g \in \mathcal{N}^*$ with $\|g\| = 1$ and $g(\xi) = \|\xi\|$. Thus,

$$\|\xi\| = \frac{g(\xi)}{\|g\|} \leq \sup_{0 \neq f \in \mathcal{N}^*} \frac{|f(\xi)|}{\|f\|} \leq \sup_{0 \neq f \in \mathcal{N}^*} \frac{\|f\|\|\xi\|}{\|f\|} = \|\xi\|.$$

DOI: 10.1201/9781003656166-12

The functional g assures that 'sup' can be replaced by 'max.' □

A family G of functionals *separates points* of a set X if, for each pair of distinct elements $\xi, \eta \in X$, there exists $f \in G$ with $f(\xi) \neq f(\eta)$; thus, by Theorem 12.1, $\mathcal{N}^*$ separates points of $\mathcal{N}$.

EXERCISE 12.1. Let $\{\xi_1, \cdots, \xi_n\} \subset \mathcal{N}$ be a linearly independent set and $\{a_1, \cdots, a_n\} \subset \mathbb{F}$. Show that there exists $f \in \mathcal{N}^*$ such that $f(\xi_j) = a_j$, for all $1 \leq j \leq n$.

Proposition 12.2. *Let X be a proper and closed subspace of $\mathcal{N}$ and $\xi \in \mathcal{N} \backslash X$. Put $\delta = d(\xi, X) := \inf_{\eta \in X} \|\xi - \eta\|$. Then there is $f \in \mathcal{N}^*$ satisfying*

$$\|f\| = 1, \quad f(\xi) = \delta, \quad \text{and} \quad f|_X = 0.$$

Proof. The distance δ between ξ and the subspace X is nonzero since X is closed. Define the linear functional $g : \mathrm{Lin}(\{\xi, X\}) \to \mathbb{F}$ by

$$g(\lambda \xi + \eta) = \lambda \delta, \quad \eta \in X, \ \lambda \in \mathbb{F}.$$

Clearly $g|_X = 0$. By the definition of δ, it follows that (for $\lambda \neq 0$)

$$|g(\lambda \xi + \eta)| = |\lambda| \, |\delta| \leq |\lambda| \, \|\xi + \eta/\lambda\| = \|\lambda \xi + \eta\|,$$

and so, $\|g\| \leq 1$. For each $\eta \in X$ one has

$$\|g\| \geq \frac{|g(\xi - \eta)|}{\|\xi - \eta\|} = \frac{\delta}{\|\xi - \eta\|};$$

it then follows that

$$\|g\| \geq \sup_{\eta \in X} \frac{\delta}{\|\xi - \eta\|} = \frac{\delta}{\inf_{\eta \in X} \|\xi - \eta\|} = \frac{\delta}{\delta} = 1.$$

Therefore $\|g\| = 1$.

Now it is enough to apply the Hahn-Banach Theorem to find an extension f of g; take $p(\zeta) = \|\zeta\|$, for all $\zeta \in \mathcal{N}$, and due to the above relation it satisfies $|g| \leq p$ for all vectors in $\mathrm{Lin}(\{\xi, X\})$ (or evoke Corollary 10.15). □

Corollary 12.3. *A vector subspace X of $\mathcal{N}$ is dense in $\mathcal{N}$ if and only if the unique element of $\mathcal{N}^*$ that vanishes in X is the null functional.*

EXERCISE 12.2. Present details of the proof of Corollary 12.3.

Proposition 12.4. *If $\mathcal{N}^*$ is separable, then $\mathcal{N}$ is separable.*

Proof. If $\mathcal{N}^*$ is separable, there is a dense sequence $(f_n)_{n=1}^\infty$ in $\mathcal{N}^*$. Choose $\xi_n \in \mathcal{N}$, so that $\|\xi_n\| = 1$ and $|f_n(\xi_n)| \geq \|f_n\|/2$, for all n, and let $X = \overline{\mathrm{Lin}(\{\xi_n\})}$, which is separable (Proposition 3.4). The goal is to show that $\mathcal{N} = X$, and to this end it is enough to show that if $f \in \mathcal{N}^*$ with $f|_X = 0$, then f is the null functional.

Consider such f with $f|_X = 0$, and take a subsequence $f_{n_j} \to f$. Since for all n_j,

$$\|f - f_{n_j}\| \geq |(f - f_{n_j})(\xi_{n_j})| = |f_{n_j}(\xi_{n_j})| \geq \|f_{n_j}\|/2,$$

it follows that

$$\|f\| \leq \|f - f_{n_j}\| + \|f_{n_j}\| \leq 3\,\|f - f_{n_j}\| \to 0, \qquad j \to \infty,$$

which shows that $f = 0$. Hence $\mathcal{N} = X$ and it is separable. $\qquad\square$

REMARK 12.5. There are examples of separable spaces whose duals are not separable. A standard example is $l^1(\mathbb{N})^* = l^\infty(\mathbb{N})$, discussed in Chapter 13.

As another application of Hahn-Banach Theorem, it will be discussed the converse of Theorem 4.5:

Proposition 12.6. *Let* $\mathcal{N}_1 \neq \{0\}$. *If the space* $\mathrm{B}(\mathcal{N}_1, \mathcal{N}_2)$ *is complete, then* $\mathcal{N}_2$ *is Banach.*

Proof. Let (η_j) be a Cauchy sequence in $\mathcal{N}_2$ and pick $\xi_0 \in \mathcal{N}_1$ with $\|\xi_0\| = 1$. By Theorem 12.1, there exists $f \in \mathcal{N}_1^*$ with $\|f\| = f(\xi_0) = 1$. The relation $\eta_j = f(\xi_0)\eta_j$ is a motivation for defining $T_j \in \mathrm{B}(\mathcal{N}_1, \mathcal{N}_2)$ by $T_j\xi = f(\xi)\eta_j$, $\xi \in \mathcal{N}_1$; it follows that (T_j) is Cauchy in $\mathrm{B}(\mathcal{N}_1, \mathcal{N}_2)$, since

$$\|(T_j - T_k)\xi\| = |f(\xi)|\,\|\eta_j - \eta_k\| \leq \|\eta_j - \eta_k\|\,\|\xi\|;$$

so there is $T \in \mathrm{B}(\mathcal{N}_1, \mathcal{N}_2)$ such that $T_j \to T$. Since

$$\|\eta_j - T\xi_0\| = \|T_j\xi_0 - T\xi_0\| \leq \|T_j - T\|\,\|\xi_0\|,$$

it follows that $\eta_j \to T\xi_0$, and so $\mathcal{N}_2$ is complete. $\qquad\square$

Since $\mathcal{N}^*$ is a Banach space, $\mathcal{N}^{**} := (\mathcal{N}^*)^*$ is well defined and it is called the *second dual* or *bidual* of $\mathcal{N}$. There is a natural way of identifying elements of $\mathcal{N}$ with elements of its second dual: to each $\xi \in \mathcal{N}$ one associates $\hat{\xi} \in \mathcal{N}^{**}$ through the relation

$$\hat{\xi}(f) := f(\xi), \qquad f \in \mathcal{N}^*.$$

Such mapping is called the *canonical map* of $\mathcal{N}$ into $\mathcal{N}^{**}$.

Proposition 12.7. *The mapping* $\hat{} : \mathcal{N} \to \mathcal{N}^{**}$ *defined above is a linear isometry.*

Proof. The linearity of the mapping $\hat{}$ is clear. It is an isometry since

$$\|\hat{\xi}\| = \sup_{\substack{f \in \mathcal{N}^* \\ \|f\| \leq 1}} \frac{|\hat{\xi}(f)|}{\|f\|} = \sup_{\substack{f \in \mathcal{N}^* \\ \|f\| \leq 1}} \frac{|f(\xi)|}{\|f\|} = \|\xi\|;$$

in the last equality Theorem 12.1 was used. $\qquad\square$

At this phase, it is possible to present (and appreciate) precisely the concept of reflexive spaces (which was mentioned in Chapter 4).

Definition 12.8. If the *canonical map* ˆ is onto, then the normed space $\mathcal{N}$ is termed *reflexive*. In other words, $\mathcal{N}$ is reflexive if it is isomorphic to $\mathcal{N}^{**}$ and the isomorphism being the canonical mapping. The relation $\hat{\mathcal{N}} := \,\hat{}\,(\mathcal{N}) \subset \mathcal{N}^{**}$ will sometimes be indicated by $\mathcal{N} \subset \mathcal{N}^{**}$.

EXERCISE **12.3.** Show that: (a) every reflexive normed space is Banach, and (b) if $\dim \mathcal{N} < \infty$, then $\mathcal{N}$ is reflexive.

REMARK **12.9.** Prominent examples of reflexive spaces are the Hilbert spaces; see Corollary 19.5.

Proposition 12.10. *Any closed subspace of a reflexive normed space is also reflexive.*

Proof. Let E be a (proper) closed subspace of a reflexive space $\mathcal{B}$; note that both E and $\mathcal{B}$ are Banach. If $f \in \mathcal{B}^*$, then its restriction $f_E := f|_E$ to the subspace E is an element of E^*. By Hahn-Banach (see Corollary 10.15), indeed $E^* = \{f_E : f \in \mathcal{B}^*\}$. Thus, for all $h \in E^{**}$ it is enough to consider $h(f_E)$, and the goal is to find $\xi_h \in E$ so that $h = \hat{\xi}_h$.

Define the linear functional $H : \mathcal{B}^* \to \mathbb{F}$, $H(f) = h(f_E)$, $f \in \mathcal{B}^*$, and since

$$|H(f)| = |h(f_E)| \leq \|h\| \|f_E\| \leq \|h\| \|f\|,$$

it follows that $H \in \mathcal{B}^{**}$, and since $\mathcal{B}$ is reflexive, there is a $\xi_h \in \mathcal{B}$ so that $H = \hat{\xi}_h$ (with ˆ denoting the canonical mapping $\mathcal{B} \mapsto \hat{\mathcal{B}}$). By construction,

$$h(f_E) = H(f) = \hat{\xi}_h(f) = f(\xi_h), \quad \forall f \in \mathcal{B}^*.$$

The next step is to show that $\xi_h \in E$. Indeed, if $\xi_h \notin E$, by Proposition 12.2, there would be an $f \in \mathcal{B}^*$ with $f(\xi_h) \neq 0$ and $f_E = 0$, but this would contradict the above relation, since it would imply that $0 \neq f(\xi_h) = h(f_E) = 0$. Therefore $\xi_h \in E$.

So, $h(f_E) = f(\xi_h) = f_E(\xi_h)$, for all $f \in \mathcal{B}^*$, and since $E^* = \{f_E : f \in \mathcal{B}^*\}$, it follows that $h(g) = g(\xi_h)$, for all $g \in E^*$, that is, $h = \hat{\xi}_h$, with ˆ denoting the canonical mapping $E \mapsto \hat{E}$. Therefore this mapping is onto and one may conclude that E is reflexive. □

Notes

A detail in the definition of reflexive space, which surely was not clear in the discussion in Chapter 4, is that the isometry between $\mathcal{N}$ and $\mathcal{N}^{**}$ must be the canonical mapping ˆ; in the work R. C. James, A Nonreflexive Banach Space Isometric with Its Second Conjugate Space, Proc. Nat. Acad. Sci. U.S.A. **37**, 174–177 (1951), it is presented an example of a nonreflexive Banach space isometric to its second dual. It is interesting to note that a vector space is algebraically reflexive (i.e., by using the action of ˆ, but forgetting about topologies) if and only if it is of finite dimension.

The term *reflexive space* was introduced by E. R. Lorch in 1939, to replace *regular space* which was the term previously used.

Due to its introductory character, some important applications of the Hahn-Banach Theorem will not be presented here; for example, the separation of convex sets by hyperplanes (also called the geometric Hahn-Banach), the Krein-Milman Theorem, Malgrange's results in Partial Differential Equations, Banach limits, etc.

For the converse of the result in Exercise 12.12 (namely, if for any $f \in \mathcal{B}^*$ there is $\xi \in \mathcal{B}$, with $\|\xi\| = 1$, such that $\|f\| = f(\xi)$, then the space is reflexive), see [James]. Finally, it is worth mentioning that the important Theorem 12.1 and Proposition 12.2 have appeared in a work of Hahn, in 1927.

Additional Exercises

EXERCISE **12.4.** Let Z be a vector subspace of $\mathcal{N}$. Show that every bounded linear functional on Z is the restriction of some element of $\mathcal{N}^*$. Conclude that $Z^* = \{f|_Z : f \in \mathcal{N}^*\}$.

EXERCISE **12.5.** Show that if $\xi \in \mathcal{N}$ is such that $f(\xi) = 0$, for all f in a dense set in $\mathcal{N}^*$, then $\xi = 0$.

EXERCISE **12.6.** Use Proposition 12.7 to show that every normed space can be completed, i.e., that every normed space is linearly isometric to a dense subset of some Banach space (see Theorem 2.6).

EXERCISE **12.7.** Show that a subset $K \subset \mathcal{B}^*$ is bounded if and only if for each $\hat{\xi} \in \hat{\mathcal{B}}$ one has $\sup_{f \in K} |\hat{\xi}(f)| < \infty$.

EXERCISE **12.8.** If $T \in B(\mathcal{N}_1, \mathcal{N}_2)$, show that

$$\|T\| = \sup_{\substack{\xi \in \mathcal{N}_1 \\ \|\xi\|=1}} \sup_{\substack{f \in \mathcal{N}_2^* \\ \|f\|=1}} |f(T\xi)|.$$

EXERCISE **12.9.** Let X be a subspace of $\mathcal{N}$ (some items ahead hold if X is just a subset of $\mathcal{N}$). The *annihilator* of X is defined by

$$X^0 := \{f \in \mathcal{N}^* : f(\xi) = 0, \ \forall \xi \in X\}.$$

In an analogous way, the *annihilator* of a subspace Λ of $\mathcal{N}^*$ is

$$\Lambda^\dagger := \{\xi \in \mathcal{N} : f(\xi) = 0, \ \forall f \in \Lambda\}.$$

(a) Show that X^0 and $\Lambda^\dagger$ are closed subspaces of $\mathcal{N}^*$ and $\mathcal{N}$, respectively.

(b) Verify that $\Lambda^\dagger = \mathcal{N} \cap \Lambda^0$ (with $\mathcal{N}$ identified with $\hat{\mathcal{N}} \subset \mathcal{N}^{**}$).

(c) Show that $X \subset (X^0)^\dagger$ and if X is closed, then $X = (X^0)^\dagger$.

(d) Show that $\Lambda \subset (\Lambda^\dagger)^0$ and, if $\mathcal{N}$ is reflexive and Λ closed, then $(\Lambda^\dagger)^0 = \Lambda$.

EXERCISE **12.10.** If E is a subspace of a normed space $\mathcal{N}$, show that its closure is given by

$$\overline{E} = \bigcap \{N(f) : f \in \mathcal{N}^*, \ E \subset N(f)\}.$$

EXERCISE **12.11.** Show that a Banach space is reflexive if and only if its dual is reflexive.

EXERCISE **12.12.** If $\mathcal{B}$ is reflexive, show that for any $f \in \mathcal{B}^*$ there is $\xi \in \mathcal{B}$, with $\|\xi\| = 1$, such that $\|f\| = f(\xi)$.

CHAPTER 13

Adjoint Operators in Normed Spaces

The Banach adjoint operator of a continuous linear operator on a normed space is introduced, establishing its connection to the solutions of several equations involving linear operators, thereby highlighting its essential role in Functional Analysis. Multiple examples are provided to illustrate the practical significance of this concept. In particular, it is shown that the Banach space $l^p(\mathbb{N})$ is reflexive when $1 < p < \infty$, while $l^1(\mathbb{N})$ is not reflexive. Finally, a version of the Hellinger-Toeplitz Theorem for Banach spaces is presented; note that the original result is for Hilbert spaces and will be discussed in Chapter 20.

Definition 13.1. Let $T \in \mathrm{B}(\mathcal{N}_1, \mathcal{N}_2)$. The operator $T^{\mathrm{a}} : \mathcal{N}_2^* \to \mathcal{N}_1^*$ given by

$$(T^{\mathrm{a}}g)(\xi) := g(T\xi), \qquad g \in \mathcal{N}_2^*, \xi \in \mathcal{N}_1,$$

is the *Banach adjoint operator* of T (or just *adjoint operator*). Note that since $T^{\mathrm{a}}(g)$ is uniquely defined for each $g \in \mathcal{N}_2^*$, then T^{a} is also uniquely defined.

Proposition 13.2. *If $T \in \mathrm{B}(\mathcal{N}_1, \mathcal{N}_2)$, then its adjoint $T^{\mathrm{a}} \in \mathrm{B}(\mathcal{N}_2^*, \mathcal{N}_1^*)$ and $\|T^{\mathrm{a}}\| = \|T\|$.*

Proof. The linearity of T^{a} is clear. Since for all $g \in \mathcal{N}_2^*$

$$\|(T^{\mathrm{a}}g)(\xi)\| = \|g(T\xi)\| \leq \|g\| \, \|T\| \, \|\xi\|, \qquad \forall \xi \in \mathcal{N}_1,$$

it follows that $\|T^{\mathrm{a}}g\| \leq \|T\| \, \|g\|$, that is, $T^{\mathrm{a}} \in \mathrm{B}(\mathcal{N}_2^*, \mathcal{N}_1^*)$ and $\|T^{\mathrm{a}}\| \leq \|T\|$. Thus, if $T = 0$, then $T^{\mathrm{a}} = 0$ and the proof finishes; suppose, then, that $T \neq 0$.

By Theorem 12.1, given $0 \neq \xi_0 \in \mathcal{N}_1$ with $T\xi_0 \neq 0$, there exists $f \in \mathcal{N}_2^*$ so that $0 \neq f(T\xi_0) = \|T\xi_0\|$ and $\|f\| = 1$. Thus,

$$\|T\xi_0\| = f(T\xi_0) = |(T^{\mathrm{a}}f)(\xi_0)| \leq \|T^{\mathrm{a}}\| \, \|f\| \, \|\xi_0\| = \|T^{\mathrm{a}}\| \, \|\xi_0\|,$$

so that $\|T\| \leq \|T^{\mathrm{a}}\|$; therefore $\|T^{\mathrm{a}}\| = \|T\|$. $\qquad\square$

Example 13.3. Given a basis of $\mathbb{F}^n$, an operator $T \in \mathrm{B}(\mathbb{F}^n)$ is represented by a matrix. Then its adjoint T^{a} is represented by the transpose of that matrix. Verify that this has a version to the case in which $T \in \mathrm{B}(\mathcal{N})$ and $\mathcal{N}$ has a Schauder basis.

•

DOI: 10.1201/9781003656166-13

EXERCISE **13.1.** Show that if $T, S \in \mathrm{B}(\mathcal{N}_1, \mathcal{N}_2)$, then $(S + T)^{\mathrm{a}} = S^{\mathrm{a}} + T^{\mathrm{a}}$, and for $\alpha \in \mathbb{F}$ one has $(\alpha T)^{\mathrm{a}} = \alpha T^{\mathrm{a}}$. Furthermore, if the product TS is well defined, then $(TS)^{\mathrm{a}} = S^{\mathrm{a}} T^{\mathrm{a}}$, and if $T^{-1} \in \mathrm{B}(\mathcal{N}_2, \mathcal{N}_1)$, then $(T^{-1})^{\mathrm{a}} = (T^{\mathrm{a}})^{-1}$.

EXERCISE **13.2.** For $T \in \mathrm{B}(\mathcal{N}_1, \mathcal{N}_2)$, show that T^{a} is injective if and only if rng T is dense in $\mathcal{N}_2$ (see also Exercise 13.5).

Proposition 13.4. $l^1(\mathbb{N})^* = l^\infty(\mathbb{N})$.

Proof. To each $f \in l^1(\mathbb{N})^*$ it will be associated an element $\alpha = (\alpha_j) \in l^\infty(\mathbb{N})$, and vice versa, and such association will be isometric and linear, so the dual of l^1 will be isomorphic to l^∞.

Let $\{e_j\}$ be the canonical basis of $l^1(\mathbb{N})$ and $f \in l^1(\mathbb{N})^*$. Thus, for $\xi = (a_j)_{j=1}^\infty = \sum_{j=1}^\infty a_j e_j$ in l^1 one has

$$f(\xi) = \sum_j a_j f(e_j) = \sum_j a_j \alpha_j,$$

with $\alpha_j = f(e_j)$; so $|\alpha_j| = |f(e_j)| \le \|f\| \|e_j\| = \|f\|$. By defining $\alpha = (\alpha_j)_{j=1}^\infty$ it follows that $\|\alpha\|_\infty \le \|f\|$ and $\alpha \in l^\infty$. On the other hand, one has

$$|f(\xi)| \le \sum_j |a_j| |\alpha_j| \le \|\alpha\|_\infty \|\xi\|_1,$$

that is, $\|f\| \le \|\alpha\|_\infty$. Therefore $\|\alpha\|_\infty = \|f\|$ and the linear mapping $f \mapsto \alpha$ defined above is an isometry between $l^1(\mathbb{N})^*$ and a subset of $l^\infty(\mathbb{N})$. Now it is enough to show that the range of this mapping is all $l^\infty(\mathbb{N})$.

If $\beta = (\beta_j) \in l^\infty$, define the linear functional g acting on l^1 by

$$g(\xi) = g\left(\sum_j a_j e_j\right) := \sum_j a_j \beta_j.$$

Since $|g(\xi)| \le \|\beta\|_\infty \|\xi\|_1$, it follows that $g \in l^1(\mathbb{N})^*$, and so the above mapping is surjective. ◻

REMARK **13.5.** By a simple variation of the above construction, it is clear that $l^1(\mathbb{N}) \subset l^\infty(\mathbb{N})^*$, but since l^∞ is not separable, it follows by Proposition 12.4, that l^1 is not the dual of l^∞ (see Example 3.7), and so l^1 is not reflexive.

Example 13.6. Let $S_r : l^1(\mathbb{N}) \hookleftarrow$ be the right shift operator

$$S_r \xi = S_r(\xi_1, \xi_2, \xi_3, \cdots) = (0, \xi_1, \xi_2, \xi_3, \cdots).$$

Then its Banach adjoint $S_r^{\mathrm{a}} : l^\infty(\mathbb{N}) \hookleftarrow$ is

$$S_l(\xi_1, \xi_2, \xi_3, \cdots) = (\xi_2, \xi_3, \xi_4, \cdots),$$

that is, the left shift operator. ●

Proposition 13.7. $l^p(\mathbb{N})^* = l^q(\mathbb{N})$ *if* $p > 1$ *and* $1/p + 1/q = 1$. *Furthermore,* l^p *is reflexive for* $1 < p < \infty$.

Proof. This proof is similar to that of Proposition 13.4, and a similar notation will be used here. To each $f \in l^p(\mathbb{N})^*$ it will be associated an element $\alpha = (\alpha_j) \in l^q(\mathbb{N})$, and vice versa, and this association will be a linear isometry, so that the dual of l^p is isomorphic to l^q.

Let $\{e_j\}$ be the canonical Schauder basis of $l^p(\mathbb{N})$ and $f \in l^p(\mathbb{N})^*$. Thus, for $\xi = (a_j)_{j=1}^\infty = \sum_{j=1}^\infty a_j e_j$ in l^p one has

$$f(\xi) = \sum_j a_j f(e_j) = \sum_j a_j \alpha_j,$$

with $\alpha_j = f(e_j)$.

By considering the sequence $\xi^{(n)} = \left(\xi_j^{(n)}\right)_{j=1}^\infty$ defined by $\xi_j^{(n)} = |\alpha_j|^q/\alpha_j$, if $\alpha_j \neq 0$ and $1 \leq j \leq n$, and $\xi_j^{(n)} = 0$ otherwise, one obtains $f(\xi^{(n)}) = \sum_{j=1}^n |\alpha_j|^q$. On the other hand,

$$f(\xi^{(n)}) \leq \|f\| \, \|\xi^{(n)}\|_p = \|f\| \left(\sum_{j=1}^n |\alpha_j|^q \right)^{1/p}.$$

Such relations imply that

$$\sum_j^n |\alpha_j|^q = f(\xi^{(n)}) \leq \|f\| \left(\sum_{j=1}^n |\alpha_j|^q \right)^{1/p},$$

i.e., $\left(\sum_{j=1}^n |\alpha_j|^q \right)^{1/q} \leq \|f\|$. Since this inequality holds for all n it follows that $\|\alpha\|_q \leq \|f\|$ and $\alpha \in l^q(\mathbb{N})$, with $\alpha = (\alpha_j)_{j=1}^\infty$.

By Hölder inequality (see Exercise 1.20), one finds

$$|f(\xi)| = \left| \sum_j a_j \alpha_j \right| \leq \|\xi\|_p \|\alpha\|_q,$$

and one concludes that $\|f\| = \|\alpha\|_q$. Therefore, the linear mapping $f \mapsto \alpha$ defined above is an isometry between $l^p(\mathbb{N})^*$ and a subset of $l^q(\mathbb{N})$. Now it is sufficient to show that the range of this mapping is all $l^q(\mathbb{N})$.

If $\beta = (\beta_j) \in l^q$, define g acting on l^p by $g(\sum_j a_j e_j) := \sum_j a_j \beta_j$. By Hölder again, one obtains that $g \in l^p(\mathbb{N})^*$, and that mapping is surjective.

Note that this representation of $l^p(\mathbb{N})^*$ shows that $l^p(\mathbb{N})^{**}$ is identified with $l^p(\mathbb{N})$ via the canonical mapping $\hat{\ }$ presented in Proposition 12.7 and so, $l^p(\mathbb{N})$ is reflexive $(1 < p < \infty)$. $\qquad\square$

REMARK 13.8. Recall that $L^p_\mu(\Omega)^* = L^q_\mu(\Omega)$ for $p > 1$ and $1/p + 1/q = 1$ (and such spaces are reflexive), and if μ is a σ-finite measure, then $L^1_\mu(\Omega)^* = L^\infty_\mu(\Omega)$, as mentioned in Chapter 4. See, in particular, Example 4.12.

Example 13.9. Consider $K : [-1, 1] \times [-1, 1] \to \mathbb{F}$ mensurable and bounded, and define the linear operator $T_K : L^p[-1, 1] \to L^q[-1, 1]$, $1 < p < \infty$, $1/p + 1/q = 1$, by

$$(T_K \psi)(t) = \int_{-1}^{1} K(t, s) \psi(s) \, ds, \qquad \psi \in L^p[-1, 1].$$

By Hölder inequality (see page 5), this operator is bounded. Since $L^q[-1, 1]^* = L^p[-1, 1]$, each element $F \in L^q[-1, 1]^*$ is represented by a unique $f \in L^p[-1, 1]$ and one has

$$F(\psi) = \int_{-1}^{1} f(t) \psi(t) \, dt, \qquad \forall \psi \in L^q[-1, 1].$$

Hence, after defining $\phi(t) = \int_{-1}^{1} K(s, t) f(s) \, ds$ and using Fubini, for $\varphi \in L^p$,

$$
\begin{aligned}
(T_K^a F)(\varphi) &= F(T_K \varphi) = \int_{-1}^{1} dt \, f(t) \int_{-1}^{1} ds \, K(t, s) \varphi(s) \\
&= \int_{-1}^{1} ds \, \varphi(s) \int_{-1}^{1} dt \, K(t, s) f(t) = \int_{-1}^{1} \varphi(s) \phi(s) \, ds,
\end{aligned}
$$

that is, ϕ (in $L^p[-1, 1]^*$) represents the functional $(T_K^a F)$ and hence,

$$(T_K^a F)(t) = \int_{-1}^{1} K(s, t) f(s) \, ds.$$

Is it possible to rule out the technical restriction $1/p + 1/q = 1$?   ●

The next result, Proposition 13.10, is a version of the Hellinger-Toeplitz's Theorem for Banach spaces (the original one is for Hilbert spaces; see Proposition 20.4). It indicates that, in the study of the adjoint of an unbounded operator, subtle domain questions are inevitable, since it implies that such operators cannot be defined on the entire Banach space.

Proposition 13.10. *Assume that the linear operators $T : \mathcal{B}_1 \to \mathcal{B}_2$ and $S : \mathcal{B}_2^* \to \mathcal{B}_1^*$ satisfy*

$$g(T\xi) = (Sg)(\xi), \qquad \forall \xi \in \mathcal{B}_1, g \in \mathcal{B}_2^*.$$

Then S and T are bounded operators and $S = T^a$.

Proof. To show that T is bounded the Closed Graph Theorem will be invoked. Suppose that $\xi_n \to \xi$ in $\mathcal{B}_1$ and $T\xi_n \to \eta$ in $\mathcal{B}_2$. Thus, for each $g \in \mathcal{B}_2^*$ one has

$$g(\eta) = \lim_{n \to \infty} g(T\xi_n) = \lim_{n \to \infty} (Sg)(\xi_n) = (Sg)(\xi) = g(T\xi),$$

so that $\eta = T\xi$ (Theorem 12.1) and the graph of T is closed. Hence T is bounded, T^a is well defined, and by the definition of adjoint operator,

$$(T^a g)(\xi) = g(T\xi) = (Sg)(\xi), \qquad \forall \xi \in \mathcal{B}_1, g \in \mathcal{B}_2^*.$$

Therefore, the functionals $T^a(g)$ and $S(g)$ coincide for all $g \in \mathcal{B}_2^*$, that is, $S = T^a$ and so S is bounded. □

Notes

Apparently it was Lagrange who introduced the concept of adjoint, but for a particular differential operator; he noted some relations between the adjoint operator and solutions of equations involving the original operator. This idea was generalized only just before 1900 by S. Pincherle. The notion of the adjoint of an operator is particularly important in Hilbert spaces; such spaces are "identical" to their duals, so that it is possible for an operator to be the same as its (Hilbert) adjoint, the so-called self-adjoint operators (see Chapters 19 and 20).

It is possible to define the adjoint of some unbounded operators (it is necessary that they are densely defined [de Oliv]); at this point there was a large contribution of J. von Neumann, around 1930.

The notation T^{a} for the Banach adjoint operator is not standard; many authors denote it by T^*, but here this latter notation is reserved to the *Hilbert adjoint* (this one is actually a standard notation), which will be discussed in Chapter 19.

For a quite general version of the Theorem of Hellinger-Toeplitz, which holds for some locally convex vector topological spaces, see [Ptak].

Additional Exercises

EXERCISE **13.3.** If $f \in \mathcal{N}^*$, determine the Banach adjoint f^{a}.

EXERCISE **13.4.** If $T \in \mathrm{B}(\mathcal{N}_1, \mathcal{N}_2)$, show that rng $T \subset \mathrm{N}(T^{\mathrm{a}})^\dagger$ (the annihilator, defined in Exercise 12.9). Discuss the consequences of this relation for the solution to the equation

$$T\xi = \eta, \qquad \eta \in \mathcal{N}_2.$$

Conclude that the adjoint can be useful for analyzing if a linear operator is invertible or not.

EXERCISE **13.5.** If $T \in \mathrm{B}(\mathcal{N}_1, \mathcal{N}_2)$, then $\mathrm{N}(T) = (\mathrm{rng}\ T^{\mathrm{a}})^\dagger$ and $\mathrm{N}(T^{\mathrm{a}}) = (\mathrm{rng}\ T)^0$ (notation as in Exercise 12.9). Conclude that T is injective if and only if $T^{\mathrm{a}}(\mathcal{N}_2^*)$ separates points of $\mathcal{N}_1$.

EXERCISE **13.6.** For $T \in \mathrm{B}(\mathcal{N}_1, \mathcal{N}_2)$, define T^{aa}, identify $\mathcal{N}_1$ and $\mathcal{N}_2$ with $\hat{\mathcal{N}}_1$ and $\hat{\mathcal{N}}_2$, respectively, and show that $T^{\mathrm{aa}}|_{\mathcal{N}_1} = T$. If $\mathcal{N}_1$ is reflexive, then $T^{\mathrm{aa}} = T$ (this is related to Exercise 25.3).

EXERCISE **13.7.** If $T \in \mathrm{B}(\mathcal{B}_1, \mathcal{B}_2)$ with T^{a} surjective, verify, via the Open Mapping Theorem, that there exists $r > 0$ so that $T^{\mathrm{a}} B_{\mathcal{B}_2^*}(0; 1) \supset B_{\mathcal{B}_1^*}(0; r)$; again by Theorem 12.1, conclude that $\|T\xi\| \geq r\|\xi\|$, for all $\xi \in \mathcal{B}_1$. From such results, show that T is invertible if and only if T^{a} is invertible.

EXERCISE **13.8.** Check that $c_0(\mathbb{N})^* = l^1(\mathbb{N})$ and conclude that c_0 is not reflexive.

EXERCISE **13.9.** Let $\mathcal{N}_p \subset l^p(\mathbb{N})$, $1 \leq p \leq \infty$, the space of sequences with a finite number of nonzero entries. Show that $\mathcal{N}_p^* = l^q(\mathbb{N})$, $1/q + 1/p = 1$.

EXERCISE **13.10.** (a) Show that if $\mathcal{B}_1$ and $\mathcal{B}_2$ are isomorphic, then $\mathcal{B}_1^*$ and $\mathcal{B}_2^*$ are isomorphic.

(b) If for some $T \in \mathrm{B}(\mathcal{B}_1, \mathcal{B}_2)$ the adjoint T^{a} is an isomorphism between $\mathcal{B}_2^*$ and $\mathcal{B}_1^*$ (onto), show that $\mathcal{B}_1$ and $\mathcal{B}_2$ are isomorphic.

EXERCISE **13.11.** For $\xi = (\xi_1, \xi_2, \xi_3, \cdots) \in l^1(\mathbb{N})$, define $T\xi$ by

$$(T\xi)_n = \sum_{j \geq n} \xi_j.$$

Show that $T \in \mathrm{B}(l^1, c_0)$ and find T^{a}.

EXERCISE **13.12.** Let E be a vector subspace of $\mathcal{N}$, and $i : E \to \mathcal{N}$ the canonical inclusion mapping. Check that i is a bounded linear operator, and show that its Banach adjoint $i^{\mathrm{a}} : \mathcal{N}^* \to E^*$ is the restriction operator $i^{\mathrm{a}}(f) = f|_E$, for all $f \in \mathcal{N}^*$.

EXERCISE **13.13.** Let $T \in \mathrm{B}(\mathcal{N}, l^\infty(\mathbb{N}))$. Show that there exists a bounded sequence $(f_j) \subset \mathcal{N}^*$ so that $T\xi = (f_j(\xi))_j$, for all $\xi \in \mathcal{N}$.

CHAPTER 14

Weak Convergence

In this chapter, the notions of weak and strong convergences of sequences in normed spaces are presented, and their basic properties are discussed. These concepts are then adapted to sequences of linear operators between normed spaces, accompanied by relevant examples, including those that converge in one sense but not in another. Finally, an example due to Schur is provided, showing that in $l^1(\mathbb{N})$ the concepts of weak and strong convergences are equivalent. The weak topologies, including the weak* topology, are covered in later chapters.

Definition 14.1. A sequence $(\xi_n) \subset \mathcal{N}$ converges weakly to $\xi \in \mathcal{N}$ if

$$\lim_{n \to \infty} f(\xi_n) = f(\xi),$$

for all $f \in \mathcal{N}^*$.

REMARK 14.2. (a) $\xi_n \rightharpoonup \xi$, $\xi_n \xrightarrow{\text{w}} \xi$ and $\text{w} \cdot \lim \xi_n = \xi$ will be used to indicate that (ξ_n) converges weakly to ξ.

(b) The convergence of (ξ_n) to ξ in the norm of $\mathcal{N}$ will be called *strong convergence* and indicated by $\xi_n \to \xi$, $\xi_n \xrightarrow{\text{s}} \xi$ and $\text{s} \cdot \lim \xi_n = \xi$.

The next result assures that this notion of weak convergence is well defined.

Proposition 14.3. *Assume that $\xi_n \rightharpoonup \xi$ in $\mathcal{N}$. Then the limit ξ is unique and the sequence (ξ_n) is bounded.*

Proof. If $\xi_n \rightharpoonup \xi$ and $\xi_n \rightharpoonup \eta$, then for all $f \in \mathcal{N}^*$ one has

$$f(\xi - \eta) = f(\xi) - f(\eta) = \lim_{n \to \infty} (f(\xi_n) - f(\xi_n)) = 0.$$

By Theorem 12.1, it follows that $\xi = \eta$ and the weak limit is unique.

To show that $(\|\xi_n\|)$ is bounded, Proposition 12.7 will be used, in particular that $\hat{\xi}_n \in \mathcal{N}^{**}$ and $\|\xi_n\| = \|\hat{\xi}_n\|$. For each $f \in \mathcal{N}^*$ one has $\hat{\xi}_n(f) = f(\xi_n)$, which is convergent and so bounded. By the Uniform Boundedness Principle it follows that $\sup_n \|\xi_n\| = \sup_n \|\hat{\xi}_n\| < \infty$. $\qquad \square$

Since $|f(\xi_n) - f(\xi)| \le \|f\| \|\xi_n - \xi\|$, the strong convergence of a sequence in a normed space implies the weak convergence and with the same limits, justifying the nomenclature. Below there are examples of sequences that converge in the weak sense, but do not converge in the strong sense.

DOI: 10.1201/9781003656166-14

EXERCISE **14.1.** If $\dim \mathcal{N} < \infty$, show that the concepts of weak and strong convergences coincide.

Proposition 14.4. $\xi_n \rightharpoonup \xi$ in $\mathcal{N}$ if and only if $f(\xi_n) \to f(\xi)$ for f in a dense subset in $\mathcal{N}^*$ and $(\|\xi_n\|)$ is a bounded set.

Proof. The necessity follows straightly by Proposition 14.3. To show the sufficiency, denote by W the set of $f \in \mathcal{N}^*$ with $f(\xi_n) \to f(\xi)$, and pick a strictly positive constant $C \geq \|\xi_n\|$, for all n.

If $g \in \mathcal{N}^*$, given $\varepsilon > 0$ there exists $f \in W$ with $\|f - g\| < \varepsilon$. Thus,

$$\begin{aligned} |g(\xi_n) - g(\xi)| &\leq |g(\xi_n) - f(\xi_n)| + |f(\xi_n) - f(\xi)| + |f(\xi) - g(\xi)| \\ &\leq \varepsilon\|\xi_n\| + |f(\xi_n) - f(\xi)| + \varepsilon\|\xi\|. \end{aligned}$$

For n large enough one has $|f(\xi_n) - f(\xi)| < \varepsilon$ and $|g(\xi_n) - g(\xi)| < \varepsilon(1 + C + \|\xi\|)$; since this holds for all $\varepsilon > 0$, $g(\xi_n) \to g(\xi)$ and therefore, $\xi_n \rightharpoonup \xi$. □

Now, different notions of convergence of bounded linear operators will be introduced and compared with each other.

Definition 14.5. Let (T_n) be a sequence of operators in $B(\mathcal{N}_1, \mathcal{N}_2)$ and $T : \mathcal{N}_1 \to \mathcal{N}_2$ linear. One says that:

(a) T_n converges *uniformly*, or *in norm,* to T if

$$\|T_n - T\| \to 0.$$

The uniform convergence will be denoted by $T_n \to T$ or $\lim_{n\to\infty} T_n = T$.

(b) T_n converges *strongly* to T if

$$\|T_n\xi - T\xi\|_{\mathcal{N}_2} \to 0, \qquad \forall \xi \in \mathcal{N}_1.$$

The strong convergence of linear operators will be denoted by $T_n \xrightarrow{s} T$ or $s \cdot \lim_{n\to\infty} T_n = T$.

(c) T_n converges *weakly* to T if

$$|f(T_n\xi) - f(T\xi)| \to 0, \qquad \forall \xi \in \mathcal{N}_1, \ f \in \mathcal{N}_2^*.$$

The weak convergence of linear operators will be denoted by $T_n \xrightarrow{w} T$, $T_n \rightharpoonup T$, or $w \cdot \lim_{n\to\infty} T_n = T$.

EXERCISE **14.2.** Show that in $B(\mathcal{N}_1, \mathcal{N}_2)$ the three kinds of limits defined above are well defined and unique (if they exist, of course). Moreover, verify that the uniform convergence $\Longrightarrow$ strong convergence $\Longrightarrow$ weak convergence, and with the same limits.

Example 14.6. Let $P_n : l^1(\mathbb{N}) \hookleftarrow$, $P_n\xi = (\xi_1, \xi_2, \cdots, \xi_n, 0, 0, \cdots)$, with $\xi = (\xi_1, \xi_2, \xi_3, \cdots)$. Since $\|P_n\xi - \xi\| = \sum_{j=n+1}^{\infty} |\xi_j|$, it is found that $P_n \xrightarrow{s} 1$. On the other hand, $\|P_n\xi - \xi\| \leq \|\xi\|$ and if (e_j) is the canonical basis of $l^1(\mathbb{N})$,

$$\|P_n e_{(n+1)} - e_{(n+1)}\| = \|e_{(n+1)}\| = 1, \qquad \forall n,$$

and so the sequence o operators (P_n) is not convergent in norm. ●

Example 14.7. Let $T_n : c_0 \to c_0$ be the linear operator

$$T_n \xi = (\underbrace{0, 0, \cdots, 0}_{n \text{ entries}}, \xi_1, \xi_2, \xi_3, \cdots).$$

Since $c_0^* = l^1$ (Exercise 13.8), each $f \in c_0^*$ is represented by some $\eta \in l^1$ of the form

$$f(T_n \xi) = \sum_{j=1}^{\infty} (T_n \xi)_j \eta_j = \sum_{j=1}^{\infty} \xi_j \eta_{(j+n)}.$$

It then follows that $|f(T_n \xi)| \le \|\xi\|_\infty \sum_{j=1}^{\infty} |\eta_{(j+n)}|$, which converges to zero as $n \to \infty$. Therefore $T_n \xrightarrow{w} 0$.

However, $\|T_n \xi\| = \|\xi\|$, for all $\xi \in c_0, n \in \mathbb{N}$, and so T_n does not converge strongly to zero. Note that T_n is a linear isometry, for all n, which converges weakly to zero!

●

EXERCISE **14.3.** Show that the sequence of operators $T_n : l^2(\mathbb{N}) \hookleftarrow$

$$T_n \xi = (\underbrace{0, 0, \cdots, 0}_{n \text{ entries}}, \xi_{n+1}, \xi_{n+2}, \xi_{n+3}, \cdots)$$

converges strongly to zero, but does not converge in norm.

EXERCISE **14.4.** Show that the sequence of operators $T_n : l^2(\mathbb{N}) \hookleftarrow$

$$T_n \xi = (\underbrace{0, 0, \cdots, 0}_{n \text{ entries}}, \xi_1, \xi_2, \xi_3, \cdots)$$

converges weakly to zero, but does not converge strongly.

As a reformulation of the Banach-Steinhaus' Theorem as presented in Exercise 7.1, one has

Proposition 14.8. *If (T_n) in $\mathrm{B}(\mathcal{B}, \mathcal{N})$ converges strongly to the linear operator $T : \mathcal{B} \to \mathcal{N}$, then $T \in \mathrm{B}(\mathcal{B}, \mathcal{N})$ and $\|T\| \le \liminf_{n \to \infty} \|T_n\|$.*

Example 14.9. The hypothesis of $\mathcal{B}$ be complete in Proposition 14.8 cannot be dispensed with. Indeed, consider the sequence (T_n) of operators with the same domain $\operatorname{dom} T_n = \{\xi \in l^1(\mathbb{N}) : \xi_j \ne 0 \text{ only for a finite number of indices}\}$,

$$T_n \xi = (\xi_1, 2\xi_2, 3\xi_3, \cdots, n\xi_n, \xi_{n+1}, \xi_{n+2}, \cdots).$$

Then (T_n) is a sequence of bounded operators, $\|T_n\| = n$, that converges strongly to T, with $\operatorname{dom} T = \operatorname{dom} T_n$ and $(T\xi)_j = j\xi_j$ for all j; however, T is not bounded. ●

EXERCISE **14.5.** Show that if (T_n) in $\mathrm{B}(\mathcal{B}, \mathcal{N})$ converges in norm to $T : \mathcal{B} \to \mathcal{N}$, then $T \in \mathrm{B}(\mathcal{B}, \mathcal{N})$ and $\|T\| = \lim_{n \to \infty} \|T_n\|$.

Similarly to Proposition 14.4, it is possible to show

Proposition 14.10. *(T_n) is strongly convergent in $\mathrm{B}(\mathcal{B}_1, \mathcal{B}_2)$ if and only if $(T_n \xi)$ is a Cauchy sequence for ξ in a total set in $\mathcal{B}_1$, and $\{\|T_n\|\}$ is a bounded set.*

EXERCISE **14.6.** Prove Proposition 14.10.

The next result, originally due to L. Schur, is rather surprising!

Example 14.11. In $l^1(\mathbb{N})$ the weak convergence of sequences is equivalent to strong convergence. •

Proof. Let (ξ^n) be a sequence in l^1 that converges weakly; since strong convergence implies weak convergence, it is enough to show that (ξ^n) converges strongly. It is possible to assume that (ξ^n) converges weakly to zero.

If (ξ^n) does not converge strongly to zero, it has a subsequence, also denoted by (ξ^n), with $\|\xi^n\|_1 \geq 5\varepsilon$, for all n, and for some $\varepsilon > 0$. Since this subsequence converges weakly to zero one has

$$\lim_{n\to\infty} \sum_{j=1}^\infty \eta_j \xi_j^n = 0, \qquad \forall \eta = (\eta_1, \eta_2, \eta_3, \cdots) \in l^1(\mathbb{N})^* = l^\infty(\mathbb{N}).$$

By replacing η by the elements e^j of the canonical basis of $l^1(\mathbb{N}) \subset l^\infty(\mathbb{N})$, one gets that $\lim_{n\to\infty} \xi_j^n = 0$ for all fixed j. Note that, for each fixed m, $\sum_{j=1}^m |\xi_j^n| < \varepsilon_0$ for large enough n, and given k one has $\sum_{j=M}^\infty |\xi_j^k| < \varepsilon_0$ for large enough M; such properties will be used in what follows.

Define $m_0 = n_0 = 1$, and inductively, construct the strictly increasing sequences (m_k) and (n_k) in the following way: n_k is the least integer greater than n_{k-1} so that

$$\sum_{j=1}^{m_{k-1}} |\xi_j^{n_k}| < \varepsilon,$$

and m_k as the least integer, greater than m_{k-1}, satisfying

$$\sum_{j=m_k}^\infty |\xi_j^{n_k}| < \varepsilon.$$

Note that in the last inequality it was explicitly used that the sequence (ξ^n) is in l^1, and that both m_k and n_k converge to infinity.

Construct $\eta \in l^\infty$ as follows: $\eta_1 = 1$ and for $m_{k-1} < j \leq m_k$ define $\eta_j = 0$ if $\xi_j^{n_k} = 0$ and $\eta_j = \overline{\xi_j^{n_k}}/|\xi_j^{n_k}|$ if $\xi_j^{n_k} \neq 0$ (the bar denotes complex conjugate); note that $\|\eta\|_\infty = 1$. Thus (by using the inequality $|a| - |b| \leq |a - b|$, $a, b \in \mathbb{F}$)

$$5\varepsilon - \left|\sum_{j=1}^\infty \eta_j \xi_j^{n_k}\right| \leq \sum_{j=1}^\infty |\xi_j^{n_k}| - \left|\sum_{j=1}^\infty \eta_j \xi_j^{n_k}\right| \leq \left|\sum_{j=1}^\infty |\xi_j^{n_k}| - \eta_j \xi_j^{n_k}\right|$$

$$= \left|\left(\sum_{j=1}^{m_{k-1}} + \sum_{m_k+1}^\infty\right)\left(|\xi_j^{n_k}| - \eta_j \xi_j^{n_k}\right)\right|$$

$$\leq \left(\sum_{j=1}^{m_{k-1}} + \sum_{m_k+1}^\infty\right)\left(|\xi_j^{n_k}| + |\eta_j \xi_j^{n_k}|\right) \leq 2\varepsilon + 2\varepsilon = 4\varepsilon.$$

Therefore, $\varepsilon \leq \left|\sum_{j=1}^\infty \eta_j \xi_j^{n_k}\right|$, for all $k > 1$, which is a contradiction with the weak convergence of the subsequence $\xi^{n_k} \xrightarrow{w} 0$; so (ξ^n) converges strongly to zero. Thereby the proof is complete. □

EXERCISE **14.7.** Why the above proof does not adapt to the spaces l^p, $1 < p < \infty$? Check that in l^p, $p > 1$, there are weakly convergent sequences that do not converge in the strong sense.

Notes

The first to introduce the concept of weak convergence was Hilbert, in 1906, for the spaces l^2, although he did not use this terminology. Hilbert has also shown that the unit ball in l^2 is weakly sequentially compact (see next chapters), a useful property that justifies the introduction of weak convergences. The extension of such results to the spaces $L^p[a, b]$ was due to F. Riesz, in his studies of solutions in L^p of certain systems of equations. The proofs that the dual of l^2 is l^2 was due to Hellinger and Toeplitz in 1906 and, in the next year, E. Landau adapted it to general l^p. By using the Landau's results, Riesz has found the dual of L^p. Example 14.11 was originally published by Schur in 1921.

Additional Exercises

EXERCISE **14.8.** If $T \in B(\mathcal{N}_1, \mathcal{N}_2)$ and $\xi_n \rightharpoonup \xi$ in $\mathcal{N}_1$, show that $T\xi_n \rightharpoonup T\xi$ in $\mathcal{N}_2$.

EXERCISE **14.9.** Assume that $\psi_n \xrightarrow{\text{w}} \psi$ in $C[a, b]$. Show that ψ_n converges pointwise to ψ.

EXERCISE **14.10.** Let $\psi \in \mathcal{B} = C[-1, 1]$ with $\psi(-1) = \psi(1) = 0$ and $\psi(0) \neq 0$. For each $n \geq 2$, define $\psi_n(t) = \psi(nt)$ if $|t| \leq 1/n$, and equal to zero otherwise. Show that for all $f \in \mathcal{B}^*$ the sequence $(f(\psi_n))$ is convergent in $\mathbb{F}$, however (ψ_n) does not converge weakly in $\mathcal{B}$ (this does not occur in reflexive spaces; see Exercise 16.8).

EXERCISE **14.11.** Verify the following version of the Uniform Boundedness Principle: If for the subset $\{T_j\}_{j \in J} \subset B(\mathcal{B}, \mathcal{N})$ one has $\sup_{j \in J} |g(T_j\xi)| < \infty, \forall \xi \in \mathcal{B}, g \in \mathcal{N}^*$, then $\sup_{j \in J} \|T_j\| < \infty$. Is this a generalization of that Principle?

EXERCISE **14.12.** The sequence $(\xi_n) \subset \mathcal{N}$ is called *weakly bounded* if $(f(\xi_n))$ is bounded for all $f \in \mathcal{N}^*$. Show that every weakly bounded sequence is bounded and that if (T_n) converges weakly in $B(\mathcal{B}, \mathcal{N})$, then $(\|T_n\|)$ is bounded.

EXERCISE **14.13.** If $\xi_n \rightharpoonup \xi$ in the Banach space $\mathcal{B}$, use Hahn-Banach's Theorem to show that there exists a sequence of linear combination of elements (ξ_n) that converges strongly to ξ. See also Proposition 19.13.

EXERCISE **14.14.** Show that if $T_n \xrightarrow{\text{s}} T$ and $S_n \xrightarrow{\text{w}} S$ in $B(\mathcal{B})$, then $S_n T_n \xrightarrow{\text{w}} ST$. Picking $T_n = \mathbf{1}$, conclude that $S_n T_n$ not necessarily converges strongly to ST.

EXERCISE **14.15.** Let $\{T_n, T\} \subset B(\mathcal{B}_1, \mathcal{B}_2)$ and $\{S_n, S\} \subset B(\mathcal{B}_2, \mathcal{B}_3)$ with $T_n \xrightarrow{\text{s}} T$ and $S_n \xrightarrow{\text{s}} S$. Show that $S_n T_n \xrightarrow{\text{s}} ST$.

EXERCISE **14.16.** For each $n \in \mathbb{N}$, let $T_n : l^2(\mathbb{N}) \hookleftarrow$ be given by

$$T_n\xi = (\underbrace{0, 0, \cdots, 0}_{n \text{ entries}}, \xi_1, \xi_2, \xi_3, \cdots),$$

which converges weakly to zero. Find the adjoint T_n^{a}, show that $T_n^{\text{a}} \xrightarrow{\text{s}} 0$ and therefore, $T_n^{\text{a}} \xrightarrow{\text{w}} 0$. Show, then, that $T_n^{\text{a}} T_n = \mathbf{1}$, for all n, and therefore the conclusions of Exercise 14.15 do not hold if strong convergence is replaced by weak convergence.

EXERCISE **14.17.** Show that the sequence $\xi^n = (\xi_1^n, \xi_2^n, \cdots)$ in $l^p(\mathbb{N})$ $(1 \leq p < \infty)$ converges weakly to $\xi^0 = (\xi_1^0, \xi_2^0, \cdots)$ in $l^p(\mathbb{N})$ if and only if $\{\|\xi^n\|_p\}$ is a bounded set and $\xi_j^n \to \xi_j^0$ for all $j \in \mathbb{N}$ (that is, pointwise convergence of ξ^n to ξ^0). Show that in l^1 such conditions coincide with strong convergence.

CHAPTER 15

Weak Topologies and Alaoglu Theorem

Now, the notion of weak* (pronounced "weak star") convergence of sequences in the dual space is introduced, with basic properties discussed, including examples (it is a "very weak" form of convergence in the dual space). Next, some weak topologies in normed spaces are defined, along with relevant topological subbases. It is underlined that if the normed space is reflexive, then the weak and weak* topologies on its dual space coincide. The chapter concludes with Alaoglu's Theorem and an application to a general characterization of normed spaces.

15.1   WEAK TOPOLOGIES

The weak convergence in a normed space $\mathcal{N}$ is defined by virtue of its dual $\mathcal{N}^*$. In a similar way there is the notion of weak convergence in $\mathcal{N}^*$ through $\mathcal{N}^{**}$. By recalling that $\hat{\mathcal{N}} \subset \mathcal{N}^{**}$ (See Chapter 12), a still weaker notion of convergence in $\mathcal{N}^*$ will be introduced, but now based on $\hat{\mathcal{N}}$.

Definition 15.1. Given a normed space $\mathcal{N}$, a sequence $(f_n) \subset \mathcal{N}^*$ *converges weakly** to $f \in \mathcal{N}^*$ if $\lim_{n \to \infty} \hat{\xi}(f_n) = \hat{\xi}(f)$, for all $\hat{\xi} \in \hat{\mathcal{N}}$ (i.e., pointwise convergence). The notation $f_n \xrightarrow{\mathrm{w}^*} f$ will indicate this kind of convergence.

Proposition 15.2. *If* $f_n \xrightarrow{\mathrm{w}^*} f$ *in* $\mathcal{N}^*$, *then the limit* f *is unique. Moreover,* $\hat{\mathcal{N}}$ *separates points of* $\mathcal{N}^*$ *and if* $\mathcal{N}$ *is Banach, then* $\{\|f_n\|\}$ *is a bounded set.*

Proof. Given $f, g \in \mathcal{N}^*$, assume that for all $\xi \in \mathcal{N}$ one has $\hat{\xi}(f) = \hat{\xi}(g)$; then $f(\xi) = g(\xi)$ for all $\xi \in \mathcal{N}$, that is, $f = g$. This proves the uniqueness of the limit, and as a corollary, that $\hat{\mathcal{N}}$ separates points of $\mathcal{N}^*$.

To prove that $\{\|f_n\|\}$ is bounded the Banach-Steinhaus Theorem will be used, so the necessity of $\mathcal{N}$ be a complete space. Since $f_n \xrightarrow{\mathrm{w}^*} f$, it follows that for all $\xi \in \mathcal{N}$, $|\hat{\xi}(f_n)| = |f_n(\xi)|$ is convergent, hence bounded. By Banach-Steinhaus, it follows that $\{\|f_n\|\}$ is bounded. $\qquad\square$

DOI: 10.1201/9781003656166-15

Proposition 15.3. $\{f_n\} \subset \mathcal{B}^*$ *is weak* convergent if and only if* $(f_n(\xi))$ *is a Cauchy sequence for* ξ *in a total set in* $\mathcal{B}$, *and* $\{\|f_n\|\}$ *is a bounded set.*

EXERCISE **15.1.** Prove Proposition 15.3.

Example 15.4. Let $\mathcal{B} = C[0,1]$ and f, f_n be the following elements of $\mathcal{B}^*$ (in Example 4.15 there is a characterization of this dual space):

$$f(\psi) = \psi(0), \qquad f_n(\psi) = n \int_0^{1/n} \psi(t)\,\mathrm{d}t, \qquad \psi \in \mathcal{B}.$$

Note that $f_n(\psi)$ is the average of ψ over the interval $[0, 1/n]$.

Consider $\psi_n(t) = 4nt(1 - nt)$ if $0 \le t \le 1/n$, and $\psi_n(t) = 0$ if $t \ge 1/n$. Then, for all n, $\|\psi_n\| = 1$, $f(\psi_n) = 0$ and $f_n(\psi_n) = 2/3$, so $\|f - f_n\| \ge 2/3$ in the norm of the space $\mathcal{B}^*$. However, for each $\psi \in \mathcal{B}$, given $\varepsilon > 0$ there exists $N \in \mathbb{N}$ so that $|\psi(t) - \psi(0)| < \varepsilon$, if $0 \le t \le 1/n$, for all $n \ge N$; thus,

$$|f_n(\psi) - f(\psi)| = \left| n \int_0^{1/n} (\psi(t) - \psi(0))\,\mathrm{d}t \right| \le \varepsilon,$$

for $n \ge N$, and one concludes that $\hat{\psi}(f_n) \to \hat{\psi}(f)$ for all $\psi \in \mathcal{B}$, that is, $f_n \xrightarrow{\mathrm{w}^*} f$ in $\mathcal{B}^*$. ●

Let τ_1 and τ_2 be two topologies on a set X; recall that τ_1 is *weaker* than τ_2 if $\tau_1 \subset \tau_2$, that is, every element of τ_1 belongs to τ_2; in this case, it is also said that τ_2 is *stronger* than τ_1. It is useful to note that the weaker the topology the "weaker the notion of convergence" (see Exercise 15.2).

Definition 15.5. The *strong topology* on $\mathcal{N}$ is the metric topology induced by the norm on $\mathcal{N}$. An (open) basis for this topology is given by the open balls $B_{\mathcal{N}}(\xi; \varepsilon)$, with $\xi \in \mathcal{N}, \varepsilon > 0$.

Definition 15.6. The *weak topology* on $\mathcal{N}$ is the topology $\tau(\mathcal{N}, \mathcal{N}^*)$ generated by the linear functionals in $\mathcal{N}^*$, that is, it is the weakest topology in $\mathcal{N}$ for which all elements of $\mathcal{N}^*$ remain continuous. An open subbase of $\tau(\mathcal{N}, \mathcal{N}^*)$ is the collection of sets

$$V(\xi; f; \varepsilon) = f^{-1} B_{\mathbb{F}}(f(\xi); \varepsilon) = \{\eta \in \mathcal{N} : |f(\xi) - f(\eta)| < \varepsilon\},$$

with $\xi \in \mathcal{N}, \varepsilon > 0$, and $f \in \mathcal{N}^*$. Recall that the family of finite intersections of elements of the subbasis is a basis for the topology.

Clearly, the weak topology in $\mathcal{N}^*$ is $\tau(\mathcal{N}^*, \mathcal{N}^{**})$; now another useful topology in the dual space $\mathcal{N}^*$ will be introduced.

Definition 15.7. The *weak* topology* on $\mathcal{N}^*$ is the topology $\tau(\mathcal{N}^*, \hat{\mathcal{N}})$ generated by the linear functionals in $\hat{\mathcal{N}}$, that is, it is the weakest topology on $\mathcal{N}^*$ so that all elements of $\hat{\mathcal{N}}$ remain continuous. A (open) subbasis of $\tau(\mathcal{N}^*, \hat{\mathcal{N}})$ is the collection

$$\begin{aligned} V^*(f; \xi; \varepsilon) &= \hat{\xi}^{-1} B_{\mathbb{F}}(\hat{\xi}(f); \varepsilon) = \left\{ g \in \mathcal{N}^* : |\hat{\xi}(f) - \hat{\xi}(g)| < \varepsilon \right\} \\ &= \left\{ g \in \mathcal{N}^* : |f(\xi) - g(\xi)| < \varepsilon \right\}, \end{aligned}$$

with $f \in \mathcal{N}^*, \xi \in \mathcal{N}$ and $\varepsilon > 0$.

A typical element of the basis generated by the above subbasis for $\tau(\mathcal{N}^*, \hat{\mathcal{N}})$ is

$$V^*(f; \xi_1, \cdots, \xi_n; \varepsilon) = \left\{ g \in \mathcal{N}^* : \max_{1 \leq j \leq n} |\hat{\xi}_j(f) - \hat{\xi}_j(g)| < \varepsilon \right\}.$$

In a similar way, one obtains the elements of the basis for $\tau(\mathcal{N}, \mathcal{N}^*)$:

$$V(\xi; f_1, \cdots, f_n; \varepsilon) = \left\{ \eta \in \mathcal{N} : \max_{1 \leq j \leq n} |f_j(\xi) - f_j(\eta)| < \varepsilon \right\}.$$

Proposition 15.8. *Let $\mathcal{N}$ be a normed space. Then:*

(i) The weak* topology on $\mathcal{N}^*$ is weaker than the weak topology on $\mathcal{N}^*$.

(ii) If $\mathcal{N}$ is reflexive, then the weak and weak* topologies in $\mathcal{N}^*$ coincide (see also Exercise 15.11).

(iii) The weak topology on $\mathcal{N}$ and weak* on $\mathcal{N}^*$ are Hausdorff.

Proof. The first two items follow straightly from the definitions. The third one follows from the fact that the respective set of functionals separates points of the underlying space (Propositions 14.3 and 15.2). Here it will be presented details for the case of weak* topology, the other one being similar.

If $f, g \in \mathcal{N}^*$ and $f \neq g$, then there exists $\xi \in \mathcal{N}$ so that

$$0 < \delta = |f(\xi) - g(\xi)| = \left| \hat{\xi}(f) - \hat{\xi}(g) \right|.$$

Thus, $V^*(f; \xi; \delta/3)$ and $V^*(g; \xi; \delta/3)$ are nonempty open neighborhoods of f and g, respectively; since they are also disjoint, then the weak* topology is Hausdorff. $\square$

EXERCISE **15.2.** A sequence (x_n) in a Hausdorff topological space (X, τ) converges to $x \in X$ if, for each (open) neighborhood $U \in \tau$ of x, one has $x_n \in U$ for all large enough n.

(a) Verify that if ξ_n converges to ξ in the topology $\tau(\mathcal{N}, \mathcal{N}^*)$, then $\xi_n \rightharpoonup \xi$ in $\mathcal{N}$.

(b) Verify that if f_n converges to f in $\tau(\mathcal{N}^*, \hat{\mathcal{N}})$, then $f_n \xrightarrow{\text{w}^*} f$ em $\mathcal{N}^*$.

With respect to the just introduced weak topologies, the next purely algebraic result is central.

Proposition 15.9. *Let $f, f_1, \cdots, f_n$ be linear functionals on the vector space X (without any topology). Then, there exist scalars $a_1, \cdots, a_n$ so that $f = a_1 f_1 + \cdots + a_n f_n$ if and only if $N_0 := \bigcap_{j=1}^n \mathrm{N}(f_j) \subset \mathrm{N}(f)$.*

Proof. The case $f = 0$ is trivial; so, suppose that $f \neq 0$. Note that if f can be written in the form $f = a_1 f_1 + \cdots + a_n f_n$, then clearly $N_0 \subset \mathrm{N}(f)$.

First consider the case of just two functionals f, g with $\mathrm{N}(g) \subset \mathrm{N}(f)$, whose simpler proof illustrates the general case. Choose $\zeta \in X \setminus \mathrm{N}(f)$; hence, $g(\zeta) \neq 0$ and for all $\xi \in X$ the vector $\eta = \xi - g(\xi)\zeta/g(\zeta)$ belongs to $\mathrm{N}(g)$, and since it also belongs to $\mathrm{N}(f)$, it follows that $0 = f(\eta) = f(\xi) - a_1 \, g(\xi)$, that is,

$$f = a_1 \, g, \qquad a_1 = \frac{f(\zeta)}{g(\zeta)}.$$

In the genral case, for each $1 \leq i \leq n$ pick

$$\xi_i \in \left[\left(\bigcap_{j \neq i} N(f_j) \right) \setminus N_0 \right];$$

if some ξ_i does not exist, it is possible to exclude f_i from the arguments (if only one functional f_k remains, pick $\xi_k \in X \setminus N(f)$). Define $\eta_i = \xi_i / f_i(\xi_i)$; then $f_j(\eta_i) = \delta_{ij}$ (Kronecker's delta). Now, for each $\xi \in X$ note that $\eta = \xi - \sum_{j=1}^n f_j(\xi) \eta_j$ belongs to N_0, hence

$$f(\eta) = 0 = f(\xi) - \sum_{j=1}^n a_j f_j(\xi),$$

with $a_j = f(\xi_i)/f_i(\xi_i)$, and so one gets $f = a_1 f_1 + \cdots + a_n f_n$. $\qquad\square$

15.2   ALAOGLU THEOREM

One of the reasons for the introduction of the weak* topology is Alaoglu's Theorem. If $\dim \mathcal{N}^* = \infty$, it is known that $\overline{B}_{\mathcal{N}^*}(0;1)$ is not compact in the usual topology on $\mathcal{N}^*$ (Theorem 2.2). But one has

Theorem 15.10 (Alaoglu). *Let $\mathcal{N}$ be a normed space. Then the closed ball $B^* := \overline{B}_{\mathcal{N}^*}(0;1) = \{f \in \mathcal{N}^* : \|f\| \leq 1\}$ is a compact Hausdorff topological space in the weak* topology.*

Proof. By Proposition 15.8, it is known that B^* with the weak* topology is Hausdorff. To show that B^* is compact, the Tychonov Theorem [Simm] will be used in order to find a compact topological space K, for which B^* is a closed subset of K, hence compact, and the induced topology coincides with the weak* topology. This will finish the proof of the theorem.

For each $\xi \in \mathcal{N}$ one associates $K_\xi = \{z \in \mathbb{F} : |z| \leq \|\xi\|\}$, which is compact in $\mathbb{F}$ and, by Tychonov Theorem, the Cartesian product K of all K_ξ is compact in the product topology. Recall that the product topology on a product space is the weakest topology so that all projections π_t are continuous (see page 129).

Each element of K is a function f which associates, to each $\xi \in \mathcal{N}$, a scalar $f(\xi)$ with $|f(\xi)| \leq \|\xi\|$; thus, the unity ball B^* is the subset of K obtained by the restriction to the functions $f \in K$ that are linear. For $f \in B^*$ consider the families $V^*(f;\xi;\varepsilon)$ (defined above) and

$$U(f;\xi;\varepsilon) := \{g \in K : |f(\xi) - g(\xi)| < \varepsilon\},$$

for ξ in $\mathcal{N}$ and $\varepsilon > 0$. Such families are local subbases of neighborhoods of $f \in B^*$ in the weak* topology and in the product topology, respectively. Since $B^* \subset K \cap \mathcal{N}^*$, it is found that

$$V^*(f;\xi;\varepsilon) \cap B^* = U(f;\xi;\varepsilon) \cap B^*$$

and the weak* topology on B^* and the induced topology on K on B^* coincide. Hence, to finish the proof, it is sufficient to show that B^* is a closed subset of K.

Let g be an element in the closure of B^* in K. By the definition of K one has $|g(\xi)| \le \|\xi\|$, so that to show that $g \in B^*$ it is enough to verify that g is linear. Every neighborhood of g in K intersects B^*; thus, given $\xi, \eta \in \mathcal{N}$ and $\varepsilon > 0$, there exists

$$h \in [B^* \cap U(g; \xi; \varepsilon/3) \cap U(g; \eta; \varepsilon/3) \cap U(g; \xi + \eta; \varepsilon/3)].$$

By the linearity of h, it follows that

$$
\begin{aligned}
|g(\xi + \eta) \quad &- \quad g(\xi) - g(\eta)| \\
&= \quad |g(\xi + \eta) - h(\xi + \eta) - g(\xi) - g(\eta) + h(\xi + \eta)| \\
&= \quad |g(\xi + \eta) - h(\xi + \eta) - g(\xi) + h(\xi) - g(\eta) + h(\eta)| \\
&\le \quad |g(\xi + \eta) - h(\xi + \eta)| + |g(\eta) - h(\eta)| + |g(\xi) - h(\xi)| < \varepsilon.
\end{aligned}
$$

Therefore, $g(\xi + \eta) = g(\xi) + g(\eta)$. Similarly, for all $\xi \in \mathcal{N}$ one obtains $g(\alpha \xi) = \alpha g(\xi)$, for all $\alpha \in \mathbb{F}$, and so g is linear. □

EXERCISE **15.3.** Show that if $\mathcal{B}$ is reflexive, then the unity ball B^* (as in the proof of Alaoglu Theorem) and the ball $\overline{B}_{\mathcal{B}}(0; 1)$ are weakly compact.

The first application of Alaoglu's Theorem is a characterization of normed spaces.

Proposition 15.11. *If $\mathcal{N}$ is a normed space, then there exists a compact Hausdorff topological space X so that $\mathcal{N}$ is isomorphic (via a linear isometric mapping) to a subspace of $C(X)$.*

Proof. By Alaoglu Theorem, it is enough to choose $X = \overline{B}_{\mathcal{N}^*}(0; 1)$ with the weak* topology and the canonical mapping^: $\mathcal{N} \to \hat{\mathcal{N}} \subset \mathcal{N}^{**}$ as the linear isometry between $\mathcal{N}$ and a subspace of $C(X)$. Note that such subspace is closed if and only if $\mathcal{N}$ is a Banach subspace. □

REMARK **15.12.** Proposition 15.11 shows how rich is the set of subspaces of $C(X)$! It will be seen that each Hilbert space is essentially some l^2 (Proposition 22.1).

Notes

Alaoglu's Theorem was published in [Alaog]; there are related proofs due to N. Bourbaki and S. Kakutani. With respect to Proposition 15.11, it is worth recalling that, by Banach-Mazur Theorem, any separable Banach space is isometric to some closed subspace of $C[-1, 1]$.

The weak and weak* topologies, as well as general topologies involving the notion of pointwise convergence, can be introduced in an elegant way by means of families of seminorms.

The strong point in Tychonov Theorem, proved in 1930, is that it holds for the Cartesian product of any family of (compact) sets, and it is closely related to Zorn Lemma; for the enumerable case, there are simpler proofs that use the Cantor diagonal process. It is possible to say that Alaoglu Theorem indicates that the natural topology on the Cartesian product is actually the product topology, weaker than the topology generated by the product of open sets. Some authors risk saying that Alaoglu's Theorem is the most important application of Tychonov's Theorem.

Additional Exercises

EXERCISE **15.4.** Consider $\mathcal{N}$ with the weak topology. If $\zeta \in \mathcal{N}$ and $0 \neq \alpha \in \mathbb{F}$, show that the maps $M_\alpha, T_\zeta : \mathcal{N} \hookleftarrow$, given by $T_\zeta \xi = \zeta + \xi$ and $M_\alpha \xi = \alpha \xi$, are homeomorphisms. Which is the analogue for $\mathcal{N}^*$ with the weak* topology?

EXERCISE **15.5.** $(\psi_n) \subset$ C$[-1,1]$ can be considered in C$[-1,1]^*$ via the mapping $\psi_n(\phi) = \int_{-1}^{1} \phi(t)\psi_n(t)\,\mathrm{d}t$, $\phi \in$ C$[-1,1]$. Show that if $\psi_n \xrightarrow{w^*} \delta$, where $\delta(\phi) = \phi(0)$, then

$$\lim_{n \to \infty} \int_{-1}^{1} \psi_n(t)\,\mathrm{d}t = 1$$

and there exists $C > 0$ so that $\int_{-1}^{1} |\psi_n(t)|\,\mathrm{d}t \leq C$ for all n.

EXERCISE **15.6.** If $\dim \mathcal{N} = \infty$, show that for any given $f_1, \cdots, f_n$ in $\mathcal{N}^*$, there exists a nontrivial vector subspace made up of solutions ξ of $f_j(\xi) = 0$, $j = 1, \cdots, n$ (see Proposition 15.9). Use this to conclude that, in case $\dim \mathcal{N} = \infty$, all nonempty open sets of $\mathcal{N}$ in the weak topology, as well as every nonempty open set of $\mathcal{N}^*$ with the weak* topology, contain elements of arbitrarily large norm. Show that, then, the respective norms are not continuous in these topologies (see also Exercise 21.5).

EXERCISE **15.7.** Based on Exercise 15.6, show that the weak topology on $\mathcal{N}$, with $\dim \mathcal{N} = \infty$, as well as the weak* topology on $\mathcal{N}^*$, is not generated by a norm.

EXERCISE **15.8.** If $\dim \mathcal{N}$ is finite, show that the weak and strong topologies in $\mathcal{N}$ coincide.

EXERCISE **15.9.** Use with the following guide to show a result that will be used in some exercises ahead: "Let X be a vector space and Y a vector space of linear functionals on X that separates points of X. Denote by Z the vector space X with the topology $\tau(X,Y)$ generated by Y. Then the dual Z^* of Z (i.e., continuous linear functionals in this topology) is precisely Y."

(a) If $f \in Z^*$ then $\{\xi \in X : |f(\xi)| < 1\}$ contains an open set of the form $U = \{\xi : |f_j(\xi)| < \varepsilon, 1 \leq j \leq n\}$, for functionals $f_j \in Y$ and some $\varepsilon > 0$, and this guarantees the existence of $C > 0$ so that $|f(\xi)| \leq C \max_{1 \leq j \leq n} |f_j(\xi)|$, for all $\xi \in X$.

(b) Conclude that $\cap_{j=1}^{n} \mathrm{N}(f_j) \subset \mathrm{N}(f)$ and, by Proposition 15.9, there are scalars $a_j, 1 \leq j \leq n$, with $f = \sum_{j=1}^{n} a_j f_j$. Since Y is a vector space it follows that $f \in Y$.

EXERCISE **15.10.** Use Exercise 15.9 to show that:
  (a) the weak topology on $\mathcal{N}$ does not introduce new continuous linear functionals, that is, the set of continuous linear functionals on $\mathcal{N}$, with the weak topology, is again $\mathcal{N}^*$;
  (b) in the weak* topology on $\mathcal{N}^*$ the set of continuous linear functionals on $\mathcal{N}^*$ is again $\hat{\mathcal{N}}$.

EXERCISE **15.11.** Use Exercise 15.9 to show that the weak and weak* topologies in $\mathcal{N}^*$ coincide if and only if $\mathcal{N}$ is reflexive (see Proposition 15.8).

EXERCISE **15.12.** (a) Show that if $E \subset \mathcal{N}^*$ is closed in the weak* topology and bounded (in the norm of $\mathcal{N}^*$), then E is weak* compact.
  (b) Show that $E \subset \mathcal{B}^*$ is weak* compact if and only if E is closed in the weak* topology and bounded (in the norm of $\mathcal{B}^*$).
  (c) If $\mathcal{B}$ is reflexive, verify that (b) holds with the weak* topology replaced by the weak topology (and $\mathcal{B}^*$ replaced by of $\mathcal{B}$, of course).

EXERCISE **15.13.** If the vector space X admits a linear functional $f : X \to \mathbb{F}$ with $\mathrm{N}(f) = \{0\}$, what is possible to say about $\dim X$?

CHAPTER 16

Reflexive Spaces and Sequential Compactness

Continuing with the applications of Alaoglu's Theorem, several important relationships between weak topologies and separable reflexive spaces are examined. A fundamental question addressed is whether, in infinite-dimensional normed spaces, sequences that converge weakly necessarily converge strongly. Additionally, the conditions under which the unit ball in the dual space can be metrized in the weak* topology are investigated. Another relevant problem considered is how the separability of a normed space can be characterized in terms of its dual space. As an application of the results presented, the Krylov-Bogolioubov Theorem, a fundamental result in Ergodic Theory, is proven.

Properties of $\mathcal{N}$ are reflected in $\hat{\mathcal{N}}$, so in the weak* topology. Recall that $\mathcal{N}$ is separable if and only if $\hat{\mathcal{N}}$ is separable.

Proposition 16.1. *If $B^* = \overline{B}_{\mathcal{N}^*}(0;1)$ is metrizable in the weak* topology, then $\mathcal{N}$ is separable.*

Proof. Since B^* is metrizable, there exists a sequence of open subsets (A_n) in B^* (in the weak* topology $\tau(\mathcal{N}^*, \hat{\mathcal{N}})$) in such way that $\cap_n A_n = \{0\}$. Each A_n contains a basic element of this topology of the form

$$V_n^*(0; J_n; \varepsilon_n) = \left\{ f \in \mathcal{N}^* : |\hat{\xi}(f)| < \varepsilon_n, \forall \xi \in J_n \right\},$$

with $J_n \subset \mathcal{N}$ finite. So $J = \cup_n J_n$ is countable and $E := \mathrm{Lin}(J)$ is a separable subspace of $\mathcal{N}$ by Proposition 3.4.

Take $g \in \mathcal{N}^*$ such that the restriction $g|_J = 0$, i.e., $g|_{\mathrm{Lin}(J)} = 0$; hence $g \in A_n$, for all n, and so $g = 0$. By Corollary 12.3, it follows that $\mathrm{Lin}(J)$ is dense in $\mathcal{N}$, showing that $\mathcal{N}$ is separable (Proposition 3.4). □

To discuss a reciprocal result to Proposition 16.1, the following result will be useful.

Proposition 16.2. *Let (X, τ) be a compact topological space. If there exists a sequence (f_n) of continuous functions, $f_n : X \to \mathbb{F}$, that separates points of X, then (X, τ) is metrizable.*

DOI: 10.1201/9781003656166-16

Proof. It is possible to assume that $\|f_n\|_\infty \leq 1$, for all n. Since (f_n) separates points of X, define a metric in X by

$$d(x,y) = \sum_{n=1}^{\infty} \frac{1}{2^n} |f_n(x) - f_n(y)|, \qquad x, y \in X,$$

whose series converges uniformly in $X \times X$, entailing that for every $x \in X$ the function $d_x : (X, \tau) \to [0, \infty)$ given by $d_x(\cdot) := d(x, \cdot)$ is continuous. Thus, the balls $B(x; r) = d_x^{-1}([0, r))$, $x \in X, r > 0$, are open sets in (X, τ) and since such balls constitute a basis for τ_d, then τ is stronger than the topology τ_d generated by that metric in X. To conclude that these topologies are equivalent, it will be shown that every closed set in (X, τ) is also closed in (X, τ_d).

Let F be a closed subset in (X, τ). Because (X, τ) is compact, then F is also compact in (X, τ). Since $\tau_d \subset \tau$, every open cover of F by elements in τ_d is also an open cover in τ and, therefore, F is also compact in (X, τ_d). Since every compact set in a metric space is closed, it follows that F is also closed in (X, τ_d), and so $\tau \subset \tau_d$. Therefore the topologies τ and τ_d coincide. $\qquad\square$

Proposition 16.3. *If $\mathcal{N}$ is separable and $S \subset \mathcal{N}^*$ is compact in the weak* topology, then S is metrizable in the weak* topology.*

Proof. Let (ξ_n) be a dense sequence in $\mathcal{N}$. By definition, $\hat{\xi}_n$ is continuous in the weak* topology; if for all n one has $\hat{\xi}_n(f) = \hat{\xi}_n(g), f, g$ in $\mathcal{N}^*$, i.e., $f(\xi_n) = g(\xi_n)$, it follows that $f = g$, since both are continuous functionals that agree on a dense set in $\mathcal{N}$. Hence $(\hat{\xi}_n)$ is a sequence of continuous functions that separates points of $\mathcal{N}^*$ and, in particular, separates points of S. By Proposition 16.2, S is metrizable in the weak* topology. $\qquad\square$

EXERCISE **16.1.** If $\mathcal{N}$ is separable and $S \subset \mathcal{N}^*$ is weak* compact, show that the convergence of a sequence in S in the associated metric (that in Proposition 16.3) is equivalent to the weak* convergence (Definition 15.1).

EXERCISE **16.2.** If $\mathcal{N}$ is separable, show that every bounded sequence (f_n) in $\mathcal{N}^*$ has a weakly* convergent subsequence. In this case it is said that every bounded subset of $\mathcal{N}^*$ is *weakly* sequentially compact* (see Exercise 16.12).

An important consequence of these results is

Theorem 16.4. *$\mathcal{N}$ is separable if and only if for any $r > 0$, the closed ball $\overline{B}_{\mathcal{N}^*}(0; r)$ is metrizable in the weak* topology.*

Proof. It is enough to consider $r = 1$. If B_1^* is metrizable in the weak* topology, then $\mathcal{N}$ is separable by Proposition 16.1. On the other hand, by Alaoglu Theorem, B_1^* is compact in this topology, and since $\mathcal{N}$ is separable, Proposition 16.3 guarantees that B_1^* is metrizable. $\qquad\square$

Related to Proposition 12.4 one has:

EXERCISE **16.3.** If $\mathcal{B}$ is reflexive and separable, show that $\mathcal{B}^*$ is separable (note that it is also reflexive).

Theorem 16.5. *If $\mathcal{B}$ is reflexive, then any bounded sequence in $\mathcal{B}$ has a weakly convergent subsequence.*

Proof. Let (ξ_n) be a bounded sequence in $\mathcal{B}$ and $E = \overline{\mathrm{Lin}((\xi_n))}$. By construction, E is separable (see Proposition 3.4) and since it is closed, it follows by Proposition 12.10 that it is also reflexive. By Exercise 16.3, E^* is separable. Furthermore, $E^{**} = \hat{E}$ is also separable.

Thus, by using the Exercise 16.2 applied to $\mathcal{N} = E^*$, it is concluded that $(\hat{\xi}_n) \subset \hat{E}$ (with $\hat{\ }$ denoting the canonical application $E \mapsto \hat{E} = E^{**}$) has a subsequence $(h_j := \hat{\xi}_{n_j})$ weakly* convergent to certain $\hat{\xi} \in \hat{E}$, that is, for any $g \in \check{E}^*$ (with $\check{\ }$ denoting the canonical application $E^* \mapsto \check{E}^* \subset E^{***}$) $g(h_j) \to \hat{\xi}$.

Since $g = \check{f}$, for a unique $f \in E^*$, one finds that

$$g(h_j) = \check{f}(h_j) = h_j(f) = \hat{\xi}_{n_j}(f) = f(\xi_{n_j})$$

is convergent to $g(\hat{\xi}) = f(\xi)$ for any $f \in E^*$. However, for all $F \in \mathcal{B}^*$ one has $F|_E \in E^*$, and so $F(\xi_{n_j}) = F|_E(\xi_{n_j}) \to F|_E(\xi) = F(\xi)$ is convergent for all $F \in \mathcal{B}^*$; in other words, (ξ_{n_j}) is weakly convergent to ξ. □

It is then said that in the case of reflexive spaces $\mathcal{B}$, all bounded subsets of $\mathcal{B}$ are *weakly sequentially compact.*

Corollary 16.6. *If $\mathcal{B}$ is reflexive and $\dim \mathcal{B} = \infty$, then there are weakly convergent sequences in $\mathcal{B}$ that are not strongly convergent.*

Proof. By Theorem 2.2, the unity ball $\overline{B}_{\mathcal{B}}(0; 1)$ is not strongly compact, so there exists a sequence $(\xi_n) \subset \overline{B}_{\mathcal{B}}(0; 1)$ which has no strongly convergent subsequence. Apply Theorem 16.5 to the bounded sequence (ξ_n). □

Note that the proof of Corollary 16.6 also implies another conclusion: If $\mathcal{B}$ is reflexive with $\dim \mathcal{B} = \infty$, then it has weakly convergent sequences that have no strongly convergent subsequences.

Example 16.7. Example 14.11 shows that there are cases of nonreflexive Banach spaces for which the conclusions of Corollary 16.6 do not hold, since in $l^1(\mathbb{N})$ the weak and strong convergences are equivalent. •

EXERCISE **16.4.** $C[a, b]$ is not reflexive (see Riesz-Markov Theorem and the Notes in Chapter 4). Show that in $C[a, b]$ there are weakly convergent sequences that are not strongly convergent.

REMARK **16.8.** If $\dim \mathcal{N} = \infty$, the weak and strong topologies are distinct (see, for example, Exercises 15.7 and 16.13), even if the weak and strong convergences are equivalent, as in $l^1(\mathbb{N})$ (Example 14.11).

Now it is discussed an application to Ergodic Theory. If Ω is a compact metric space, say that a Borel measure μ in Ω is *invariant* under the continuous mapping $A : \Omega \hookleftarrow$ if

$$\int_\Omega \psi \, d\mu = \int_\Omega (\psi \circ A) \, d\mu, \qquad \forall \psi \in C(\Omega).$$

The measure μ is said to be a *probability measure* if $\mu(\Omega) = 1$.

Theorem 16.9 (Krylov-Bogolioubov). *Let Ω be a compact metric space and $A : \Omega \hookleftarrow$ continuous. Then A has an invariant probability measure.*

Proof. Set $\mathcal{B} = C(\Omega)$. By Riesz-Markov $\mathcal{B}^* = M(\Omega)$, with $\mu \in M(\Omega)$, $\hat{\psi}(\mu) = \mu(\psi) = \int_\Omega \psi d\mu$, $\psi \in C(\Omega)$. By recalling that $\mathcal{B}$ is separable one has, by Alaoglu Theorem and Proposition 16.3, that $\overline{B}_{\mathcal{B}^*}(0; r)$ is compact and metrizable in the weak* topology, for all $r > 0$; so it is sequentially compact.

For $\xi \in \Omega$, let δ_ξ be the measure $\delta_\xi(\psi) = \psi(\xi)$; the sequence

$$\mu_n = \frac{1}{n} \sum_{j=0}^{n-1} \delta_{A^j(\xi)}$$

is contained in $\overline{B}_{\mathcal{B}^*}(0; r)$ for some $r > 0$ (see the proof of Proposition 16.2) and, thus, it has a weakly* convergent subsequence to some $\nu \in \overline{B}_{\mathcal{B}^*}(0; r)$; such subsequence will also be denoted by μ_n. Hence, for any $\psi \in C(\Omega)$ one has

$$\hat{\psi}(\nu) = \int_\Omega \psi \, d\nu = \lim_{n\to\infty} \hat{\psi}(\mu_n) = \lim_{n\to\infty} \frac{1}{n} \sum_{j=0}^{n-1} \psi(A^j(\xi));$$

now,

$$\int_\Omega \psi \circ A \, d\mu_n - \int_\Omega \psi \, d\mu_n = \frac{1}{n} \left[\psi \circ A^n(\xi) - \psi(\xi) \right],$$

and since ψ is bounded one finds, for $n \to \infty$, that $\frac{1}{n}[\psi \circ A^n(\xi) - \psi(\xi)] \to 0$ and, therefore, $\int_\Omega \psi \circ A \, d\nu = \int_\Omega \psi \, d\nu$, i.e., ν is invariant under A. By choosing the constant function $\psi = 1$ it is found that $\nu(\Omega) = 1$. $\qquad \square$

Notes

The Hilbert spaces are important examples of reflexive spaces (see Corollary 19.5), and they have the advantage of the notion of orthogonality. Such advantage is clear, for instance, in a less abstract version of Corollary 16.6; see Exercise 21.5. It is worth mentioning an interesting result that was not discussed here: $\mathcal{B}$ is reflexive if and only if $\overline{B}_{\mathcal{N}}(0; 1)$ is compact in the weak topology. A short proof of this result can be found in [Oberlin].

Theorem 16.5 is due to Pettis in 1938. Sometimes even in reflexive spaces the weak closure cannot be defined in terms of weak convergence; indeed, in 1929 von Neumann presented cases for which a sequence has a unique accumulation point in the weak topology, however it has no subsequence that converges weakly to that point (see Example 22.4)! The search for invariant measures by virtue of arithmetic means, similar to that in the proof of Theorem 16.9, was initiated in the work [KryBog].

Additional Exercises

EXERCISE **16.5.** Let $\psi_n(t) = e^{int}$ in $L^1[-\pi, \pi]$. Show that although $\psi_n \rightharpoonup 0$, it is not strongly convergent.

EXERCISE **16.6.** Show that on a compact topological space (X, τ), any metric topology weaker than τ is, in fact, equivalent to τ.

EXERCISE **16.7.** Let $\delta_n \in l^2(\mathbb{N})^* = l^2(\mathbb{N})$ be defined by $\delta_n(\xi_1, \xi_2, \xi_3, \cdots) = \xi_n$. Show that $\delta_n \xrightarrow{w^*} 0$, but (δ_n) does not converge in $l^2(\mathbb{N})$.

EXERCISE **16.8.** Use the following guide to show that any reflexive space $\mathcal{B}$ is *weakly sequentially complete*, i.e., if $(\xi_n) \subset \mathcal{B}$ is a sequence for which $(f(\xi_n))$ is convergent in $\mathbb{F}$ for all $f \in \mathcal{B}^*$, then (ξ_n) is weakly convergent (see Exercise 14.10 for the case of a nonreflexive space that is not weakly sequentially complete):

(a) Show that under such conditions there is $R > 0$ such that $\|\xi_n\| \leq R$, for all n (see Exercise 14.12).

(b) Since $\mathcal{B}$ is reflexive, use Alaoglu's Theorem to conclude that $\overline{B}_\mathcal{B}(0; R)$ is weakly compact, so (ξ_n) has a weak accumulation point $\xi \in \overline{B}_\mathcal{B}(0; R)$ (definition left to the reader).

(c) Since $(f(\xi_n))$ is convergent, conclude that $\xi_n \xrightarrow{w} \xi$.

EXERCISE **16.9.** Show that the dual of a reflexive Banach space is weakly sequentially complete (see definition in Exercise 16.8).

EXERCISE **16.10.** Verify that in the so-called *Hilbert cube*

$$\mathcal{C} = \{\xi = (\xi_1, \xi_2, \cdots) \in l^p(\mathbb{N}) : |\xi_j| \leq 1/j, \ \forall j\}, \qquad 1 \leq p < \infty,$$

the weak convergence is equivalent to strong convergence.

EXERCISE **16.11.** Examine the following expression: "If in a bounded and closed subset X, of a reflexive Banach space, the weak convergence of sequences is equivalent to the strong convergence, then X is strongly compact. Therefore the Hilbert cube (Exercise 16.10) in $l^2(\mathbb{N})$ is sequentially compact."

EXERCISE **16.12.** Verify that the sequence $(f_n) \subset l^\infty(\mathbb{N})^*$, $f_n(\xi) = \xi_n$, is bounded but does not have any weak* convergent subsequence; this illustrates Proposition 16.1 (or, better, that the hypothesis of separability of $\mathcal{N}$ cannot be removed in Exercise 16.2).

EXERCISE **16.13.** From the basis for the weak topology on $\mathcal{N}$, show that if $\dim \mathcal{N} = \infty$, then 0 belongs to the weak closure of $S_\mathcal{N}(0; 1)$. Use this to argument that the weak and strong topologies are distinct in infinite-dimensional spaces. What is the weak closure of $S_\mathcal{N}(0; 1)$?

CHAPTER 17

Hilbert Spaces

Hilbert spaces are regarded as the most important class of Banach spaces. In addition to the norm, the notion of the *inner product*, which serves as a generalization of the scalar product on $\mathbb{R}^3$, is also included. The study of the spaces l^2 and L^2 was primarily conducted by Hilbert himself. However, it was around 1930 that the abstract definition of Hilbert space was introduced by von Neumann. This definition was required for the mathematical formulation of Quantum Mechanics. In this chapter, the definition of Hilbert space is provided, and several of their fundamental properties are discussed. Finally, the concept of orthogonality is introduced and analyzed in detail.

17.1 INNER PRODUCT

Definition 17.1. An *inner product* on the vector space X is a functional $(\xi, \eta) \mapsto \langle \xi, \eta \rangle$, defined on $X \times X \to \mathbb{F}$, so that for any $\xi, \eta, \zeta \in X$ and $\alpha \in \mathbb{F}$,

(i) $\langle \alpha\xi + \eta, \zeta \rangle = \bar{\alpha}\langle \xi, \zeta \rangle + \langle \eta, \zeta \rangle$

(ii) $\langle \xi, \eta \rangle = \overline{\langle \eta, \xi \rangle}$

(iii) $\langle \xi, \xi \rangle \geq 0$ and $\langle \xi, \xi \rangle = 0$ if and only if $\xi = 0$.

REMARK **17.2.** As before, the bar $\bar{\alpha}$ denotes the complex conjugate of the scalar α. It will also be used the notation $\|\xi\| = \sqrt{\langle \xi, \xi \rangle}$, since later on it will be seen that this function is a norm. In case it is necessary to make explicit the space X on which the inner product is defined, then the notation $\langle \xi, \eta \rangle_X$ will be invoked.

Note that for each $\xi \in X$ the function $\eta \mapsto \langle \xi, \eta \rangle$ is linear, while for each $\eta \in X$ the function $\xi \mapsto \langle \xi, \eta \rangle$ is *conjugate linear*, the latter is also called *antilinear* (recall that $S : X \to Y$ is antilinear if, for all ξ, η in the vector space X and $\alpha \in \mathbb{F}$, one has $S(\alpha\xi + \eta) = \bar{\alpha}S(\xi) + S(\eta)$). Together such properties are referred to as *sesquilinear* ("sesqui" means one and a half).

EXERCISE **17.1.** Check that if $\xi = 0$ then $\langle \xi, \eta \rangle = \langle \eta, \xi \rangle = 0$, for all $\eta \in X$.

A vector space X with inner product $\langle \cdot, \cdot \rangle$ is called a *inner product space* or *pre-Hilbertian space* and usually, denoted by the pair $(X, \langle \cdot, \cdot \rangle)$ or just by X, when it is clear which is the inner product in question.

DOI: 10.1201/9781003656166-17

Example 17.3. On C$[a,b]$ the function $\langle \psi, \phi \rangle = \int_a^b \overline{\psi(t)} \phi(t)\, dt$ is an inner product.
•

Example 17.4. $\langle \xi, \eta \rangle = \sum_{j=1}^n \overline{\xi_j} \eta_j$ is an inner product on $\mathbb{C}^n$. Similarly, on $\mathbb{R}^n$ (without the complex conjugate, of course). •

Example 17.5. $\langle \xi, \eta \rangle = \sum_{j=1}^\infty \overline{\xi_j} \eta_j$ is an inner product on $l^2(\mathbb{N})$. This series is well defined, namely, it is absolutely convergent, since $l^2(\mathbb{N})^* = l^2(\mathbb{N})$; that is, since $|\overline{\xi_j} \eta_j| \leq 1/2(|\xi_j|^2 + |\eta_j|^2)$, then $\sum_{j=1}^\infty |\overline{\xi_j} \eta_j| \leq 1/2(\|\xi\|^2 + \|\eta\|^2) < \infty$. It is shown in Example 18.2 that l^p is not an inner product space (compatible with the norm) if $p \neq 2$. •

Example 17.6. $\langle \psi, \phi \rangle = \int_\Omega \overline{\psi(t)} \phi(t)\, d\mu(t)$ is an inner product on $L^2_\mu(\Omega)$. This integral is well defined (either by Hölder inequality or) since

$$|\overline{\psi(t)}\, \phi(t)| \leq \frac{1}{2} \left(|\psi(t)|^2 + |\phi(t)|^2 \right),$$

and so $\overline{\psi}\phi \in L^1_\mu(\Omega)$. In Example 18.3 it is shown that $L^p_\mu(\Omega)$ is not an inner product space (compatible with the norm) if $p \neq 2$. •

Proposition 17.7. *Let* $(X, \langle \cdot, \cdot \rangle)$ *be an inner product space. Then for all elements* $\xi, \eta \in X$, *one has:*

(i) (Cauchy-Schwarz inequality) $|\langle \xi, \eta \rangle| \leq \|\xi\| \|\eta\|$; *the equality occurs if and only if* $\{\xi, \eta\}$ *is linearly dependent.*

(ii) (triangle inequality) $\|\xi + \eta\| \leq \|\xi\| + \|\eta\|$; *the equality occurs if and only if* $\xi = 0$ *or* $\eta = t\xi$ *for some* $t \geq 0$.

Proof. *(i)* If $\langle \xi, \eta \rangle = 0$ it is immediate. If $\langle \xi, \eta \rangle \neq 0$, then $\eta \neq 0$ and for $t \in \mathbb{F}$ one has

$$0 \leq \langle \xi - t\eta, \xi - t\eta \rangle = \|\xi\|^2 - t\langle \xi, \eta \rangle - \bar{t}\langle \eta, \xi \rangle + |t|^2 \|\eta\|^2;$$

by choosing $t = \langle \eta, \xi \rangle / \|\eta\|^2$, it follows that $0 \leq \|\xi\|^2 - |\langle \xi, \eta \rangle|^2 / \|\eta\|^2$ and, therefore, $\|\xi\|^2 \|\eta\|^2 \geq |\langle \xi, \eta \rangle|^2$, with the equality holding if and only if there exists $t \in \mathbb{F}$ with $\xi - t\eta = 0$, i.e., $\xi = t\eta$.
    (ii) Item *(i)* yields

$$
\begin{aligned}
\|\xi + \eta\|^2 &= \|\xi\|^2 + \langle \xi, \eta \rangle + \langle \eta, \xi \rangle + \|\eta\|^2 \\
&\leq \|\xi\|^2 + 2|\langle \xi, \eta \rangle| + \|\eta\|^2 \\
&\leq \|\xi\|^2 + 2\|\xi\|\, \|\eta\| + \|\eta\|^2 = (\|\xi\| + \|\eta\|)^2,
\end{aligned}
$$

and the triangle inequality is proven. The equality occurs if and only if $2\|\xi\|\, \|\eta\| = \langle \xi, \eta \rangle + \langle \eta, \xi \rangle = 2\mathrm{Re}\,\langle \xi, \eta \rangle$. Assuming this, by item *(i)* one has

$$|\langle \xi, \eta \rangle| \geq \mathrm{Re}\,\langle \xi, \eta \rangle = \|\xi\|\, \|\eta\| \geq |\langle \xi, \eta \rangle|,$$

that is, $|\langle \xi, \eta \rangle| = \|\xi\|\, \|\eta\| = \mathrm{Re}\,\langle \xi, \eta \rangle = \langle \xi, \eta \rangle \geq 0$. Such conditions include the equality in Cauchy-Schwarz; hence either $\xi = 0$ or there exists $t \in \mathbb{F}$ so that $\eta = t\xi$. Again by the above conditions one gets $0 \leq \langle \xi, \eta \rangle = \langle \xi, t\xi \rangle = t\|\xi\|^2$, and so $t \geq 0$. $\square$

REMARK **17.8.** The condition $t \geq 0$ for the equality in the triangle inequality becomes more natural if one recalls that complex numbers with nonzero imaginary part also "rotate" vectors after multiplication. Negative real numbers "rotate vectors by an angle of 180^0"; for instance, for $\xi \neq 0$ and $t = -1$ one has $0 = \|\xi - \xi\| \neq \|\xi\| + \|\xi\| = 2\|\xi\|$.

At this point, the following consequence is clear:

Corollary 17.9. *On an inner product space* $(X, \langle \cdot, \cdot \rangle)$ *the function*

$$\xi \mapsto \sqrt{\langle \xi, \xi \rangle}, \qquad \xi \in X,$$

is a norm, the so-called induced norm by the inner product. If no norm is mentioned, it is assumed that this is the norm on an inner product space.

EXERCISE **17.2.** Give the details of the proof of Corollary 17.9.

EXERCISE **17.3.** If $\xi_n \to \xi$ and $\eta_n \to \eta$ in an inner product space, show that $\langle \xi_n, \eta_n \rangle \to \langle \xi, \eta \rangle$, that is, the inner product is continuous (under strong convergence).

Definition 17.10. A *Hilbert space* is an inner product space which is complete with respect to the induced norm. In this book $\mathcal{H}, \mathcal{H}_1, \mathcal{H}_2, \cdots$, always denote Hilbert spaces.

Example 17.11. l^2, L^2_μ, $\mathbb{F}^n$ with the norm $\|\cdot\|_2$ (including $\mathbb{F}$) are Hilbert spaces (if no explicit norm is given, it is understood that this is the norm on $\mathbb{F}^n$). $C[a, b]$ with the inner product of Example 17.3 is not a Hilbert space, since it is not complete; check this! ●

EXERCISE **17.4.** Prove the *polarization identity*

$$\langle \eta, \xi \rangle = \frac{1}{4} \left(\|\xi + \eta\|^2 - \|\xi - \eta\|^2 \right),$$

$$\langle \eta, \xi \rangle = \frac{1}{4} \left(\|\xi + \eta\|^2 - \|\xi - \eta\|^2 + i\|\xi + i\eta\|^2 - i\|\xi - i\eta\|^2 \right);$$

the first one for a real inner product space and the second one for a complex inner product space. This distinguished identity shows that the inner product can be recovered from the norm.

17.2   ORTHOGONALITY

Since the inner product is a generalization of the scalar product on $\mathbb{R}^3$, it can be used to define angles between vectors; however, here only the notion of *orthogonality* will be introduced. This concept will appear in several arguments and proofs in this book.

Definition 17.12. Two elements ξ, η in an inner product space X are *orthogonal* if $\langle \xi, \eta \rangle = 0$. The symbol $\xi \perp \eta$ will indicate that such vectors are orthogonal. If E, F are subsets of X, then $E \perp F$ indicates that $\xi \perp \eta$ for all $\xi \in E$ and $\eta \in F$; if also E and F are subspaces, one says that they are orthogonal. $E^\perp$ denotes the set of all vectors in X orthogonal to the set E, namely, $E^\perp := \{\xi \in X : \langle \xi, \eta \rangle = 0, \forall \eta \in E\}$.

EXERCISE **17.5.** (a) Show that if $\xi \perp \eta$ (both nonzero), then $\{\xi, \eta\}$ is a linearly independent set. Note that the null vector is the unique vector orthogonal to any other vector (check this!).

(b) Verify that if $\xi \perp \eta$, then $\|\xi + \eta\|^2 = \|\xi\|^2 + \|\eta\|^2$ (*Pythagoras*).

(c) If $\xi_n \perp \eta$, for all n, and $\xi_n \to \xi$, conclude that $\xi \perp \eta$.

The next characterization of orthogonality between vectors will be used in the text; interpret it geometrically.

Lemma 17.13. *In an inner product space, one has $\xi \perp \eta$ if and only if*

$$\|\xi + t\eta\| \geq \|\xi\|, \qquad \forall t \in \mathbb{F}.$$

Proof. If $\eta = 0$ such inequality is evident. Assume that $\eta \neq 0$. Clearly

$$0 \leq \|\xi + t\eta\|^2 = \|\xi\|^2 + 2\mathrm{Re}\,(t\langle \xi, \eta \rangle) + |t|^2 \|\eta\|^2.$$

If $\xi \perp \eta$, then for all $t \in \mathbb{F}$, $\|\xi + t\eta\|^2 = \|\xi\|^2 + |t|^2 \|\eta\|^2 \geq \|\xi\|^2$, that is, $\|\xi + t\eta\| \geq \|\xi\|$.

On the other hand, if $\|\xi + t\eta\| \geq \|\xi\|$ for all $t \in \mathbb{F}$, taking the square of such expression and then choosing the particular value $t = -\langle \eta, \xi \rangle / \|\eta\|^2$, one obtains $0 \leq -|\langle \xi, \eta \rangle|^2$, so that $\xi \perp \eta$. □

Next, it is pointed out that the completion of an inner product space is well defined as a Hilbert space.

Definition 17.14. A linear operator $U : (X, \langle \cdot, \cdot \rangle) \to (Y, [\cdot, \cdot])$, between two inner product spaces, is *unitary* if it is surjective on Y and $\langle \xi, \eta \rangle = [U\xi, U\eta]$ for all $\xi, \eta \in X$. If there is such unitary operator, then the spaces X and Y are *unitarily equivalent*. This is the natural notion of isomorphism between inner product spaces.

Note that in case of real spaces, it is also used the term *orthogonal operator*, as synonym of unitary operator.

EXERCISE **17.6.** (a) Use the polarization identity to show that every isometric and surjective linear operator between inner product spaces is unitary. This is a typical application of polarization.

(b) Show that every unitary operator is an isometry, so invertible, and that its inverse is also unitary. If U is unitary, check that $UU^{-1} = \mathbf{1} = U^{-1}U$.

Theorem 17.15. *If $(X, \langle \cdot, \cdot \rangle)$ is an inner product space, then it is unitarily equivalent to a dense subspace of a Hilbert space $\mathcal{H}$; such $\mathcal{H}$ is said to be a completion of X. Furthermore, any two completions of X are unitarily equivalent.*

Proof. By Theorem 2.6 and its notation, consider $\mathcal{H} = \tilde{X}$, and so it is only left to show the compatibility of the inner product on X with the one on $\mathcal{H}$, that is, taking into account the proof of Theorem 2.6 it is needed to properly define the inner product on the Banach space $\tilde{X}$.

By the continuity of the inner product (Exercise 17.3), it is possible to define an inner product on $\tilde{X}$ by

$$\langle \tilde{\xi}, \tilde{\eta} \rangle := \lim_{n \to \infty} \langle \xi_n, \eta_n \rangle,$$

which induces the norm on $\tilde{X}$. By the polarization identity (Exercise 17.4), it is found that κ is a unitary operator between X and $\kappa(X)$, both considered as inner product spaces. □

Notes

As already mentioned, David Hilbert has particularly studied the spaces l^2 and L^2. Around 1906, E. Fischer and F. Riesz, independently, showed that $l^2(\mathbb{Z})$ is isomorphic to $L^2[a, b]$, what has certainly motivated the abstract definition of Hilbert spaces by von Neumann around 1930, although he included the condition of separability in his original definition. In 1934 Rellich, Riesz, and Löwig showed that many results hold without the separability condition, and so reaching the current definition of Hilbert space.

It is interesting to note that any Hilbert space is unitarily equivalent to some l^2 (Proposition 22.1). Based on results like Lemma 17.13, there are attractive proposals to extend the notion of orthogonality to Banach spaces; the interested readers are referred, for instance, to [Saidi].

Additional Exercises

EXERCISE **17.7.** Show that Cauchy-Schwarz inequality still holds if the condition $\langle \xi, \xi \rangle = 0 \Rightarrow \xi = 0$, in the definition of inner product, is removed.

EXERCISE **17.8.** (a) If for all ζ in an inner product space one has $\langle \xi, \zeta \rangle = \langle \eta, \zeta \rangle$, show that $\xi = \eta$.
(b) Show that the condition $\|\xi\| = \|\eta\|$, in a real inner product space, implies that $\langle \xi + \eta, \xi - \eta \rangle = 0$, that is, $(\xi - \eta)$ and $(\xi + \eta)$ are orthogonal.

EXERCISE **17.9.** Verify that in an inner product space $\|\xi\| = \max_{\|\eta\|=1} |\langle \xi, \eta \rangle|$.

EXERCISE **17.10.** How many different inner products can induce a given norm on a vector space?

EXERCISE **17.11.** Let $V : \mathcal{N} \to \mathcal{H}$ be linear, isometric and surjective. Show that $\mathcal{N}$ is a Hilbert space.

EXERCISE **17.12.** Let ξ_1, ξ_2, ξ_3 be vectors in an inner product space. Show that the set of these vectors is linearly independent if and only if the determinant of the matrix $(\langle \xi_k, \xi_j \rangle)_{1 \le k, j \le 3}$ is nonzero.

EXERCISE **17.13.** Show that in a Hilbert space the sequence (ξ_n) converges to ξ if and only if $\|\xi_n\| \to \|\xi\|$ and $\langle \xi_n, \xi \rangle \to \|\xi\|^2$.

EXERCISE **17.14.** Given a subset E of an inner product space, show that $E^\perp$ is a closed vector subspace, and if E is also a vector subspace, then $E \cap E^\perp = \{0\}$.

EXERCISE **17.15.** If E is a subspace of an inner product space, show that $\eta \in E^\perp$ if and only if $\|\eta - \xi\| \ge \|\eta\|$, for all $\xi \in E$.

EXERCISE **17.16.** Let $S : \mathcal{H} \hookleftarrow$ with rng $S = \mathcal{H}$ and $\langle S(\xi), S(\eta) \rangle = \langle \xi, \eta \rangle$, for all $\xi, \eta \in \mathcal{H}$. Show that there exists the inverse operator S^{-1}, which satisfies $\langle S^{-1}(\xi), S^{-1}(\eta) \rangle = \langle \xi, \eta \rangle$, and also $\langle S(\xi), \eta \rangle = \langle \xi, S^{-1}(\eta) \rangle$, for all $\xi, \eta \in \mathcal{H}$. Use such results to conclude that S is linear and so, a unitary operator.

EXERCISE **17.17.** Denote by X the space $C[-1, 1]$ with the inner product of Example 17.3, and by Y the space $C[-1, 1]$ with the usual norm $\| \cdot \|_\infty$. Show that the mappings $\mathbf{1} : X \to Y$ (identity), and $S : X \to \mathbb{F}$, $(S\psi)(t) = \psi(0)$ are not continuous, while $T : X \to Y$, $(T\psi)(t) = \int_0^t \psi(s)\, ds$ is continuous. Find the norm of T.

CHAPTER 18

Orthogonal Projection

Two significant aspects related to inner product spaces are examined in detail. First, the norms induced by inner products are characterized through the so-called parallelogram law, which enables the presentation of a broad range of examples. In particular, it is shown that the space $l^p(\mathbb{N})$, equipped with the usual norm, forms an inner product space if and only if $p = 2$. Second, orthogonal projection operators are introduced, and examples are provided to illustrate this fundamental concept, which plays an essential role in a version of the spectral theorem that will be discussed at the end of this book.

18.1 PARALLELOGRAM LAW

EXERCISE **18.1.** Check that in an inner product space $(X, \langle \cdot, \cdot \rangle)$ the *parallelogram law*

$$\|\xi + \eta\|^2 + \|\xi - \eta\|^2 = 2\|\xi\|^2 + 2\|\eta\|^2, \qquad \forall \xi, \eta \in X,$$

holds. Give a geometric interpretation.

Theorem 18.1. *A norm* $\| \cdot \|$ *on a normed space* $\mathcal{N}$ *is induced by an inner product if and only if it satisfies the parallelogram law.*

Proof. If the norm is induced by an inner product then, by Exercise 18.1, it satisfies the parallelogram law. Now, assume that the parallelogram law holds in $\mathcal{N}$; this law will be used to show, initially in the real case, that

$$f(\xi, \eta) := \frac{1}{2} \left(\|\xi + \eta\|^2 - \|\xi\|^2 - \|\eta\|^2 \right), \qquad \xi, \eta \in \mathcal{N},$$

defines an inner product which induces the norm $\| \cdot \|$. The motivation for such f comes from the expansion of $\|\xi + \eta\|^2$ and then solving for $\langle \xi, \eta \rangle$.

Clearly $f(\xi, \xi) = \|\xi\|^2$ and $f(\xi, \xi) = 0 \iff \xi = 0$; also, $f(\xi, \eta) = f(\eta, \xi)$. Therefore, in order to show that f defines an inner product it is enough to check the bilinearity; note that the parallelogram law was not used yet. The following easily checked properties will be used: (a) $f(0, \eta) = 0$, for all η, and (b) if $\xi_n \to \xi$, then $f(\xi_n, \eta) \to f(\xi, \eta)$.

DOI: 10.1201/9781003656166-18

Writing out $f(\xi, \zeta) + f(\eta, \zeta)$, and using the parallelogram law for the two pairs of vectors $(\xi + \zeta), (\eta + \zeta)$ and ξ, η, successively, one obtains

$$
\begin{aligned}
f(\xi, \zeta) + f(\eta, \zeta) &= \frac{1}{4}\left(\|\xi + \eta + 2\zeta\|^2 - \|\xi + \eta\|^2 - 4\|\zeta\|^2\right) \\
&= 2f\left(\frac{1}{2}(\xi + \eta), \zeta\right).
\end{aligned}
$$

By picking $\eta = 0$, it is found that $f(\xi, \zeta) = 2f(\xi/2, \zeta)$, for all ξ, ζ, and so

$$
f(\xi, \zeta) + f(\eta, \zeta) = f(\xi + \eta, \zeta).
$$

Note that, by choosing $\eta = -\xi$, one gets $f(\xi, \zeta) = -f(-\xi, \zeta)$. The above relations implies that $f(n\xi, \eta) = nf(\xi, \eta)$, for all $n \in \mathbb{Z}$. To finish the proof in the real case, it is then enough to check that the generalization of this expression for all scalars in $\mathbb{R}$ holds.

If $0 \neq m \in \mathbb{Z}$ one has (for $n \in \mathbb{Z}$)

$$
f\left(\frac{n}{m}\xi, \eta\right) = f\left(n\frac{\xi}{m}, \eta\right) = nf\left(\frac{\xi}{m}, \eta\right) = \frac{n}{m}mf\left(\frac{\xi}{m}, \eta\right) = \frac{n}{m}f(\xi, \eta).
$$

For $\alpha \in \mathbb{R}$, choose a sequence of rational numbers (q_n) converging to α, and use property (b) to get

$$
f(\alpha\xi, \eta) = \lim_{n\to\infty} f(q_n\xi, \eta) = \lim_{n\to\infty} q_n f(\xi, \eta) = \alpha f(\xi, \eta),
$$

and the theorem is proved in the real case.

For the complex case define

$$
F(\xi, \eta) = f(\xi, \eta) - if(\xi, i\eta), \qquad \xi, \eta \in \mathcal{N},
$$

with f denoting the function employed above (see Lemma 11.2 for the motivation of the definition of F). The aim is to show that F defines an inner product which induces the norm on the complex space. Note that, for all $\xi, \eta \in \mathcal{N}$, the following properties hold: $F(\xi, \xi) = \|\xi\|^2$, $f(i\xi, i\eta) = f(\xi, \eta)$,

$$
F(\xi, i\eta) = f(\xi, i\eta) - if(\xi, -\eta) = i(f(\xi, \eta) - if(\xi, i\eta)) = iF(\xi, \eta).
$$

Now it is sufficient to show that $F(\xi, \eta) = \overline{F(\eta, \xi)}$, and this also shows that the (sesqui)linearity extends to the complex numbers. The other properties then follow in a way similar to the real case discussed above.

Now, the following chain of equalities,

$$
\begin{aligned}
F(\xi, \eta) &= f(\xi, \eta) - if(\xi, i\eta) = f(\eta, \xi) - if(i\eta, \xi) \\
&= f(\eta, \xi) - if(-\eta, i\xi) = f(\eta, \xi) + if(\eta, i\xi) \\
&= \overline{F(\eta, \xi)},
\end{aligned}
$$

completes the proof of the theorem. $\qquad\qquad\qquad\qquad\qquad\qquad\square$

Example 18.2. $l^p(\mathbb{N})$ is an inner product space if and only if $p = 2$. •

Proof. These spaces are Banach, but the parallelogram law does not hold if $p \neq 2$. Indeed, pick $\xi = e_j$ and $\eta = e_{2j}$, with $\{e_j\}$ being the canonical basis of $l^p(\mathbb{N})$ (in the case $p = \infty$ it is not a Schauder basis!); thus, $\|\xi\|^2 = \|\eta\|^2 = 1$ and $\|\xi - \eta\|^2 = \|\xi + \eta\|^2 = 2^{2/p}$, if $p \neq \infty$, and equals to 1 if $p = \infty$. □

Example 18.3. Let $(\Omega, \mathcal{A}, \mu)$ be a measure space with two disjoint subsets $A, B \in \mathcal{A}$, $A \cap B = \emptyset$, $0 < \mu(A) \neq \mu(B) < \infty$. Then $L^p_\mu(\Omega)$ is an inner product space if and only if $p = 2$. Note that such conditions are quite general. •

Proof. Only the case $1 \leq p < \infty$ will be discussed, while $p = \infty$ will be left as an exercise. It is enough to verify that the functions $\psi = \chi_A$ and $\phi = \chi_B$ (the characteristic functions of A and B, respectively) satisfy $\|\psi\|^2_p = \mu(A)^{2/p}$, $\|\phi\|^2_p = \mu(B)^{2/p}$, $\|\psi + \phi\|^2_p = \|\psi - \phi\|^2_p = (\mu(A) + \mu(B))^{2/p}$, and so the parallelogram law does not hold if $p \neq 2$. □

REMARK 18.4. Note that $\mathbb{F}^n$ with the corresponding norms $\|\cdot\|_p$ (similar to those on $l^p(\mathbb{N})$) is an inner product space if and only if $p = 2$, although such norms are all equivalent.

Example 18.5. $C[-1, 1]$ is not an inner product space. •

Proof. The elements $\psi(t) = t$ and $\phi(t) = 1$ satisfy $\|\psi\| = \|\phi\| = 1$, while $\|\psi - \phi\|^2 = \|\psi + \phi\|^2 = 4$, and the parallelogram law does not hold in this space. □

18.2 ORTHOGONAL PROJECTION

Definition 18.6. A vector space X is the *direct sum* of two of its subspaces X_1 and X_2, which is denoted by $X = X_1 \oplus X_2$, if every $\xi \in X$ has a unique representation in the form

$$\xi = \xi_1 + \xi_2, \qquad \xi_1 \in X_1, \xi_2 \in X_2.$$

Theorem 18.7 (Orthogonal Projection). *If E is a closed vector subspace of a Hilbert space $\mathcal{H}$, then*

$$\mathcal{H} = E \oplus E^\perp.$$

Hence $E^\perp$ is called the orthogonal complement of the subspace E in $\mathcal{H}$ (Exercise 18.6 complements the geometric view of this result.)

Proof. Let $\xi \in \mathcal{H}$, $\delta := \inf_{\zeta \in E} \|\xi - \zeta\|$ and $(\eta_n) \subset E$ so that $\|\xi - \eta_n\| \to \delta$. By the parallelogram law, one has

$$2\|\eta_n - \xi\|^2 + 2\|\eta_k - \xi\|^2 = \|\eta_n - \eta_k\|^2 + \|\eta_n + \eta_k - 2\xi\|^2,$$

and since $(\eta_n + \eta_k)/2 \in E$, it follows that

$$\begin{aligned}
\|\eta_n - \eta_k\|^2 &= 2\|\eta_n - \xi\|^2 + 2\|\eta_k - \xi\|^2 - 4\|(\eta_n + \eta_k)/2 - \xi\|^2 \\
&\leq 2\|\eta_n - \xi\|^2 + 2\|\eta_k - \xi\|^2 - 4\delta^2,
\end{aligned}$$

which shows that (η_n) is a Cauchy sequence in E, and so it converges to some η in E, since E is complete (it is closed in $\mathcal{H}$). By the continuity of the norm, it is found that $\|\xi - \eta\| = \delta$.

Since $(t\zeta - \eta) \in E$, for all $\zeta \in E$ and $t \in \mathbb{F}$, one obtains

$$\|(\xi - \eta) + t\zeta\| = \|\xi + (t\zeta - \eta)\| \geq \delta = \|\xi - \eta\|,$$

and therefore $(\xi - \eta) \in E^\perp$ by Lemma 17.13. Thus, one gets the decomposition

$$\xi = \eta + (\xi - \eta), \qquad \eta \in E, (\xi - \eta) \in E^\perp.$$

Now it is only needed to show the uniqueness of such decomposition. Suppose that $\xi = \eta' + \zeta'$, with $\eta' \in E$ and $\zeta' \in E^\perp$; then

$$\eta' + \zeta' = (\xi - \eta) + \eta \Longrightarrow \zeta' - (\xi - \eta) = (\eta - \eta') \in E \cap E^\perp,$$

so both are zero; hence $\zeta' = (\xi - \eta)$ and $\eta' = \eta$. □

Corollary 18.8. *Let $\mathcal{H}$ be a Hilbert space.*

(i) If E is a closed subspace of $\mathcal{H}$, then $(E^\perp)^\perp = E$. So, in this case, E is the orthogonal complement of $E^\perp$.

(ii) If M s a subset of $\mathcal{H}$, then

$$\overline{\mathrm{Lin}(M)} = \mathcal{H} \Longleftrightarrow M^\perp = \{0\}.$$

Proof. (i) Clearly $E \subset E^{\perp\perp} := (E^\perp)^\perp$, and since

$$E \oplus E^\perp = \mathcal{H} = E^{\perp\perp} \oplus E^\perp,$$

then, by uniqueness of the direct sum, it follows that $E = E^{\perp\perp}$ (recall that $E^\perp$ is closed).

(ii) Let $N = \mathrm{Lin}(M)$ and $\overline{N}$ be its closure. It then follows that $M^\perp = N^\perp = \overline{N}^\perp$ (verify the last equality!). Thus,

$$\mathcal{H} = \overline{N} \oplus \overline{N}^\perp = \overline{N} \oplus M^\perp,$$

so that $\overline{N} = \mathcal{H}$ if and only if $M^\perp = \{0\}$. □

REMARK 18.9. The direct sum of vector subspaces does not assume, in principle, the orthogonality between the subspaces under consideration; however, for the case of direct sum of the subspaces E and F of an inner product space X, in the notation $X = E \oplus F$ it will be tacitly assumed that $E \perp F$. In many instances it is also used $F = X \ominus E$ to indicate that F is the orthogonal complement of E in X. Chapter 23 will discuss, in a deeper way, the concept of direct sums of subspaces.

EXERCISE 18.2. Find the orthogonal complement, in $L^2[-1,1]$, of the subspaces generated by: (a) polynomials with null constant term; (b) polynomials with only even powers.

The decomposition $\mathcal{H} = E \oplus E^\perp$, for each closed subspace $E \subset \mathcal{H}$, naturally defines the *orthogonal projection operator* P_E onto E by

$$P_E : \mathcal{H} \to E, \qquad P_E \xi := \xi_E,$$

with $\xi = \xi_E + \xi_{E^\perp}$, $\xi_E \in E$, $\xi_{E^\perp} \in E^\perp$. One says that $(\xi_E + \xi_{E^\perp})$ is the *orthogonal decomposition* of ξ with respect to E, and $P_E \xi = \xi_E$ is the *orthogonal component* of ξ in E, which is also called the *orthogonal projection* of ξ onto E. Similarly, $\xi_{E^\perp}$ is the orthogonal projection of ξ onto $E^\perp$.

EXERCISE **18.3.** Prove the following basic properties of the orthogonal projection operator P_E onto the closed subspace $\{0\} \neq E \subset \mathcal{H}$:

(a) P_E is linear and bounded with $\|P_E\| = 1$.

(b) P_E is surjective, that is, rng $P_E = E$.

(c) $P_E^2 = P_E$ (idempotent) and $(P_E)|_E = \mathbf{1}$ (identity on E).

(d) $E^\perp = N(P_E)$ and $P_{E^\perp} = \mathbf{1} - P_E$ ($\mathbf{1}$ is the identity on $\mathcal{H}$).

(e) $P_E P_{E^\perp} = P_{E^\perp} P_E = 0$ (null operator).

The notion of orthogonal projection is meaningful only on inner product spaces; a *projection operator* onto a general vector space is simply defined as an idempotent operator (see the examples below, and Theorem 20.11 for the case of orthogonal projections). If $\mathcal{R}$ is a projection operator defined on the vector space X, then the reader is invited to check, in Exercise 18.16, that any $\xi \in X$ can be uniquely written in the form $\xi = \eta + \zeta$, $\eta \in$ rng $\mathcal{R}$ and $\zeta \in N(\mathcal{R})$.

Example 18.10. [Nonorthogonal Projection] If $\{\zeta, \eta\}$ is a nonorthogonal basis of $\mathbb{F}^2$, and $\eta^\perp \perp \eta$ with $\langle \eta^\perp, \zeta \rangle = 1$, then the operator $\mathcal{R} : \mathbb{F}^2 \hookleftarrow$, $\mathcal{R}\xi := \langle \eta^\perp, \xi \rangle \zeta$, for all $\xi \in \mathbb{F}^2$, is linear, idempotent, rng $\mathcal{R} = \mathrm{Lin}(\{\zeta\})$, $N(\mathcal{R}) = \mathrm{Lin}(\{\eta\})$ and $\|\mathcal{R}\| = \|\zeta\| \, \|\eta^\perp\|$. Under which conditions one has $\|\mathcal{R}\| = 1$? •

Example 18.11. [Unbounded Projection] The condition $\mathcal{R}^2 = \mathcal{R}$ that defines a projection operator (see also Theorem 20.11) is not enough to guarantee that $\mathcal{R}$ is bounded. Let $0 \leq \varphi \in L^2(\mathbb{R})$ so that $\|\varphi\|_1 = \int_\mathbb{R} \varphi(t) \, dt = 1$. Put dom $\mathcal{R} = L^2(\mathbb{R}) \cap L^1(\mathbb{R})$, and the action in $L^2(\mathbb{R})$

$$(\mathcal{R}\psi)(t) := \psi(t) - \varphi(t) \int_\mathbb{R} \psi(s) \, ds, \qquad \psi \in \text{dom } \mathcal{R}.$$

A direct verification shows that $\mathcal{R}^2 = \mathcal{R}$ on dom $\mathcal{R}$, and $\mathcal{R}\psi = \psi$ if and only if $\int_\mathbb{R} \psi(t) \, dt = 0$, that is, $\mathcal{R}$ is the projection onto the subspace of $L^2(\mathbb{R})$ of functions with vanishing integrals. Now choose $\psi_n(t) = \chi_{[0,n]}(t)/\sqrt{n}$, so that $\psi_n \in$ dom $\mathcal{R}$, $\|\psi_n\|_2 = 1$, and $\|\psi_n\|_1 = \sqrt{n}$, for all n, and

$$
\begin{aligned}
\|\mathcal{R}\psi_n\|_2^2 &= \int_\mathbb{R} \left| \psi_n(t) - \varphi(t)\sqrt{n} \right|^2 dt \\
&= \|\psi_n\|_2^2 + n\|\varphi\|_2^2 - 2 \int_\mathbb{R} \varphi(t) \chi_{[0,n]}(t) \, dt \\
&\geq 1 + n\|\varphi\|_2^2 - 2 \int_\mathbb{R} \varphi(t) \, dt = 1 + n\|\varphi\|_2^2 - 2.
\end{aligned}
$$

Hence, $\|\mathcal{R}\psi_n\|_2 \to \infty$ as $n \to \infty$, and the projection operator $\mathcal{R}$ is not bounded. ●

The notion of orthogonal projection, and orthogonality in general, makes simpler several technical arguments and proofs in Hilbert space theory, at least in comparison with the corresponding arguments for general Banach spaces.

Notes

The first proof of Theorem 18.1 was by Jordan and von Neumann in 1935. There are more recent and elaborate sufficient conditions for a norm to be induced by an inner product, with applications to certain partial differential equations and variational problems.

Fourier, Poisson and others had used orthogonality relations and projections in specific function spaces (trigonometric series, in particular), but without this language, long before the abstract formulation discussed here. Contribution by Sturm and Liouville were important for a more general formulation of such results, culminating with the famous "Sturm-Liouville problem;" in fact, those authors have presented rather general solutions to this "problem" (see, for instance, [LevSar]). The Sturm-Liouville problem was one of the seeds of the spectral theory of operators acting in Hilbert spaces.

In the paper [Clarks], there is a discussion about the existence and uniqueness of the minimum distance in Exercise 18.6 for uniformly convex Banach spaces (which include $L^p(\mathbb{R}^n)$, $1 < p < \infty$).

Additional Exercises

EXERCISE **18.4.** Let (ξ_j) be an orthogonal sequence, that is, $\xi_j \perp \xi_k$, for all $j \neq k$.

(a) Verify the Pythagoras relation

$$\left\| \sum_{j=1}^n \xi_j \right\|^2 = \sum_{j=1}^n \|\xi_j\|^2, \qquad \forall n \in \mathbb{N}.$$

(b) If $\xi = \sum_{j=1}^\infty a_j \xi_j$ (convergent in the norm), show that

$$\|\xi\|^2 = \sum_{j=1}^\infty |a_j|^2 \|\xi_j\|^2.$$

EXERCISE **18.5.** Given a subset J with at least two elements, show that $l^p(J)$ is a Hilbert space if and only if $p = 2$ (see Example 1.3).

EXERCISE **18.6.** Let E be a closed subspace of $\mathcal{H}$ and $\xi \in \mathcal{H}$. Thus, from the decomposition $\mathcal{H} = E \oplus E^\perp$, one has $\xi = \zeta + \eta$, $\zeta \in E, \eta \in E^\perp$. Show that ζ and η are the unique elements of E and $E^\perp$, respectively, whose distances to ξ are minimal. This exercise complements the geometric view of the decomposition stated in Theorem 18.7.

EXERCISE **18.7.** In $L^2[0,\infty)$, consider $E = \text{Lin}(\{e^{-t}, te^{-t}\})$ and $\psi(t) = (1 + t^2)e^{-t}$. Find the decomposition $\psi = \psi_E + \psi_{E^\perp}$, with $\psi_E \in E$ and $\psi_{E^\perp} \in E^\perp$.

EXERCISE **18.8.** Consider $\mathbb{R}^2$ with the norm $\|\|\xi\|\| = |\xi_1| + |\xi_2|$ (in this case $\mathbb{R}^2$ is not an inner product space), with $\xi = (\xi_1, \xi_2)$, and the subspace $E = \{(\xi_1, \xi_1) : \xi_1 \in \mathbb{R}\}$ of $\mathbb{R}^2$. If $\eta = (-1, 1)$, show that

$$d(\eta, E) = \inf_{\xi \in E} \|\|\eta - \xi\|\| = 2,$$

and that there exist infinite many elements $\zeta \in \mathbb{R}^2$ with $d(\eta, E) = \|\|\zeta - \eta\|\|$ (draw the spheres centered at η). Compare with Exercise 18.6.

EXERCISE **18.9.** If $E = \{\xi \in l^2(\mathbb{N}) : \xi_{2j} = 0, \forall j \in \mathbb{N}\}$, show that E is closed and determine $E^\perp$.

EXERCISE **18.10.** Let $\iota : \mathcal{N} \times \mathcal{N} \setminus \{0, 0\} \to \mathbb{R}$ be given by

$$\iota(\xi, \eta) = \frac{1}{2} \frac{\|\xi + \eta\|^2 + \|\xi - \eta\|^2}{\|\xi\|^2 + \|\eta\|^2}.$$

Verify that $(1/2) \inf_{(\xi, \eta)} \iota(\xi, \eta) \le 1 \le \sup_{(\xi, \eta)} \iota(\xi, \eta) \le 2$. In terms of the mapping ι, when is the norm of $\mathcal{N}$ derived from an inner product?

EXERCISE **18.11.** Set $E = \{\psi \in \mathrm{L}^2[-1, 1] : \int_{-1}^{1} \psi(t)\, dt = 0\}$. Determine $E^\perp$.

EXERCISE **18.12.** A subset M of a vector space is *convex* if, for all $\xi, \eta \in M$, one has $t\xi + (1 - t)\eta$ in M for any $t \in [0, 1]$. Show that in a Hilbert space any nonempty closed and convex subset has a unique element with minimal norm.

EXERCISE **18.13.** Let E, F be subsets of $\mathcal{H}$. Show that:

(a) If $E \subset F$, then $F^\perp \subset E^\perp$ and $E^{\perp\perp\perp} = E^\perp$.

(b) $E^{\perp\perp}$ is the smallest closed vector subspace that contains E. Conclude that if E is a vector subspace, then $\overline{E} = E^{\perp\perp}$.

(c) E is a closed subspace of $\mathcal{H}$ if and only if $E = E^{\perp\perp}$.

EXERCISE **18.14.** Show that, for real Hilbert spaces, every operator $S : \mathcal{H} \hookleftarrow$ that preserves distance between vectors, that is, $\|S(\xi) - S(\eta)\| = \|\xi - \eta\|$, for all $\xi, \eta \in \mathcal{H}$, has the form $S(\xi) = S(0) + V\xi$, for some linear isometry $V \in \mathrm{B}(\mathcal{H})$.

EXERCISE **18.15.** If E is a closed subspace of $\mathcal{H}$ and $T \in \mathrm{B}(\mathcal{H})$, then E is invariant under T if $T(E) \subset E$. Show that E is invariant under T if and only if $TP_E = P_E T P_E$, with P_E denoting the orthogonal projection onto E.

EXERCISE **18.16.** Let X be a vector space. If $\mathcal{R} : \mathrm{dom}\,\mathcal{R} \subset X \to X$ is a projection operator, show that any $\xi \in \mathrm{dom}\,\mathcal{R}$ can be uniquely written in the form $\xi = \eta + \zeta$, with $\eta \in \mathrm{rng}\,\mathcal{R}$ and $\zeta \in \mathrm{N}(\mathcal{R})$.

CHAPTER 19

Riesz Representation on Hilbert Spaces

The well-known Riesz Representation Theorem for Hilbert spaces is presented, establishing a natural identification of every Hilbert space with its dual. Important applications of this result are considered. First, it is shown that every Hilbert space is reflexive, a property widely utilized in applications. Second, a characterization of bounded sesquilinear forms on Hilbert spaces is provided. Third, the Hilbert adjoint of a bounded linear operator is introduced. Additionally, the Lax-Milgram Theorem, regarded as a variation of the Riesz Representation Theorem, is discussed. Finally, it is shown that if a sequence in a Hilbert space weakly converges to a vector, then its "arithmetic mean" is guaranteed to strongly converge to the same vector.

19.1 RIESZ REPRESENTATION

Theorem 19.1 (Riesz Representation). *Let $\mathcal{H}^*$ denote the dual of a Hilbert space and $\mathcal{H}^*$. The mapping $\gamma : \mathcal{H} \to \mathcal{H}^*$, $\gamma(\xi) = f_\xi$, for $\xi \in \mathcal{H}$, with action*

$$\gamma(\xi)(\eta) = f_\xi(\eta) = \langle \xi, \eta \rangle, \qquad \forall \eta \in \mathcal{H},$$

is an antilinear and surjective isometry on $\mathcal{H}^*$.

REMARK **19.2.** Hence each element of $\mathcal{H}^*$ is identified with a unique $\xi \in \mathcal{H}$, through f_ξ, and $\|f_\xi\| = \|\xi\|$; it is said that ξ *represents* f_ξ. Two notations were employed for this mapping: $\gamma(\xi)$ and f_ξ, what will be convenient below.

Proof. $f_\xi = 0$ clearly corresponds to $\xi = 0$. For $\xi \in \mathcal{H}$, the mapping $f_\xi(\cdot) = \langle \xi, \cdot \rangle$ is a linear functional and $|f_\xi(\eta)| = |\langle \xi, \eta \rangle| \leq \|\xi\| \|\eta\|$, so that $f_\xi \in \mathcal{H}^*$ with $\|f_\xi\| \leq \|\xi\|$. Since $\|\xi\|^2 = f_\xi(\xi) \leq \|f_\xi\| \|\xi\|$, then $\|f_\xi\| \geq \|\xi\|$. Thus $\|f_\xi\| = \|\xi\|$, and γ is an antilinear isometry. In this way, it is enough to show that every $f \in \mathcal{H}^*$ is of the form f_ξ, for some $\xi \in \mathcal{H}$. If $f = 0$, then $f = f_\xi$ for $\xi = 0$. If $f \neq 0$, since the kernel $\mathrm{N}(f)$ is a proper closed vector subspace (because f is continuous) of $\mathcal{H}$, by Theorem 18.7,

$$\mathcal{H} = \mathrm{N}(f) \oplus \mathrm{N}(f)^\perp,$$

DOI: 10.1201/9781003656166-19

and there exists $\zeta \in N(f)^{\perp}$ with $\|\zeta\| = 1$. Since the vector $(f(\eta)\zeta - f(\zeta)\eta) \in N(f)$, for all $\eta \in \mathcal{H}$ (this easy remark is important here), one gets that

$$\langle \zeta, f(\eta)\zeta - f(\zeta)\eta \rangle = 0, \qquad \forall \eta \in \mathcal{H},$$

that is, $f(\eta) = \langle \overline{f(\zeta)}\zeta, \eta \rangle$, and so $f = \gamma(\overline{f(\zeta)}\zeta)$. □

EXERCISE 19.1. What is the dimension of $N(f)^{\perp}$, for each If $f \in \mathcal{H}^*$?

Example 19.3. For the subspace $\mathcal{N}$ of $l^2(\mathbb{N})$ whose elements have just a finite number of nonzero entries, take $f : \mathcal{N} \to \mathbb{F}$, $f(\eta) = \sum_{j=1}^{\infty} \eta_j / j$, which is an element of $\mathcal{N}^*$, but there is no $\xi \in \mathcal{N}$ with $f = f_\xi$, since the vector $(1, 1/2, 1/3, \cdots) \notin \mathcal{N}$. The conclusion is that the Riesz Representation may fail in chase the inner product space is not complete. ●

Lemma 19.4. $\mathcal{H}^*$ *is a Hilbert space with the inner product*

$$\langle f_\xi, f_\eta \rangle_{\mathcal{H}^*} := \langle \eta, \xi \rangle_{\mathcal{H}}, \qquad f_\xi, f_\eta \in \mathcal{H}^*.$$

Proof. Since γ is an isometry, the parallelogram law is satisfied in $\mathcal{H}^* = \gamma(\mathcal{H})$, so, by Theorem 18.1, $\mathcal{H}^*$ is an inner product space, and since it is complete, it is in fact a Hilbert space. By polarization and the antilinearity of γ, it follows that the inner product is the one in the statement of the lemma. □

Corollary 19.5. *Every Hilbert space is reflexive.*

Proof. One must show that $\hat{\mathcal{H}} = \mathcal{H}^{**}$, i.e., if $g \in \mathcal{H}^{**}$, then there exists $\xi \in \mathcal{H}$ so that $g = \hat{\xi}$. Since every element of $\mathcal{H}^*$ is of the form $\gamma(\eta) = f_\eta$, $\eta \in \mathcal{H}$, it is enough to consider $g(f_\eta)$.

By the Riesz Representation Theorem, given $g \in \mathcal{H}^{**}$ (recall that $\mathcal{H}^*$ and $\mathcal{H}^{**}$ are Hilbert spaces), there is a unique $f_\xi \in \mathcal{H}^*$, for some $\xi \in \mathcal{H}$, so that

$$g(f_\eta) = \langle f_\xi, f_\eta \rangle_{\mathcal{H}^*}, \qquad \forall \eta \in \mathcal{H}.$$

By this expression and Lemma 19.4, for all $\eta \in \mathcal{H}$ one has

$$g(f_\eta) = \langle \eta, \xi \rangle_{\mathcal{H}} = f_\eta(\xi) = \hat{\xi}(f_\eta),$$

that is, $g = \hat{\xi}$. □

19.2 HILBERT ADJOINT AND LAX-MILGRAM

Definition 19.6. A *sesquilinear form* on two normed spaces $\mathcal{N}_1, \mathcal{N}_2$ is a mapping $b : \mathcal{N}_1 \times \mathcal{N}_2 \to \mathbb{F}$, linear on the second variable and antilinear on the first one. b is bounded if its *norm* $\|b\|$ is finite, that is,

$$\|b\| := \sup_{\substack{0 \neq \xi_1 \in \mathcal{N}_1 \\ 0 \neq \xi_2 \in \mathcal{N}_2}} \frac{|b(\xi_1, \xi_2)|}{\|\xi_1\| \|\xi_2\|} < \infty.$$

Proposition 19.7. *If $b : \mathcal{H}_1 \times \mathcal{H}_2 \to \mathbb{F}$ is a bounded sesquilinear form, then there exists a unique operator $T_b \in \mathrm{B}(\mathcal{H}_1, \mathcal{H}_2)$ so that*

$$b(\xi_1, \xi_2) = \langle T_b\xi_1, \xi_2\rangle, \qquad \forall \xi_1 \in \mathcal{H}_1, \xi_2 \in \mathcal{H}_2.$$

Furthermore, $\|T_b\| = \|b\|$.

Proof. For each $\xi_1 \in \mathcal{H}_1$ the functional $L_{\xi_1} : \mathcal{H}_2 \to \mathbb{F}$, $L_{\xi_1}(\xi_2) = b(\xi_1, \xi_2)$ is linear, and since

$$|L_{\xi_1}(\xi_2)| = |b(\xi_1, \xi_2)| \leq \|b\| \, \|\xi_1\| \, \|\xi_2\|,$$

it follows that $\|L_{\xi_1}\| \leq \|b\| \, \|\xi_1\|$ and so $L_{\xi_1} \in \mathcal{H}_2^*$.

By Riesz's Representation Theorem, there exists a unique $\eta_2 \in \mathcal{H}_2$ so that $L_{\xi_1}(\xi_2) = \langle \eta_2, \xi_2\rangle$, for all $\xi_2 \in \mathcal{H}_2$. Define $T_b : \mathcal{H}_1 \to \mathcal{H}_2$ by $T_b\xi_1 = \eta_2$, which is linear and

$$b(\xi_1, \xi_2) = \langle T_b\xi_1, \xi_2\rangle, \qquad \forall \xi_1 \in \mathcal{H}_1, \xi_2 \in \mathcal{H}_2.$$

Note that $T_b = 0$ if and only if b is the null form (the definition should be immediate).

Thus, if $b \neq 0$, since $\langle T_b\xi_1, T_b\xi_1\rangle = b(\xi_1, T_b\xi_1)$, for all $\xi_1 \in \mathcal{H}_1$, one has

$$
\begin{aligned}
\|T_b\| &= \sup_{\substack{0 \neq \xi_1 \\ T_b\xi_1 \neq 0}} \frac{\|T_b\xi_1\|}{\|\xi_1\|} = \sup_{\substack{0 \neq \xi_1 \\ T_b\xi_1 \neq 0}} \frac{|\langle T_b\xi_1, T_b\xi_1\rangle|}{\|\xi_1\| \, \|T_b\xi_1\|} \leq \|b\| \\
&= \sup_{\substack{0 \neq \xi_1 \\ 0 \neq \xi_2}} \frac{|\langle T_b\xi_1, \xi_2\rangle|}{\|\xi_1\| \, \|\xi_2\|} \leq \sup_{\substack{0 \neq \xi_1 \\ 0 \neq \xi_2}} \frac{\|T_b\xi_1\| \, \|\xi_2\|}{\|\xi_1\| \, \|\xi_2\|} = \|T_b\|,
\end{aligned}
$$

showing that $T_b \in \mathrm{B}(\mathcal{H}_1, \mathcal{H}_2)$ and $\|T_b\| = \|b\|$.

The uniqueness of the operator follows from the relation $\langle T_b\xi_1, \xi_2\rangle = \langle S\xi_1, \xi_2\rangle$, for any ξ_1, ξ_2, and so the operators S and T_b coincide. $\qquad\square$

The standard example of bounded sesquilinear form is the inner product on a Hilbert space, and in this case (with $\mathcal{H}_1 = \mathcal{H}_2$) $T_b = \mathbf{1}$.

The above results allow one to define the Hilbert adjoint operator for each $T \in \mathrm{B}(\mathcal{H}_1, \mathcal{H}_2)$, which will be denoted by T^* in order to distinguish it from the Banach adjoint operator T^{a}. In Chapter 20 the relation between these operators will be discussed.

Given $T \in \mathrm{B}(\mathcal{H}_1, \mathcal{H}_2)$, the relation

$$b_T(\xi_2, \xi_1) := \langle \xi_2, T\xi_1\rangle_{\mathcal{H}_2}, \qquad \xi_1 \in \mathcal{H}_1, \xi_2 \in \mathcal{H}_2,$$

defines a sesquilinear form $b_T : \mathcal{H}_2 \times \mathcal{H}_1 \to \mathbb{F}$, and since $|b_T(\xi_2, \xi_1)| \leq \|T\| \, \|\xi_1\| \, \|\xi_2\|$, then b_T is bounded with $\|b_T\| \leq \|T\|$. By following the proof of Proposition 19.7, one concludes that $\|b_T\| = \|T\|$. Also by this proposition, there exists a unique $T^* \in \mathrm{B}(\mathcal{H}_2, \mathcal{H}_1)$ so that

$$\langle T^*\xi_2, \xi_1\rangle = b_T(\xi_2, \xi_1) = \langle \xi_2, T\xi_1\rangle, \qquad \xi_1 \in \mathcal{H}_1, \xi_2 \in \mathcal{H}_2,$$

and, in addition, $\|T^*\| = \|b_T\| = \|T\|$.

Definition 19.8. If $T \in \mathrm{B}(\mathcal{H}_1, \mathcal{H}_2)$, the unique operator $T^* \in \mathrm{B}(\mathcal{H}_2, \mathcal{H}_1)$ just defined is named as *Hilbert adjoint operator* of T. Unless otherwise stated to the contrary, when Hilbert spaces are considered, the term "adjoint" refers to the Hilbert adjoint operator.

Definition 19.9. An operator $T \in \mathrm{B}(\mathcal{H})$ is *normal* if $T^*T = TT^*$. If $T = T^*$, it is called *self-adjoint* (i.e., $\langle T\xi, \eta \rangle = \langle \xi, T\eta \rangle$, for all ξ, η), a particular case of normal operators.

EXERCISE **19.2.** Show that an operator $U \in \mathrm{B}(\mathcal{H}_1, \mathcal{H}_2)$ is unitary if and only if $U^{-1} = U^*$. Note that a unitary operator is also normal.

EXERCISE **19.3.** Show that if $T \in \mathrm{B}(\mathcal{H}_1, \mathcal{H}_2)$, then $T^{**} := (T^*)^* = T$.

There is a variation, due to Lax and Milgram, of Riesz Representation that is useful for studies of certain partial differential equations.

Theorem 19.10 (Lax-Milgram). *If $b : \mathcal{H} \times \mathcal{H} \to \mathbb{F}$ is a bounded sesquilinear form, and there is $c > 0$ so that $|b(\eta, \eta)| \geq c\|\eta\|^2$, for all $\eta \in \mathcal{H}$, then for each $f \in \mathcal{H}^*$ there is a unique $\xi_f \in \mathcal{H}$ so that*

$$f(\eta) = b(\xi_f, \eta), \qquad \forall \eta \in \mathcal{H}.$$

Proof. By Proposition 19.7, there exists a unique $T_b \in \mathrm{B}(\mathcal{H})$ so that $b(\zeta, \eta) = \langle T_b\zeta, \eta \rangle$, for all $\zeta, \eta \in \mathcal{H}$. Since

$$c\|\eta\|^2 \leq |b(\eta, \eta)| = |\langle T_b\eta, \eta \rangle| \leq \|\eta\| \, \|T_b\eta\| \leq \|\eta\|^2 \|T_b\|,$$

that is, $c\|\eta\| \leq \|T_b\eta\| \leq \|\eta\| \, \|T_b\|$ for all $\eta \in \mathcal{H}$, it follows that T_b is invertible, that rng T_b is closed in $\mathcal{H}$, and T_b^{-1} is bounded (see Exercise 19.14).

Let $\zeta \in (\mathrm{rng}\, T_b)^{\perp}$; thus

$$0 = |\langle T_b\zeta, \zeta \rangle| = |b(\zeta, \zeta)| \geq c\|\zeta\|^2,$$

which implies that $\zeta = 0$, and so rng T_b is dense in $\mathcal{H}$. Therefore rng $T_b = \mathcal{H}$ and dom $T_b^{-1} = \mathcal{H}$. By Riesz Representation, $f = f_\xi$ for a unique $\xi \in \mathcal{H}$, and for all $\eta \in \mathcal{H}$, $f(\eta) = \langle \xi, \eta \rangle = b(T_b^{-1}\xi, \eta)$. Select $\xi_f = T_b^{-1}\xi$ to finish the proof. □

Now it is presented a simple and useful technical result, although it is restricted to complex inner product spaces, as illustrated by Example 19.12.

Proposition 19.11. *Let $(X, \langle \cdot, \cdot \rangle)$ be a complex inner product space. If $T : X \hookleftarrow$ is a linear operator and $\langle T\xi, \xi \rangle = 0$, for all $\xi \in X$, then $T = 0$.*

Proof. For all $\alpha \in \mathbb{C}$ and all $\xi, \eta \in X$, one has

$$0 = \langle T(\alpha\xi + \eta), \alpha\xi + \eta \rangle = \bar{\alpha}\langle T\xi, \eta \rangle + \alpha\langle T\eta, \xi \rangle.$$

By picking out, successively, $\alpha = 1$ and $\alpha = -i$ one obtains

$$\langle T\xi, \eta \rangle + \langle T\eta, \xi \rangle = 0 \quad \text{and} \quad \langle T\xi, \eta \rangle - \langle T\eta, \xi \rangle = 0,$$

whose unique solution is $\langle T\xi, \eta \rangle = 0$, for all $\xi, \eta \in X$, that is, T is the zero operator. □

Example 19.12. Consider the rotation R by a right angle on the real Hilbert space $\mathbb{R}^2$ (see Example 20.6). Then $R \neq 0$ whereas $\langle R\xi, \xi \rangle = 0$, for all $\xi \in \mathbb{R}^2$ (cf., Proposition 19.11). If $T, S \in \mathrm{B}(\mathcal{H})$ and $\langle T\xi, \xi \rangle = \langle S\xi, \xi \rangle$ for all $\xi \in \mathcal{H}$, is it true that $T = S$? •

The next result should be compared with Exercise 14.13.

Proposition 19.13. *If* $\xi_n \xrightarrow{w} \xi$ *in* $\mathcal{H}$, *then there exists a subsequence* (ξ_{n_j}) *such that its "arithmetic mean" strongly converges to* ξ, *that is,*

$$\lim_{M\to\infty} \frac{1}{M} \sum_{j=1}^{M} \xi_{n_j} = \xi.$$

Proof. To construct the subsequence, first observe that for all $\eta \in \mathcal{H}$ one has, as $n \to \infty$,

$$f_{\xi_n - \xi}(\eta - \xi) = \langle \xi_n - \xi, \eta - \xi \rangle \to 0,$$

since $\xi_n \xrightarrow{w} \xi$. Put $\xi_{n_1} = \xi_1$. Now pick n_2 such that $|f_{\xi_{n_2}-\xi}(\xi_{n_1} - \xi)| < 1$. Suppose that $n_1, \cdots, n_j$ have been successively selected; then use the above observation to pick n_{j+1} so that $|f_{\xi_{n_{j+1}}-\xi}(\xi_{n_k} - \xi)| < 1/(2j)$, for $k = 1, 2, \cdots, j$. For this subsequence (ξ_{n_j}) one has (let $L = \sup_n \|\xi_n - \xi\|^2 < \infty$; see Proposition 14.3)

$$\left\| \frac{1}{M} \sum_{j=1}^{M} \xi_{n_j} - \xi \right\|^2 = \frac{1}{M^2} \left(\sum_{j=1}^{M} \|\xi_{n_j} - \xi\|^2 + \sum_{k \neq j} f_{\xi_{n_j}-\xi}(\xi_{n_k} - \xi) \right)$$

$$\leq \frac{1}{M^2} \left(ML + 2 \sum_{j=2}^{M} \sum_{k=1}^{j-1} |f_{\xi_{n_j}-\xi}(\xi_{n_k} - \xi)| \right)$$

$$\leq \frac{1}{M^2} \left(ML + 2 \sum_{j=2}^{M} \sum_{k=1}^{j-1} \frac{1}{2(j-1)} \right) \leq \frac{1}{M}(L+1),$$

which vanishes as $M \to \infty$. □

Notes

The Riesz Representation Theorem was first proved in 1907, independently, by F. Riesz and M. Fréchet; so sometimes it is called Riesz-Fréchet Theorem. It was used by von Neumann in an elegant proof of Radon-Nikodym Theorem; see details in [Rudin2].

It is possible to show that any Hilbert space is reflexive (Corollary 19.5) from the theory of uniformly convex spaces, without the explicit use of the Riesz Representation Theorem (see [Brezis]).

Additional Exercises

EXERCISE **19.4.** Show that all bounded sequences in a Hilbert space $\mathcal{H}$ have a weakly convergent subsequence.

EXERCISE **19.5.** Show that if $\dim \mathcal{H} = \infty$, then there are sequences in $\mathcal{H}$ which are weakly convergent but have no strongly convergent subsequences (some specific examples will appear during the discussion of orthonormal bases in Chapter 21).

EXERCISE **19.6.** State and prove an analogous result to the Riesz Representation Theorem for bounded antilinear functionals on $\mathcal{H}$.

EXERCISE **19.7.** Show that if $T, S \in B(\mathcal{H}_1, \mathcal{H}_2)$, then $(S + T)^* = S^* + T^*$, and for $\alpha \in \mathbb{F}$ one has $(\alpha T)^* = \bar{\alpha} T^*$; if $T^{-1} \in B(\mathcal{H}_2, \mathcal{H}_1)$, then $(T^{-1})^* = (T^*)^{-1}$. Moreover, if the product TS is defined then $(TS)^* = S^* T^*$, and therefore, $(T^n)^* = (T^*)^n$, for $T \in B(\mathcal{H})$ and $n \in \mathbb{N}$.

EXERCISE **19.8.** Show that there exists an isometric and bijective mapping between $B(\mathcal{H})$ and the set of bounded sesquilinear forms $b(\cdot, \cdot)$ on $\mathcal{H} \times \mathcal{H}$.

EXERCISE **19.9.** If $T \in B(\mathcal{H})$, show that $N(T) = (\text{rng } T^*)^{\perp}$, and also that $N(T^*) = (\text{rng } T)^{\perp}$. Thus, the adjoint operator may be useful to check if a linear operator is invertible or not.

EXERCISE **19.10.** Use (and check this!) the following variation of the polarization identity

$$4\langle \xi, \eta \rangle = \langle \xi + \zeta, \varrho + \eta \rangle - \langle \xi - \zeta, \varrho - \eta \rangle + i\big[\langle \xi + i\zeta, \varrho + i\eta \rangle - \langle \xi - i\zeta, \varrho - i\eta \rangle\big],$$

for any $\xi, \eta, \zeta, \varrho$ in the complex $\mathcal{H}$, to present another proof of Proposition 19.11.

EXERCISE **19.11.** Let T and S be self-adjoint operators on a Hilbert space. Show that TS is self-adjoint if and only if such operators commute.

EXERCISE **19.12.** Use Riesz Representation's Theorem to show that if $f, g \in \mathcal{H}^*$ have the same kernel, then there exists $a \neq 0$ so that $f = ag$. Recall that, by Proposition 15.9, this result holds with very few structure.

EXERCISE **19.13.** Verify the following assertions:

(a) The set of unitary operators on $\mathcal{H}$ is a group with respect to the product of operators.

(b) Given a subspace $E \subset \mathcal{H}$ and a unitary operator $U : \mathcal{H} \hookleftarrow$, then $U(E^{\perp}) = (UE)^{\perp}$.

(c) If $U : \mathcal{H} \hookleftarrow$ is unitary and $U^2 = \mathbf{1}$, then it is self-adjoint. In addition, any unitary and self-adjoint operator $V \in B(\mathcal{H})$ satisfies $V^2 = \mathbf{1}$.

EXERCISE **19.14.** Show that T_b in the proof of Theorem 19.10 is invertible, rng T_b is closed in $\mathcal{H}$ and T_b^{-1} is bounded.

CHAPTER **20**

Self-Adjoint Operators

First, some examples of Hilbert adjoints of bounded linear operators in Hilbert spaces are presented. Then, several additional results regarding adjoint operators (self-adjoint operators, in particular) are discussed, along with a very useful characterization of orthogonal projection operators, which also helps clarify the concept. The traditional Hellinger-Toeplitz result is presented, and the relationship between the two types of adjoint operators, the "normed adjoint" T^{a} and the "Hilbert adjoint" T^*, when defined on Hilbert spaces, is clarified.

In Chapter 19, the concept of Hilbert adjoint of elements of $B(\mathcal{H})$ was introduced; in the following, simple examples involving this concept are presented.

Example 20.1. Let $\phi \in L_\mu^\infty(\Omega)$ and consider the multiplication operator $\mathcal{M}_\phi : L_\mu^2(\Omega) \hookleftarrow$, discussed in Examples 3.13 and 4.2,

$$(\mathcal{M}_\phi \psi)(t) = \phi(t)\psi(t), \qquad \psi \in L_\mu^2(\Omega).$$

It was shown that $\mathcal{M}_\phi$ is bounded and that $\|\mathcal{M}_\phi\| = \text{ess-sup}|\phi|$. The adjoint of this operator is $\mathcal{M}_\phi^* = \mathcal{M}_{\bar\phi}$, so that it is a normal operator; thus $\mathcal{M}_\phi$ is self-adjoint if and only if ϕ is a real-valued function (μ-a.e., of course). •

Proof. For all $\varphi, \psi \in L_\mu^2(\Omega)$ one has

$$\begin{aligned} \langle \varphi, \mathcal{M}_\phi \psi \rangle &= \int_\Omega \overline{\varphi(t)}\, \phi(t)\psi(t)\, \mathrm{d}\mu(t) \\ &= \int_\Omega \overline{\overline{\phi(t)}\varphi(t)}\, \psi(t)\, \mathrm{d}\mu(t) = \langle \mathcal{M}_{\bar\phi}\varphi, \psi \rangle, \end{aligned}$$

and so $\mathcal{M}_\phi^* = \mathcal{M}_{\bar\phi}$, which is normal since $\mathcal{M}_\phi^*\mathcal{M}_\phi = \mathcal{M}_{\bar\phi\phi} = \mathcal{M}_{\phi\bar\phi} = \mathcal{M}_\phi\mathcal{M}_\phi^*$. Also, $\mathcal{M}_\phi$ is self-adjoint if and only if $\phi = \bar\phi$. $\square$

Example 20.2. Given a basis of $\mathbb{F}^n$, an operator $T \in B(\mathbb{F}^n)$ is represented by a matrix. Then its Hilbert adjoint operator T^* is represented by the complex conjugate of the corresponding transpose matrix. •

Example 20.3. For each fixed $0 \neq s \in \mathbb{R}$, the operator $T_s \in B(L^2(\mathbb{R}))$, $(T_s\psi)(t) = \frac{1}{2s}[\psi(t+s) + \psi(t-s)]$, $\psi \in L^2(\mathbb{R})$, is self-adjoint. •

DOI: 10.1201/9781003656166-20

EXERCISE **20.1.** Find the adjoint operator of the left shift operator $S_l : l^2(\mathbb{N}) \hookleftarrow$ given by $S_l(\xi_1, \xi_2, \cdots) = (\xi_2, \xi_3, \cdots)$.

EXERCISE **20.2.** Show that every eigenvalue of a linear self-adjoint operator is real.

The next result is due to Hellinger-Toeplitz, published in 1910. It has important consequences in the mathematical formulation of many problems. The proof of the Banach space version (Proposition 13.10) can be adapted here, however an alternative proof, which explicitly uses the Uniform Boundedness Principle rather than the Closed Graph Theorem, will be presented.

Proposition 20.4 (Hellinger-Toeplitz). *Let $T : \mathcal{H} \hookleftarrow$ be a linear operator so that*

$$\langle T\eta, \xi \rangle = \langle \eta, T\xi \rangle, \qquad \forall \eta, \xi \in \mathcal{H}.$$

Then $T \in \mathrm{B}(\mathcal{H})$ and it is self-adjoint.

Proof. For each $\eta \in \mathcal{H}$ with $\|\eta\| = 1$ one has, by Cauchy-Schwarz, that the linear functional $\mathrm{L}_\eta : \mathcal{H} \to \mathbb{F}$, given by $\mathrm{L}_\eta(\xi) = \langle T\eta, \xi \rangle = \langle \eta, T\xi \rangle$ is bounded, since $|\mathrm{L}_\eta(\xi)| \leq \|T\eta\| \, \|\xi\|$. Again by Cauchy-Schwarz, one has

$$|\mathrm{L}_\eta(\xi)| = |\langle \eta, T\xi \rangle| \leq \|T\xi\|, \qquad \forall \|\eta\| = 1.$$

By the Uniform Boundedness Principle, there is $C > 0$ so that $\|\mathrm{L}_\eta\| \leq C$, for all η with $\|\eta\| = 1$. Thus

$$\|T\eta\|^2 = |\mathrm{L}_\eta(T\eta)| \leq C\|T\eta\|, \quad \|\eta\| = 1,$$

and T is bounded. By this boundedness and the hypotheses, it follows straightly that T is also self-adjoint. $\qquad\qquad\square$

EXERCISE **20.3.** If $T : \mathcal{H}_1 \to \mathcal{H}_2$ is linear and there exists $S : \mathcal{H}_2 \to \mathcal{H}_1$, not necessarily linear, so that $\langle T\xi, \eta \rangle = \langle \xi, S(\eta) \rangle$ for all $\xi \in \mathcal{H}_1$ and $\eta \in \mathcal{H}_2$, conclude that S is linear, T and S are bounded, and finally, that $T^* = S$.

Proposition 20.5. *Let $\mathcal{H}$ be a complex Hilbert space and $T \in \mathrm{B}(\mathcal{H})$. Then T is self-adjoint if and only if $\langle T\xi, \xi \rangle \in \mathbb{R}$ for all $\xi \in \mathcal{H}$.*

Proof. If T is self-adjoint, then for all $\xi \in \mathcal{H}$, one has

$$\overline{\langle T\xi, \xi \rangle} = \langle \xi, T\xi \rangle = \langle T\xi, \xi \rangle;$$

hence $\langle T\xi, \xi \rangle$ is real. If $\langle T\xi, \xi \rangle$ is real for all $\xi \in \mathcal{H}$, then

$$\begin{aligned}
\langle (T - T^*)\xi, \xi \rangle &= \langle T\xi, \xi \rangle - \langle T^*\xi, \xi \rangle = \overline{\langle \xi, T\xi \rangle} - \langle T^*\xi, \xi \rangle \\
&= \langle \xi, T\xi \rangle - \langle T^*\xi, \xi \rangle = \langle T^*\xi, \xi \rangle - \langle T^*\xi, \xi \rangle = 0.
\end{aligned}$$

Therefore, by Proposition 19.11, $T = T^*$. $\qquad\qquad\square$

Example 20.6. The example in which R denotes a rotation by a right angle on $\mathbb{R}^2$, that is, Example 19.12, shows that, in the case of real Hilbert spaces, it may occur that $\langle R\xi, \xi \rangle$ is (obviously!) real for all vectors ξ, although R is not self-adjoint. Indeed, one has the representations

$$R = \begin{pmatrix} 0 & -1 \\ 1 & 0 \end{pmatrix} \quad \text{and} \quad R^* = \begin{pmatrix} 0 & 1 \\ -1 & 0 \end{pmatrix},$$

obtained by means of the canonical basis of $\mathbb{R}^2$. ●

Proposition 20.7. *If $T \in B(\mathcal{H})$, then $\|TT^*\| = \|T^*T\| = \|T\|^2$. Therefore:*

(i) $T^*T = 0$ if and only if $T = 0$.

(ii) If T is normal, then $\|T^2\| = \|T\|^2$.

Proof. If $T \in B(\mathcal{H})$ one has

$$\begin{aligned} \|T\|^2 \quad &= \quad \sup_{\|\xi\|=1} \|T\xi\|^2 = \sup_{\|\xi\|=1} \langle T\xi, T\xi \rangle = \sup_{\|\xi\|=1} \langle T^*T\xi, \xi \rangle \\ &\leq \quad \sup_{\|\xi\|=1} \|T^*T\| \, \|\xi\|^2 = \|T^*T\| \leq \|T^*\| \, \|T\| = \|T\|^2, \end{aligned}$$

and so $\|T^*T\| = \|T\|^2$. By adapting the roles of T and T^* in this relation one obtains $\|TT^*\| = \|T\|^2$. Then *i)* is immediate from such relation.

Now, if T is normal, it commutes with its adjoint, and so for all $\xi \in \mathcal{H}$ one has $\|T^*T\xi\|^2 = \langle T^*T\xi, T^*T\xi \rangle = \langle T^2\xi, T^2\xi \rangle = \|T^2\xi\|^2$; hence $\|T^2\| = \|T^*T\| = \|T\|^2$. □

REMARK **20.8.** As already mentioned, the unitary operators are the natural isomorphisms between Hilbert spaces. It is possible to extend this notion to define the equivalence of bounded operators: The operators $T \in B(\mathcal{H}_1)$ and $S \in B(\mathcal{H}_2)$ are said to be *unitarily equivalent* if there exists a unitary operator $U : \mathcal{H}_1 \to \mathcal{H}_2$ so that $T = U^*SU$. The Spectral Theorem for self-adjoint operators (considered here only for the compact case; see Chapter 30) states that every bounded self-adjoint operator is unitarily equivalent to some $\mathcal{M}_\phi$ (see Example 20.1), with ϕ a bounded real function (details and references can be found in [de Oliv]).

Example 20.9. The right shift operator $S_r : l^2(\mathbb{N}) \hookleftarrow$, given by $S_r(\xi_1, \xi_2, \xi_3, \cdots) = (0, \xi_1, \xi_2, \cdots)$, is a linear isometry between Hilbert spaces, preserve the inner product (that is, $\langle S_r\xi, S_r\eta \rangle = \langle \xi, \eta \rangle$, for all ξ, η), but it is not unitary, since it is not surjective. ●

Other two results will be discussed in this chapter. The first one clarifies the relation between the two definitions of adjoint operators on Hilbert spaces, namely, T^a and T^*, the Banach and Hilbert adjoint, respectively. The second one is an important characterization of orthogonal projection operators (see page 102) in terms of self-adjointness.

Proposition 20.10. *Let $\gamma_1 : \mathcal{H}_1 \to \mathcal{H}_1^*$ and $\gamma_2 : \mathcal{H}_2 \to \mathcal{H}_2^*$ be the antilinear isometries taking place in the Riesz Representation Theorem in $\mathcal{H}_1$ and $\mathcal{H}_2$, respectively. If $T \in \mathrm{B}(\mathcal{H}_1, \mathcal{H}_2)$, then*

$$T^* = \gamma_1^{-1} \circ T^{\mathrm{a}} \circ \gamma_2.$$

Note that the operator on the right hand side is linear.

Proof. This proof reduces to properly writing some definitions; thus, it is worth recalling them: $\gamma_j(\xi_j)(\eta_j) = f_{\xi_j}(\eta_j) = \langle \xi_j, \eta_j \rangle$, for all $\xi_j, \eta_j \in \mathcal{H}_j$, for $j = 1, 2$, and $T^{\mathrm{a}} : \mathcal{H}_2^* \to \mathcal{H}_1^*$, $(T^{\mathrm{a}} f)(\xi_1) = f(T\xi_1)$, for all $\xi_1 \in \mathcal{H}_1, f \in \mathcal{H}_2^*$.

For all $\xi_1 \in \mathcal{H}_1, \xi_2 \in \mathcal{H}_2$, if one denotes $\eta_1 = \gamma_1^{-1}(T^{\mathrm{a}} f_{\xi_2})$, then

$$
\begin{aligned}
\langle \xi_2, T\xi_1 \rangle &= f_{\xi_2}(T\xi_1) = (T^{\mathrm{a}} f_{\xi_2})(\xi_1) = f_{\eta_1}(\xi_1) = \langle \eta_1, \xi_1 \rangle \\
&= \left\langle \gamma_1^{-1}(T^{\mathrm{a}} f_{\xi_2}), \xi_1 \right\rangle = \left\langle (\gamma_1^{-1} \circ T^{\mathrm{a}} \circ \gamma_2)(\xi_2), \xi_1 \right\rangle.
\end{aligned}
$$

It is then immediate that $T^* = \gamma_1^{-1} \circ T^{\mathrm{a}} \circ \gamma_2$. $\qquad\square$

Theorem 20.11. *Let $P \in \mathrm{B}(\mathcal{H})$. Then P is an orthogonal projection operator (i.e., there exists a closed subspace $E \subset \mathcal{H}$ so that $P = P_E$) if and only if $P^2 = P$ and P is self-adjoint.*

Proof. If $P = P_E$ for some closed subspace $E \subset \mathcal{H}$, then every $\xi \in \mathcal{H}$ can be written in the form $\xi = \xi_E + \xi_{E^\perp}$, $\xi_E \in E$ and $\xi_{E^\perp} \in E^\perp$ (this notation will be used again in this proof). Since $\xi_E = P_E \xi$ one has

$$P^2 \xi = P(P\xi) = P\xi_E = \xi_E = P\xi,$$

and so $P^2 = P$. In order to check that in this case P is also self-adjoint, pick two arbitrary vectors $\xi = \xi_E + \xi_{E^\perp}$ and $\eta = \eta_E + \eta_{E^\perp}$ in $\mathcal{H}$. Since,

$$\langle P\xi, \eta \rangle = \langle \xi_E, \eta_E + \eta_{E^\perp} \rangle = \langle \xi_E, \eta_E \rangle = \langle \xi, \eta_E \rangle = \langle \xi, P\eta \rangle,$$

it follows that P is self-adjoint.

Assume now that P is self-adjoint with $P^2 = P$. Define

$$E := \{ \xi \in \mathcal{H} : P\xi = \xi \} = \mathrm{N}(1 - P).$$

Since $P \in \mathrm{B}(\mathcal{H})$, then E is a closed subspace of $\mathcal{H}$. It will be shown that $P = P_E$. Note that every $\xi \in \mathcal{H}$ can be written in the form

$$\xi = P\xi + (1 - P)\xi,$$

with $P\xi \in E$, since $P(P\xi) = P^2\xi = P\xi$, and $(1 - P)\xi \in E^\perp$; indeed, for all $\eta \in E$ one has $\langle \eta, (1 - P)\xi \rangle = \langle P\eta, (1 - P)\xi \rangle = \langle \eta, (P - P^2)\xi \rangle = 0$. Then, by uniqueness of the direct sum decomposition $\mathcal{H} = E \oplus E^\perp$, it is found that $E = \mathrm{rng}\, P$ (with $P\xi = \xi_E$) and $E^\perp = \mathrm{rng}\, (1 - P)$ (with $(1 - P)\xi = \xi_{E^\perp}$).

In addition, $P\xi_E = P(P\xi) = P^2\xi = P\xi = \xi_E$ and so one concludes that $P|_E = 1|_E$, while $P\xi_{E^\perp} = P(1 - P)\xi = 0$, hence $P|_{E^\perp} = 0$. In summary, $P = P_E$. $\qquad\square$

Notes

It is interesting to note that the set of bounded self-adjoint operators is a real Banach subspace of B($\mathcal{H}$). As described in Exercise 20.4, every operator in B($\mathcal{H}$) is a linear combination of two self-adjoint operators. It is possible to show that any self-adjoint operator in B($\mathcal{H}$) is a linear combination of two unitary operators (see Section 9.5 of [de Oliv]); therefore, any operator in B($\mathcal{H}$) is a linear combination of four unitary operators.

An intriguing analogy: maybe the main difference between operators and functions is that operators may be noncommuting. In the world of Hilbert space operators, the self-adjoint ones would correspond to the real functions, the normal ones to complex functions (real and imaginary parts commute, see Exercise 20.4), and the unitary ones to the bijective functions taking values in the circle $\{z \in \mathbb{C} : |z| = 1\}$. The projection operators would correspond to the characteristic functions, while the simple functions to the linear combinations $\sum_{j=1}^{n} \alpha_n P_{E_n}$.

Additional Exercises

EXERCISE 20.4. Let $T \in$ B($\mathcal{H}$), with complex $\mathcal{H}$. Show that there are unique self-adjoint operators T_R and T_I so that $T = T_R + iT_I$ and $T^* = T_R - iT_I$. Verify that T is normal if and only if T_R and T_I commute; show also that it is unitary if and only if T_R and T_I commute and $T_R^2 + T_I^2 = \mathbf{1}$.

EXERCISE 20.5. Let $\mathcal{H}$ be a Hilbert space and (T_n) a sequence of self-adjoint operators in B($\mathcal{H}$) with $T_n \xrightarrow{w} T \in$ B($\mathcal{H}$). Show that T is also self-adjoint. Verify that this conclusion also holds if weak convergence of operators is replaced by strong or uniform convergence.

EXERCISE 20.6. By Exercise 13.7, show that $T \in$ B($\mathcal{H}_1, \mathcal{H}_2$) is invertible if and only if T^* is invertible. Is it possible to conclude this straightly form the definitions?

EXERCISE 20.7. An operator $T \in$ B($\mathcal{H}$) is *greater than* $\lambda \in \mathbb{R}$ if, for all $\xi \in \mathcal{H}$, $\langle (T - \lambda\mathbf{1})\xi, \xi \rangle \geq 0$. Show that any such operator, acting on a complex $\mathcal{H}$, is self-adjoint. In the particular case in which $\lambda = 0$ the operator T is said to be a *positive operator*.

EXERCISE 20.8. Adapt the proof of Proposition 13.10 to prove the Hellinger-Toeplitz result for the particular case of Hilbert spaces, that is, Proposition 20.4.

EXERCISE 20.9. [Normal Operators] Let $\mathcal{H}$ be a complex Hilbert space and $T \in$ B($\mathcal{H}$). Verify the following items:

(a) T is normal if and only if $\|T\xi\| = \|T^*\xi\|$, for all $\xi \in \mathcal{H}$. Use this to give another proof that $\|T^*T\| = \|TT^*\| = \|T\|^2$, if T is a normal operator.

(b) If T is normal, then N(T) = N(T^*) = (rng T)$^\perp$; hence, if $\alpha \in \mathbb{C}$ is an eigenvalue of the normal operator T, then $\bar{\alpha}$ is an eigenvalue of T^*.

(c) If $\alpha \neq \beta$ are distinct eigenvalues of a normal operator T, then the corresponding eigenspaces are orthogonal.

(d) If $T^2 = T$ and T is normal, then T is self-adjoint and an orthogonal projection operator.

EXERCISE 20.10. Let $T \in$ B($\mathcal{H}$). Show that N(T) = N(T^*T) and conclude that $\overline{\text{rng } T^*} = \overline{\text{rng } T^*T}$.

EXERCISE 20.11. Let P_E and P_F be orthogonal projection operators on closed subspaces $E, F \subset \mathcal{H}$, respectively. Show that if rng $P_E \cap$ N(P_F) $\neq \{0\}$, then $\|P_E - P_F\| = 1$. When is $(P_F - P_E)$ an orthogonal projection operator?

EXERCISE 20.12. Let E, F be closed subspaces of $\mathcal{H}$ and P_E, P_F the corresponding orthogonal projection operators. Prove that the following assertions are equivalent:

(a) The operator $(P_F - P_E)$ is positive (see Exercise 20.7).

(b) For all $\xi \in \mathcal{H}$ one has $\|P_E\xi\| \leq \|P_F\xi\|$.

(c) $E \subset F$.

(d) $P_F P_E = P_E P_F = P_E$.

EXERCISE **20.13.** Let E, F be closed subspaces of $\mathcal{H}$ and P_E, P_F the corresponding orthogonal projection operators. Show that $E \perp F$ if and only if $P_E P_F = P_F P_E = 0$.

EXERCISE **20.14.** Let $T : l^2(\mathbb{Z}) \to l^2(\mathbb{Z})$ be given by $(T\xi)_n = (\xi_{n+1} - \xi_{n-1})$, with $\xi = (\cdots, \xi_{-2}, \xi_{-1}, \xi_0, \xi_1, \xi_2, \cdots)$. Show that T is bounded. Find $\|T\|$ and T^*.

EXERCISE **20.15.** Let $T : l^2(\mathbb{Z}) \to l^2(\mathbb{Z})$ be given by $(T\xi)_n = (\xi_{n+1} + \xi_{n-1})$, with $\xi = (\cdots, \xi_{-2}, \xi_{-1}, \xi_0, \xi_1, \xi_2, \cdots)$. Show that T is bounded and self-adjoint. Find $\|T\|$.

EXERCISE **20.16.** Show that the set of bounded self-adjoint operators is a real Banach subspace of B($\mathcal{H}$).

CHAPTER 21

Orthonormal Bases in Hilbert Spaces

A convenient feature of Hilbert spaces is the existence of orthonormal sets, which can be used to decompose vectors. This allows the concept of "orthogonal coordinates" to be introduced, in a manner similar to the coordinates used in Euclidean spaces. Such a framework brings numerous technical simplifications in both arguments and proofs, while also establishing important and interesting relations between vector norms and these coordinates. Among the key technical relations discussed are the Bessel Inequality, the concept of Hilbert dimension of a Hilbert space, the Fourier coefficients of a vector with respect to an orthonormal set, and a very useful characterization of the conditions under which an orthonormal set constitutes an orthonormal basis of the space.

Definition 21.1. A family $\{\xi_\alpha\}_{\alpha \in J}$ in $\mathcal{H}$ is *orthonormal* if $\|\xi_\alpha\| = 1$, for all $\alpha \in J$, and $\xi_\alpha \perp \xi_\beta$, for all $\alpha \neq \beta \in J$. An *orthonormal basis* of $\mathcal{H}$ is a total orthonormal set $\{\xi_\alpha\}_{\alpha \in J}$, i.e., $\overline{\mathrm{Lin}(\{\xi_\alpha\}_{\alpha \in J})} = \mathcal{H}$.

EXERCISE **21.1.** Show that an orthonormal family is linearly independent.

Example 21.2. The canonical basis $\{e_j\}_{j=1}^n$ of $\mathbb{F}^n$ is a standard example of orthonormal basis. •

Example 21.3. The canonical basis of $l^2(\mathbb{N})$ is an orthonormal basis. •

Example 21.4. $\{e^{int}/\sqrt{2\pi}\}_{n \in \mathbb{Z}}$ is an orthonormal set in $\mathrm{L}^2[0, 2\pi]$. It will be shown that it is also an orthonormal basis (Theorem 22.5). •

Orthonormal bases simplify many approaches and proofs in Hilbert spaces. Given a linearly independent sequence (ξ_n) in $\mathcal{H}$, there exist orthonormal sequences that generate the same vector subspace as (ξ_n), built through the Gram-Schmidt orthonormalization process (see ahead); as a byproduct, this process shows the existence of orthonormal bases for separable $\mathcal{H}$. For general Hilbert spaces, the existence of such bases is proved by applying Zorn Lemma.

Proposition 21.5. *Every nontrivial Hilbert space has an orthonormal basis.*

DOI: 10.1201/9781003656166-21

Proof. Let $\mathcal{H} \neq \{0\}$. Denote by $\mathcal{O}$ the set of all orthonormal families in $\mathcal{H}$, which is not empty since if $0 \neq \xi \in \mathcal{H}$, then $\{\xi/\|\xi\|\}$ is an orthonormal set. Set inclusion defines a partial ordering in $\mathcal{O}$; now, given a total ordered family in $\mathcal{O}$, the union of the members of this family is its upper bound (which belongs to $\mathcal{O}$). By Zorn Lemma, there exists a maximal element M in $\mathcal{O}$. It will be shown that M is an orthonormal basis of $\mathcal{H}$.

If M is not an orthonormal basis, then, by Theorem 18.7, there exists a vector $0 \neq \eta \in \overline{\text{Lin}(M)}^\perp$, and so $M \cup \{\eta/\|\eta\|\}$ would be an orthonormal family, denying the maximally of the set M in $\mathcal{O}$. Therefore M is an orthonormal basis of $\mathcal{H}$. $\qquad\square$

Proposition 21.6 (Gram-Schmidt). *Let $\{\xi_n\}$ be a countable collection of linearly independent vectors in $\mathcal{H}$. Then there exists an orthonormal collection $\{e_n\}$ in $\mathcal{H}$, with the same cardinality of $\{\xi_n\}$, so that, for all m, $\{\xi_1, \cdots, \xi_m\}$ and $\{e_1, \cdots, e_m\}$ generate the same vector subspace.*

Proof. Note, initially, that given an orthonormal set $\{\eta_1, \cdots, \eta_k\}$, the vector $\left(\eta - \sum_{j=1}^k \langle \eta_j, \eta \rangle \eta_j\right)$ is orthogonal to each η_j, and so it is orthogonal to the subspace generated by $\{\eta_1, \cdots, \eta_k\}$.

The sequence (e_n) will be constructed by recurrence. Define $e_1 = \xi_1/\|\xi_1\|$; so clearly e_1 and ξ_1 generate the same vector subspace. Suppose that $e_1, \cdots, e_k$ were constructed; then, define

$$\xi'_{k+1} = \xi_{k+1} - \sum_{j=1}^k \langle e_j, \xi_{k+1} \rangle e_j, \qquad e_{k+1} = \xi'_{k+1}/\|\xi'_{k+1}\|.$$

Since $\{\xi_1, \cdots, \xi_{k+1}\}$ is linearly independent, the set $\{e_1, \cdots, e_k, \xi_{k+1}\}$ is also linearly independent, because $\{\xi_1, \cdots, \xi_k\}$ and $\{e_1, \cdots, e_k\}$ generate the same subspace. Thus, by the observation at the beginning of this proof, e_{k+1} is orthogonal to $\text{Lin}(\{e_1, \cdots, e_k\})$. By construction, the sets $\{\xi_1, \cdots, \xi_{k+1}\}$ and $\{e_1, \cdots, e_{k+1}\}$ generate the same vector subspace, and the latter set is orthonormal. $\qquad\square$

Proposition 21.7. *Let $\{e_\alpha\}_{\alpha \in J}$ be an orthonormal set in $\mathcal{H}$. Then the following two assertions are equivalent:*

(i) $\{e_\alpha\}_{\alpha \in J}$ is a basis of $\mathcal{H}$.

(ii) If $\xi \in \mathcal{H}$ satisfies $\xi \perp e_\alpha$, for all $\alpha \in J$, then $\xi = 0$.

Proof. $(i) \Rightarrow (ii)$ Let $\{e_\alpha\}$ be an orthonormal basis and $\xi \perp e_\alpha$, for all $\alpha \in J$. By definition, for each $\varepsilon > 0$ there exists a finite linear combination $\sum_{j=1}^n a_{\alpha_j} e_{\alpha_j}$ (a_{α_j} are scalars) that satisfies

$$\|\xi\|^2 + \sum_{j=1}^n |a_{\alpha_j}|^2 = \left\|\xi - \sum_{j=1}^n a_{\alpha_j} e_{\alpha_j}\right\|^2 < \varepsilon,$$

implying that $\|\xi\|^2 < \varepsilon$. Therefore $\xi = 0$.

$(ii) \Rightarrow (i)$ If $M = \text{Lin}(\{e_\alpha\}_{\alpha \in J})$, since $\overline{M}^\perp = (\{e_\alpha\}_\alpha)^\perp$, by Corollary 18.8, one has

$$\mathcal{H} = \overline{M} \oplus (\{e_\alpha\}_{\alpha \in J})^\perp.$$

Now (ii) implies that $(\{e_\alpha\}_{\alpha \in J})^\perp = \{0\}$, and so $\{e_\alpha\}_{\alpha \in J}$ is a maximal orthonormal set, that is, $\overline{M} = \mathcal{H}$ and $\{e_\alpha\}_{\alpha \in J}$ is an orthonormal basis. □

Proposition 21.8 (Bessel Inequality). *If $\{\xi_\alpha\}_{\alpha \in J}$ is an orthonormal set in $\mathcal{H}$, then for each $\xi \in \mathcal{H}$ one has*

$$\sum_{\alpha \in J} |\langle \xi_\alpha, \xi \rangle|^2 \leq \|\xi\|^2.$$

In particular, $\langle \xi_\alpha, \xi \rangle \neq 0$ only for a countable number of indices $\alpha \in J$.

Proof. Consider, initially, a countable orthonormal family $\{\xi_j\}$. Given $\xi \in \mathcal{H}$, write $\eta_n = \xi - \sum_{j=1}^n \langle \xi_j, \xi \rangle \xi_j$, which is orthogonal to all ξ_j with $1 \leq j \leq n$. Note that

$$\|\xi\|^2 = \|\eta_n\|^2 + \sum_{j=1}^n |\langle \xi_j, \xi \rangle|^2 \geq \sum_{j=1}^n |\langle \xi_j, \xi \rangle|^2.$$

Thus, if J is finite the proof terminates. If J is not finite, from this inequality it follows that

$$\|\xi\|^2 \geq \sum_j |\langle \xi_j, \xi \rangle|^2,$$

for all countable orthonormal sets $\{\xi_j\}$. For each $m \geq 1$, denote by $J_m = \{\alpha \in J : |\langle \xi_\alpha, \xi \rangle| \geq 1/m\}$. From the above relation it follows that J_m is finite for all m. Since

$$\{\alpha \in J : \langle \xi_\alpha, \xi \rangle \neq 0\} = \bigcup_{m=1}^\infty J_m,$$

one concludes that for each $\xi \in \mathcal{H}$ the set of indices for which $\langle \xi_\alpha, \xi \rangle \neq 0$ is countable. □

EXERCISE **21.2.** Show that a Hilbert space is separable if and only if it has a countable orthonormal basis.

Corollary 21.9. *All orthonormal bases of a given Hilbert space have the same cardinality.*

Proof. Let $\mathcal{H}$ be a Hilbert space. If $\mathcal{H}$ has a finite orthonormal basis, the proof is left as an exercise. Suppose then that $\mathcal{H}$ has an infinite orthonormal basis. Let $U = \{\xi_\alpha\}_{\alpha \in J}$ and $V = \{\eta_\beta\}_{\beta \in K}$ be two orthonormal bases of $\mathcal{H}$. For each $\alpha \in J$, by Bessel inequality the set

$$V_\alpha := \{\eta_\beta \in V : \langle \eta_\beta, \xi_\alpha \rangle \neq 0\}$$

is countable and, by Proposition 21.7, each η_β belongs at least to one V_α; thus, $V = \bigcup_{\alpha \in J} V_\alpha$.

Since V_α is countable (recall that $\aleph_0$ is the "smallest infinite cardinality"), it follows that [cardinality of V] $\leq$ [cardinality of U]. In a similar way one shows the reverse inequality. Therefore the bases U and V have the same cardinality. □

This last result allows one to associate a specific cardinality to each Hilbert space.

Definition 21.10. The *Hilbert dimension*, or simply *dimension*, of a Hilbert space is the cardinality of an orthonormal basis of this space.

EXERCISE **21.3.** Select properly sets J to construct Hilbert spaces $l^2(J)$ of arbitrary Hilbert dimension.

EXERCISE **21.4.** Show that if there exists a unitary operator $U : \mathcal{H}_1 \to \mathcal{H}_2$, then the range, under U, of an orthonormal basis of $\mathcal{H}_1$ is an orthonormal basis of $\mathcal{H}_2$. Conclude then that these spaces have the same Hilbert dimension.

It is possible to use an orthonormal basis of a Hilbert space to define vector coordinates. In this context, such coordinates are also termed Fourier coefficients.

Definition 21.11. Given an orthonormal set $\{\xi_\alpha\}_{\alpha \in J}$ in a Hilbert space $\mathcal{H}$, the family $\{\langle \xi_\alpha, \xi \rangle\}_{\alpha \in J}$ is called the *Fourier coefficients* of $\xi \in \mathcal{H}$, and

$$\sum_{\alpha \in J} \langle \xi_\alpha, \xi \rangle \xi_\alpha$$

is called the *Fourier series* of ξ with respect to $\{\xi_\alpha\}_{\alpha \in J}$.

With reference to Fourier series, given an orthonormal family, it is a natural step to give criteria for equality in Bessel inequality, and also criteria for this family to be a basis. Such criteria coincide.

Theorem 21.12. *Let $\{\xi_\alpha\}_{\alpha \in J}$ be an orthonormal set in $\mathcal{H}$. Then the following assertions are equivalent:*

(i) $\{\xi_\alpha\}_{\alpha \in J}$ is an orthonormal basis of $\mathcal{H}$.

(ii) If $\xi \in \mathcal{H}$, then the Fourier series of ξ, with respect to $\{\xi_\alpha\}_{\alpha \in J}$, converges to ξ (and independently of the sum order), that is,

$$\xi = \sum_{\alpha \in J} \langle \xi_\alpha, \xi \rangle \xi_\alpha, \qquad \forall \xi \in \mathcal{H}.$$

(iii) [Parseval Identity] For all $\xi \in \mathcal{H}$ one has

$$\|\xi\|^2 = \sum_{\alpha \in J} |\langle \xi_\alpha, \xi \rangle|^2.$$

Proof. If $\xi \in \mathcal{H}$, then $\langle \xi_\alpha, \xi \rangle \neq 0$ only for a countable subset of indices α, which here will be supposed to be $\mathbb{N}$ or $\{1, 2, \cdots, n\}$ in case it is finite; these choices are only for notational convenience.

(i)$\Rightarrow$(ii) By Bessel inequality, $\sum_{j \in J} |\langle \xi_j, \xi \rangle|^2$ is convergent, thus

$$\left\| \sum_{j=n}^{m} \langle \xi_j, \xi \rangle \xi_j \right\|^2 = \sum_{j=n}^{m} |\langle \xi_j, \xi \rangle|^2 \to 0, \quad \text{as } n, m \to \infty,$$

so that $\sum_{j\in J}\langle\xi_j,\xi\rangle\xi_j$ is convergent, since the sequence of partial sums (of nonzero terms) forms a Cauchy sequence.

By defining $\eta = \xi - \sum_j\langle\xi_j,\xi\rangle\xi_j$ and using the continuity of the inner product, it follows that $\langle\xi_\alpha,\eta\rangle = 0$ for all $\alpha \in J$, and since $\{\xi_\alpha\}_{\alpha\in J}$ is an orthonormal basis, by Proposition 21.7, one finds that $\eta = 0$, i.e.,

$$\xi = \sum_{\alpha\in J}\langle\xi_\alpha,\xi\rangle\xi_\alpha, \qquad \forall\xi \in \mathcal{H}.$$

(ii)⇒*(iii)* From the notation in the above proof, it follows that for $n < \infty$

$$\left|\|\xi\|^2 - \sum_{j=1}^{n}|\langle\xi_j,\xi\rangle|^2\right| = \left\|\xi - \sum_{j=1}^{n}\langle\xi_j,\xi\rangle\xi_j\right\|^2,$$

which converges to zero as $n \to \infty$, because $\|\eta\| = 0$.

(iii)⇒*(i)* Suppose that Parseval Identity holds and that $\langle\xi_\alpha,\xi\rangle = 0$, for all $\alpha \in J$. Then $\xi = 0$, and by Proposition 21.7, it follows that $\{\xi_\alpha\}_{\alpha\in J}$ is an orthonormal basis of $\mathcal{H}$. ☐

Notes

Many terms in Mathematics are in honor of personalities that gave notable contributions to the development of a theory; for example, the term "Fourier coefficients" in this chapter is due to their resemblance with the trigonometric series studied by Fourier. Something similar occurs with the Bessel Inequality and the Parseval Identity, discovered by these researchers in the particular setting of trigonometric series, around 1800.

As it frequently happens, a technique used in a specific case that becomes applied to various problems is then termed "method." This is the case of the Gram-Schmidt process. In 1883, J. P. Gram introduced a generalization of the minimum square process developed by Legendre, Gauss and Tchebychev, and used, apparently for the first time, the orthonormalization discussed here; of course there were contributions by Erhard Schmidt, one of Hilbert's students. The Gram-Schmidt process is also a powerful tool in some numerical calculations.

In optimization and approximation theories, barely touched in Theorem 18.7 and Exercises 18.6 and 21.8, there is remarkable a result of 1914 by Muntz, which states that given a sequence $(n_j) \subset \mathbb{N}$, then $\mathrm{Lin}((t^{n_j}))$ is dense in $L^2[0,1]$ if and only if $\sum_j 1/n_j = \infty$. Observe that if this series diverges, then it is possible to omit any finite number of terms (and sometimes even infinitely many of them) of (t^{n_j}) that the closure of the generated subspace is still $L^2[0,1]$; this strongly contrasts with orthonormal subsets, for which it is not possible to omit any element without getting a proper generated linear subspace.

Those who had the opportunity to read a book on Quantum Mechanics for physicists can recognize the so-called Dirac notation in the relation dealt with in Exercise 21.9; for this it is enough to ignore the "vectors" $'\langle\xi'$ and $'\eta\rangle'$ in that relation and replace $\langle\xi,\eta\rangle$ by the identity operator. With somewhat Functional Analysis experience (and maybe also in Physics) it is possible, from such relations, to appreciate Dirac's motivation to introduce his "δ function," which was fully rigorously treated in Distribution Theory, many years later (whose general approach was initiated in the 1940s by L. Schwartz).

Additional Exercises

EXERCISE **21.5.** (a) Show that every orthonormal sequence in a Hilbert space converges weakly to zero, but has no strongly convergent subsequence. Compare with the results related to reflexive Banach spaces in Chapter 16.

(b) In spite of item a), show that if $\xi_n \xrightarrow{w} \xi$ in $\mathcal{N}$, then $\|\xi\| \le \liminf_{n \to \infty} \|\xi_n\|$. Item a) shows that in some cases the equality does not occur.

EXERCISE **21.6.** If $\psi \in L^2[0, 2\pi]$, use Bessel inequality to prove Riemann-Lebesgue Lemma in this space:

$$\lim_{n \to \infty} \int_0^{2\pi} e^{-int} \psi(t) \, dt = 0.$$

EXERCISE **21.7.** By using that $L^2[0, 2\pi]$ is dense in $L^1[0, 2\pi]$, show that Riemann-Lebesgue Lemma (Exercise 21.6) holds for $\psi \in L^1[0, 2\pi]$.

EXERCISE **21.8.** Let $\{\xi_j\}$ be an orthonormal sequence in the Hilbert space $\mathcal{H}$ and $E_n = \mathrm{Lin}(\{\xi_1, \cdots, \xi_n\})$. Given $\xi \in \mathcal{H}$, show that the minimum of $d(\xi, E_n) = \inf_{\eta \in E_n} \|\xi - \eta\|$ is attained, and only for $\eta = \sum_{j=1}^n \langle \xi_j, \xi \rangle \xi_j$.

EXERCISE **21.9.** Show that the orthonormal set $\{\xi_\alpha\}_{\alpha \in J}$ is an orthonormal basis of $\mathcal{H}$ if and only if for any $\xi, \eta \in \mathcal{H}$, on has $\langle \xi, \eta \rangle = \sum_{\alpha \in J} \langle \xi, \xi_\alpha \rangle \langle \xi_\alpha, \eta \rangle$.

EXERCISE **21.10.** Use Theorem 21.12 to show that the transformation $\gamma : \mathcal{H} \to \mathcal{H}^*$, in Riesz Representation 19.1, is surjective. Note that this approach is independent of that result by Riesz.

EXERCISE **21.11.** In each case below, apply the Gram-Schmidt orthonormalization process to the first three elements that appear.

(a) $\mathcal{H} = L^2[-1, 1]$, and $\xi_j(t) = t^j$, $j = 0, 1, 2, 3, \cdots$. The resulting orthonormal polynomials are called *Legendre polynomials*.

(b) $\mathcal{H} = L^2(\mathbb{R})$, and $\xi_j(t) = t^j e^{-t^2/2}$, $j = 0, 1, 2, 3, \cdots$. The resulting orthonormal polynomials are called *Hermite polynomials*.

(c) $\mathcal{H} = L^2[0, \infty)$, and $\xi_j(t) = t^j e^{-t/2}$, $j = 0, 1, 2, 3, \cdots$. The resulting orthonormal polynomials are called *Laguerre polynomials*.

EXERCISE **21.12.** For general normed spaces, there is no notion of orthogonality; however, there exists a version of the Gram-Schmidt process for such spaces. Given an infinite-dimensional and separable normed space $\mathcal{N}$, based on the construction in the Gram-Schmidt process 21.6 and in Riesz Lemma 2.1, construct, for each $0 < \alpha < 1$, a sequence of vectors (ξ_n), with $\|\xi_n\| = 1$, for all n, such that $\|\xi_j - \xi_k\| \ge \alpha$, for all $j \ne k$, and with $\mathrm{Lin}(\{\xi_n\})$ dense in $\mathcal{N}$.

CHAPTER 22

Fourier Series

First, it is shown that each Hilbert space is unitarily equivalent to some l^2. In other words, abstractly, the collection of all l^2 spaces reproduces all Hilbert spaces, establishing a foundational result in Functional Analysis. Second, the classical Fourier Series are discussed within their natural setting, namely $L^2[a, b]$, which brings an important example of an orthonormal basis. This discussion provides insight into the role of abstract Fourier Series. At the conclusion of the chapter, the Riesz Representation Theorem is applied to present an integration theory in Hilbert spaces.

22.1 FOURIER SERIES

Proposition 22.1. *Given a Hilbert space $\mathcal{H}$, there exists a set J so that $\mathcal{H}$ is unitarily equivalent to $l^2(J)$.*

Proof. Let $\{\xi_j\}_{j \in J}$ be an orthonormal basis of $\mathcal{H}$; so the cardinality of J is the Hilbert dimension of $\mathcal{H}$. Define the operator $U : \mathcal{H} \to l^2(J)$ by $(U\xi)_j = \langle \xi_j, \xi \rangle$, $j \in J$, which is linear and isometric, since by Parseval

$$\|U\xi\|_{l^2}^2 = \sum_{j \in J} |\langle \xi_j, \xi \rangle|^2 = \|\xi\|_{\mathcal{H}}^2.$$

To finish the proof it is enough to show that U is surjective on $l^2(J)$.

The set $\{e_j := U\xi_j\}_{j \in J}$ is the canonical basis of $l^2(J)$, since $\|e_j\| = 1$ and their components are $(e_j)_k = \langle e_k, e_j \rangle = \delta_{j,k}$ (Kronecker's δ). Thus, rng U contains $\mathrm{Lin}(\{e_j\}_{j \in J})$, which is a dense set in $l^2(J)$, and since U is isometric, it follows that rng U is a closed set. Therefore, $U(\mathcal{H}) = l^2(J)$ and U is a unitary operator. $\square$

EXERCISE 22.1. Use the following arguments to show that rng $U = l^2(J)$ in the proof of Proposition 22.1. By Pythagoras, show that for $f = (f_j) \in l^2(J)$, the partial sums of the series $\sum_j f_j \xi_j$ form a Cauchy sequence in $\mathcal{H}$, hence convergent to some $\xi \in \mathcal{H}$; conclude that $U\xi = f$.

Corollary 22.2. *Two Hilbert spaces are unitarily equivalent if and only if they have the same Hilbert dimension.*

DOI: 10.1201/9781003656166-22

Proof. Let $\mathcal{H}_1$ and $\mathcal{H}_2$ be Hilbert spaces. If there is a unitary operator $U : \mathcal{H}_1 \to \mathcal{H}_2$, then the image of an orthonormal basis of $\mathcal{H}_1$ under U is an orthonormal basis of $\mathcal{H}_2$. Since U is bijective, it follows that these Hilbert spaces have the same dimension.

Suppose now that $\mathcal{H}_1$ and $\mathcal{H}_2$ have the same Hilbert dimension; let J be a set whose cardinality coincides with such dimension. By Proposition 22.1, both Hilbert spaces are unitarily equivalent to $l^2(J)$ and, therefore, $\mathcal{H}_1$ is unitarily equivalent to $\mathcal{H}_2$ (it is enough to perform a composition of unitary operators, which is also unitary). $\qquad\square$

REMARK **22.3.** In Proposition 22.1 as well as in Corollary 22.2, it is necessary to distinguish the real and complex cases. For each cardinality, "there are just two Hilbert spaces, a real and a complex one." In particular, for each finite dimension n "there are only the Hilbert spaces $\mathbb{R}^n$ and $\mathbb{C}^n$."

EXERCISE **22.2.** Show, without using the Hahn-Banach Theorem, that in a Hilbert space the weak and weak* limits of sequences are unique (if they exist).

Example 22.4. It will be shown that every infinite-dimensional Hilbert space has a sequence (ξ_n) for which zero is a weak accumulation point, but (ξ_n) has no weak convergent subsequences! By Proposition 22.1, it is possible to consider $\mathcal{H} = l^2$ and, in fact, $\mathcal{H} = l^2(\mathbb{N})$ (the general case can be easily adapted from this one).

The sequence $\mathcal{S} = (\xi^n)_{n \geq 1}$, with $\xi^n_m = \sqrt{n}\,\delta_{m,n}$ (Kronecker's delta), of elements of $l^2(\mathbb{N})$ has no bounded subsequence, so no weak convergent subsequence. Now, any weak neighborhood of zero has an open set of the form

$$V(0;\mathcal{C};\varepsilon) = \left\{ \xi \in l^2(\mathbb{N}) : \max_{1 \leq j \leq k} \left\{ |f_{\eta^j}(\xi)| = |\langle \eta^j, \xi \rangle| \right\} < \varepsilon \right\},$$

with $\mathcal{C} = \{\eta^1, \cdots, \eta^k\} \subset l^2(\mathbb{N})$ and $\varepsilon > 0$. The goal is to show that for any open set of this form one has $\mathcal{S} \cap V(0;\mathcal{C};\varepsilon) \neq \emptyset$, so that zero is an accumulation point of $\mathcal{S}$ in the weak topology.

If $\mathcal{S} \cap V(0;\mathcal{C};\varepsilon) = \emptyset$, then no ξ^n belongs to $V(0;\mathcal{C};\varepsilon)$. Therefore, for some j one has that $\sqrt{n_r}\,|\eta^j_{n_r}| = |\langle \eta^j, \xi^{n_r} \rangle| \geq \varepsilon$ for infinitely many indices n_r and with $\sum_r 1/n_r = \infty$, since there are finitely many possibilities for j and the harmonic series is divergent. But this implies that

$$\|\eta^j\|_2^2 \geq \varepsilon^2 \sum_r 1/n_r = \infty,$$

which is a contradiction with $\eta^j \in l^2(\mathbb{N})$; therefore $\mathcal{S} \cap V(0;\mathcal{C};\varepsilon) \neq \emptyset$. ●

Now one of the chief historical examples of orthonormal basis of an infinite-dimensional Hilbert space is discussed: the standard Fourier series in $\mathrm{L}^2[a,b]$.

Theorem 22.5. *The enumerable set* $\{\psi_n\}_{n \in \mathbb{Z}}$, *with*

$$\psi_n(t) = \frac{1}{\sqrt{b-a}}\, e^{\frac{i 2\pi n(t-a)}{(b-a)}},$$

is an orthonormal basis of the complex Hilbert space $\mathrm{L}^2[a,b]$.

Proof. Clearly $\{\psi_n\}_{n \in \mathbb{Z}}$ is an orthonormal set in $L^2[a,b]$. Put

$$E := \mathrm{Lin}(\{\psi_n\}_{n \in \mathbb{Z}}) \quad \text{and} \quad S := \{\psi \in C[a,b] : \psi(a) = \psi(b)\}.$$

Note initially that the identity mapping

$$(C[a,b], \|\cdot\|_\infty) \xrightarrow{1} (C[a,b], \|\cdot\|_2)$$

is continuous; thus there exists $C > 0$ so that $\|\psi\|_2 \le C\|\psi\|_\infty$, for all $\psi \in C[a,b]$, and so uniform convergence implies convergence in $L^2[a,b]$.

Two well-known facts will be used: 1) from integration theory the set $(C[a,b], \|\cdot\|_2)$ is dense in $L^2[a,b]$, and 2) Fejér theorem (discussed in texts on Fourier series in $C[a,b]$), which implies that $(E, \|\cdot\|_\infty)$ is dense in $(S, \|\cdot\|_\infty)$.

Now, $(S, \|\cdot\|_2)$ is dense in $(C[a,b], \|\cdot\|_2)$. Indeed, if $\phi \in C[a,b]$, for each $n \in \mathbb{N}$ define

$$\phi_n(t) = \begin{cases} \phi(t), & a \le t \le b - \frac{1}{n} \\ \phi(a) + n(b-t)\left(\phi(b - \frac{1}{n}) - \phi(a)\right), & b - \frac{1}{n} < t \le b \end{cases},$$

which belongs to S, for all n, and $\|\phi - \phi_n\|_2 \to 0$ as $n \to \infty$.

In summary, E is dense in S (with the norm $\|\cdot\|_\infty$), which in its turn is dense in $(C[a,b], \|\cdot\|_2)$, and the latter is dense in $L^2[a,b]$; this, together with the continuity of the identity mapping above, assure that E is dense in $L^2[a,b]$, and the theorem is proved. □

Corollary 22.6. *The mapping* $\mathcal{F} : L^2[a,b] \to l^2(\mathbb{Z})$ *given by*

$$(\mathcal{F}\psi)_n = \int_a^b \overline{\psi_n(t)}\psi(t)\,dt, \qquad \psi \in L^2[a,b],$$

with ψ_n as in Theorem 22.5, is a unitary operator and

$$\psi(t) = \sum_{n \in \mathbb{Z}} (\mathcal{F}\psi)_n\,\psi_n(t) \quad in \quad L^2[a,b].$$

Proof. By Theorems 21.12 and 22.5, it is found that

$$\psi(t) = \sum_{n \in \mathbb{Z}} \langle \psi_n, \psi \rangle\,\psi_n(t) \quad in \quad L^2[a,b],$$

with the inner product given by $\langle \psi_n, \psi \rangle = (\mathcal{F}\psi)_n$, for all indices $n \in \mathbb{Z}$. To complete the proof it is enough to note that $\mathcal{F}$ plays the role of the unitary operator U in the proof of Proposition 22.1, after identifying J with $\mathbb{Z}$. □

REMARK **22.7.** $\mathcal{F}$ is called the *Fourier transform* on $L^2[a,b]$. Note that its action is also defined on $L^p[a,b]$, however it is meaningless to speak about unitarity if $p \ne 2$.

22.2 INTEGRATION IN HILBERT SPACES

It is possible to develop an integration theory of functions taking values in a separable Banach space, but by restricting to Hilbert spaces one can use, in an elegant and concise way, the Riesz Representation Theorem to define the integral in some cases. This will be the procedure in this short section.

If $\mathcal{H}$ is separable, so with a countable orthonormal basis (Exercise 21.2), then $\psi : (\Omega, \mathcal{A}, \mu) \to \mathcal{H}$ is said to be measurable if for every $\xi \in \mathcal{H}$ the function $\Omega \to \mathbb{F}, t \mapsto \langle \xi, \psi(t) \rangle$ is measurable. Such ψ is integrable if it is measurable and $\int_\Omega \|\psi(t)\| \, d\mu(t) < \infty$ (see Exercise 22.3).

EXERCISE 22.3. Use Parseval and polarization identities to show that if $\psi, \phi : (\Omega, \mathcal{A}, \mu) \to \mathcal{H}$ are mensurable, then the mapping $t \mapsto \langle \phi(t), \psi(t) \rangle$ is measurable in case $\mathcal{H}$ is separable. Verify also that the set of such mensurable mappings is a vector space.

Proposition 22.8. *Let $\psi : (\Omega, \mathcal{A}, \mu) \to \mathcal{H}$ be an integrable mapping. Then there exists a unique $\xi \in \mathcal{H}$ so that*

$$\langle \xi, \eta \rangle = \int_\Omega \langle \psi(t), \eta \rangle \, d\mu(t), \qquad \forall \eta \in \mathcal{H}.$$

Furthermore, $\|\xi\| \le \int_\Omega \|\psi(t)\| \, d\mu(t) < \infty$.

REMARK 22.9. Such vector ξ is denoted by $\int_\Omega \psi(t) \, d\mu(t)$ and called the integral of ψ with respect to the measure μ.

Proof. By Exercise 22.3, $t \mapsto \|\psi(t)\|$ is measurable. Define the linear functional $f : \mathcal{H} \to \mathbb{F}$ by $f(\eta) = \int_\Omega \langle \psi(t), \eta \rangle \, d\mu(t)$. By Cauchy-Schwarz, one has

$$|f(\eta)| \le \int_\Omega \|\psi(t)\| \|\eta\| \, d\mu(t) \le \left(\int_\Omega \|\psi(t)\| \, d\mu(t) \right) \|\eta\|,$$

so that $f \in \mathcal{H}^*$. By Riesz Representation Theorem, there exists a unique $\xi \in \mathcal{H}$ with $f = f_\xi$, i.e., for all $\eta \in \mathcal{H}$ one has $f(\eta) = f_\xi(\eta) = \langle \xi, \eta \rangle$. Finally, again by Riesz Representation, it follows that $\|\xi\| = \|f_\xi\| = \|f\| \le \int_\Omega \|\psi(t)\| \, d\mu(t)$. $\square$

Notes

Around 1820, the traditional Fourier series was introduced by Joseph Fourier, in his studies of heat conduction. The original idea was a way to represent complicated periodic phenomena in terms of simpler ones, which was greatly generalized in the concept of an orthonormal basis of infinite-dimensional Hilbert spaces. Note that Fourier's work was not mathematically rigorous, but due to the importance of related questions and applications, several mathematicians (as Cauchy and Dirichlet) were motivated to develop a solid base of the mathematical analysis.

In 1926, Kolmogorov presented an example of $\psi \in L^1[a, b]$ whose Fourier series is divergent at all points in this interval $[a, b]$; only in 1966 Carleson [Carles] showed that if $\psi \in L^p[a, b], 1 < p \le \infty$, then its Fourier series converges in a set of total Lebesgue measure in $[a, b]$.

In the 1980s, an approach called "wavelet transform," that resembles Fourier transform, but more adequate to the analysis of phenomena presenting fast oscillations and different scales, was introduced. Currently, the wavelet transform is a branch of Mathematics with applications to many areas, including Fractal Theory, Image Processing, etc. There are optimized numerical methods for the wavelet transform. The interested readers may consult, for instance, [Holsch].

Additional Exercises

EXERCISE **22.4.** Use Proposition 22.1 to present an alternative solution to Exercise 21.9.

EXERCISE **22.5.** Let $U : \mathrm{L}^2(\mathbb{R}) \hookleftarrow$, $(U\psi)(t) = \psi(-t)$. Show that U is unitary and $U^2 = 1$. Determine the orthogonal complement of $E = \{\psi \in \mathrm{L}^2(\mathbb{R}) : U\psi = \psi\}$.

EXERCISE **22.6.** Let $\{\xi_j\}_{j=1}^\infty$ be an orthonormal basis of the separable Hilbert space $\mathcal{H}$, so that $\xi = \sum_{j=1}^\infty a_j \xi_j$, for all $\xi \in \mathcal{H}$. Show that the sequences of operators $T_n \xi = a_n \xi_n$ and $S_n \xi = \sum_{j=1}^\infty a_j \xi_{j+n}$ belongs to $\mathrm{B}(\mathcal{H})$, for all n, and also that $T_n \xrightarrow{\mathrm{s}} 0$, but it does not converge in norm, while $S_n \xrightarrow{\mathrm{w}} 0$, but it does not converge strongly.

EXERCISE **22.7.** (a) Show that if (U_n) is a sequence of isometric operators in $\mathcal{H}$ and $U_n \xrightarrow{\mathrm{s}} U$, then U is isometric.

(b) Show that the sequence of unitary operators (check) $U_n : l^2(\mathbb{N}) \hookleftarrow$,

$$U_n \xi = (\xi_n, \xi_1, \xi_2, \cdots, \xi_{n-1}, \xi_{n+1}, \xi_{n+2}, \cdots)$$

converges strongly to an isometry that is not unitary.

(c) Give an example of a sequence of isometric operators, in some Hilbert space, that converges weakly to an operator that is not isometric.

EXERCISE **22.8.** Let $\mathcal{N}$ be the vector space of trigonometric polynomials $p : \mathbb{R} \to \mathbb{C}$, $p(t) = \sum_{j=1}^n a_j e^{itr_j}$, $a_j \in \mathbb{C}$ and $r_j \in \mathbb{R}$, for all j. Show that

$$\langle p, q \rangle := \lim_{T \to \infty} \frac{1}{2T} \int_{-T}^{T} \overline{p(t)}\, q(t)\, \mathrm{d}t, \qquad p, q \in \mathcal{N},$$

is an inner product and that $\{\psi_r = e^{itr}\}_{r \in \mathbb{R}}$ is an orthonormal set in $\mathcal{N}$. Conclude that this space (closely related to the almost periodic functions) is not separable.

EXERCISE **22.9.** Let U be a unitary operator on $\mathcal{H}$ and $E = \{\xi \in \mathcal{H} : U\xi = \xi\}$. Show that $E = \{\xi \in \mathcal{H} : U^*\xi = \xi\}$, and by considering the operator $(1 - U)$ and Exercise 20.9, show that the sequence of bounded operators

$$T_n = \frac{1}{n+1} \sum_{j=0}^{n} U^j$$

strongly converges to the orthogonal projection operator P_E (this result is known as the "Mean Ergodic Theorem," originally due to von Neumann.)

EXERCISE **22.10.** Let $\psi \in \mathrm{L}^2[0, 2\pi]$ so that $(n(\mathcal{F}\psi)_n)_{n \in \mathbb{Z}}$ belongs to $l^2(\mathbb{Z})$. Use Cauchy-Schwarz to show that the Fourier series of ψ

$$\sum_{n \in \mathbb{Z}} (\mathcal{F}\psi)_n \frac{1}{\sqrt{2\pi}} e^{int}$$

converges absolutely and uniformly to ψ. Conclude that ψ is continuous and $\psi(0) = \psi(2\pi)$.

EXERCISE **22.11.** If $\psi \in \mathrm{C}^1[0, 2\pi]$ with $\psi(0) = \psi(2\pi)$, verify that $(\mathcal{F}\psi')_n = in(\mathcal{F}\psi)_n$. Use this to show that if ψ is continuously differentiable, its Fourier series converges uniformly (the $'$ indicates derivative).

EXERCISE **22.12.** (a) Verify the linearity and $\int_{A \cup B} \cdot = \int_A \cdot + \int_B \cdot$, if $A \cap B = \emptyset$, for the integral defined in Remark 22.9.

(b) Prove the following version of the Dominated Convergence Theorem. If $\psi_j : (\Omega, \mathcal{A}, \mu) \to \mathcal{H}$ (separable) is integrable with $\psi_j \xrightarrow{\mu\text{a.e.}} \psi$, and there is a function $g \in \mathrm{L}^1_\mu(\Omega)$ with $\|\psi_j(t)\| \le |g(t)|$, μ-a.e., for all j, then ψ is integrable and $\int_\Omega \psi_j \, \mathrm{d}\mu \to \int_\Omega \psi \, \mathrm{d}\mu$ as $j \to \infty$.

EXERCISE **22.13.** Let $T : \mathbb{R} \to \mathrm{B}(\mathcal{H})$ be continuous, considering $\mathbb{R}$ with Lebesgue measure and $\mathcal{H}$ separable.

(a) Show that for all $\xi \in \mathcal{H}$ the mapping $t \mapsto T(t)\xi$ is measurable.

(b) Assume that $\mathrm{s} \cdot \lim_{t \to \infty} T(t) = S$ in $\mathrm{B}(\mathcal{H})$. Show that, for all $\delta > 0$,

$$\int_0^\infty e^{-\delta t} T(t)\xi \, \mathrm{d}t, \qquad \forall \xi \in \mathcal{H},$$

is well defined and $S\xi = \lim_{\delta \to 0^+} \delta \int_0^\infty e^{-\delta t} T(t)\xi \, \mathrm{d}t$, for all $\xi \in \mathcal{H}$.

EXERCISE **22.14.** Let M denote the set of mensurable $\psi : (\Omega, \mathcal{A}, \mu) \to \mathcal{H}$ (separable) satisfying $\int_\Omega \|\psi(t)\|^2 \, \mathrm{d}\mu(t) < \infty$. Show that M is a vector space and that

$$[\psi, \phi] := \int_\Omega \langle \psi(t), \phi(t) \rangle_{\mathcal{H}} \, \mathrm{d}\mu(t), \qquad \psi, \phi \in M,$$

is an inner product on M. Is this pre-Hilbertian space complete? Note that in the case $\mathcal{H} = \mathbb{F}$ one has $M = \mathrm{L}^2_\mu(\Omega)$.

EXERCISE **22.15.** The following steps indicate how Fejér's Theorem follows from Stone-Weierstrass' (see [Simm]). Let E and S be as in the proof of Theorem 22.5. Note that $(S, \| \cdot \|_\infty)$ can be identified with $(\mathrm{C}(\mathrm{I}), \| \cdot \|_\infty)$, where I denotes the interval $[a, b]$ with extremes a, b identified (a circumference). Show that E is a complex subspace of $\mathrm{C}(\mathrm{I})$, which is an algebra invariant under complex conjugation, which also separates points of I and contains the constant functions. Using Stone-Weierstrass, conclude that E is dense in $\mathrm{C}(\mathrm{I})$.

CHAPTER 23

Operations on Banach Spaces

There are methods for constructing new Banach spaces from existing ones. Two of these constructions, the direct sum and quotient spaces, will be introduced here. A direct sum of Banach spaces is a construction that enables the combination of multiple Banach spaces into a larger one. A quotient space of a Banach space, on the other hand, involves partitioning a Banach space by a closed subspace; it can be understood as the formation of a new Banach space by "collapsing" the elements of the subspace to a point. As a notable application of the quotient space construction, the Open Mapping Theorem will be derived from the Closed Graph Theorem, so concluding the equivalence of these two important technical results.

23.1 DIRECT SUM

To fix notation it is convenient to recall the concept of a Cartesian product of sets. The Cartesian product of the sets $\{X_t\}_{t \in J}$, denoted by $\prod_{t \in J} X_t$, is the set of functions

$$\psi : J \to \bigcup_{t \in J} X_t$$

whose components are $\pi_t \psi := \psi(t) \in X_t$, for all $t \in J$; recall that each ψ is also called a *choice function*. By the Axiom of Choice, it follows that the Cartesian product of any family of nonempty sets is nonempty. It is tacitly assumed that every X_t is nonempty. If $X_t = X$ for all $t \in J$, it is also denoted by $\prod_{t \in J} X_t = X^J$, or X^n in case $J = \{1, 2, \cdots, n\}$, which is identified with the n-tuples of elements of X. In case all X_t are vector spaces, note that the mapping $\pi_t : \prod_{j \in J} X_j \to X_t$, the so-called *projection* onto the tth factor, is linear.

Let $\{(\mathcal{N}_t, \|\cdot\|_t)\}_{t \in J}$ be a family of normed spaces; then $\prod_{t \in J} \mathcal{N}_t$ becomes a vector space with pointwise linear operations. For each $1 \le p < \infty$, one defines the norm

$$\|\|\psi\|\|_p := \left(\sum_{t \in J} \|\pi_t \psi\|_t^p \right)^{\frac{1}{p}}$$

on the vector space $\bigoplus_p \mathcal{N}_t = \bigoplus_p (\mathcal{N}_t)_{t \in J}$ of the $\psi \in \prod_{t \in J} \mathcal{N}_t$ which satisfy $\pi_t \psi \neq 0$

DOI: 10.1201/9781003656166-23

only for a countable set of indices $t \in J$ and with $\|\|\psi\|\|_p < \infty$. For $p = \infty$, one introduces the norm

$$\|\|\psi\|\|_\infty := \sup_{t \in J} \|\pi_t \psi\|_t$$

on the vector subspace $\bigoplus_\infty \mathcal{N}_t = \bigoplus_\infty (\mathcal{N}_t)_{t \in J}$ of the ψ in $\prod_{t \in J} \mathcal{N}_t$ with $\|\|\psi\|\|_\infty < \infty$.

Proposition 23.1. *For* $1 \le p \le \infty$, $\bigoplus_p \mathcal{N}_t$ *is a normed space, and for all* $t \in J$, π_t *is a bounded linear operator with* $\|\pi_t\| \le 1$. *Furthermore,* $\bigoplus_p \mathcal{N}_t$ *is a Banach space if all* $\mathcal{N}_t$ *are Banach.*

Proof. Only $1 \le p < \infty$ will be dealt with; the case $p = \infty$ is left as an exercise. Clearly $\bigoplus_p \mathcal{N}_t$ is a normed space with $\|\| \cdot \|\|_p$, π_t is linear and $\|\pi_t \psi\|_t \le \|\|\psi\|\|_p$, which shows that $\|\pi_t\| \le 1$ for all $t \in J$.

Suppose that all $\mathcal{N}_t$ are Banach. Let $(\psi^n)_{n=1}^\infty$ be a Cauchy sequence in $\bigoplus_p \mathcal{N}_t$. The inequality

$$\|\pi_t \psi^n - \pi_t \psi^m\|_t \le \|\|\psi^n - \psi^m\|\|_p, \qquad \forall n, m, t,$$

implies that, for every $t \in J$, $(\pi_t \psi^n)$ is Cauchy in $\mathcal{N}_t$ and so, converges to some $\psi_t \in \mathcal{N}_t$. Note that for $\psi := \{\psi_t\}_{t \in J}$ one has $\psi_t \ne 0$ only for a countable set of indices $t \in J$.

Given $\varepsilon > 0$, there exists $N(\varepsilon)$ with $\|\|\psi^n - \psi^m\|\|_p < \varepsilon$ if $n, m \ge N(\varepsilon)$. Thus, for every finite set of indices $f \subset J$, one has

$$\left(\sum_{t \in f} \|\pi_t \psi^n - \pi_t \psi^m\|_t^p \right)^{\frac{1}{p}} \le \|\|\psi^n - \psi^m\|\|_p < \varepsilon, \qquad n, m \ge N(\varepsilon).$$

Taking $m \to \infty$ it is found that $\sum_{t \in f} \|\pi_t \psi^n - \psi_t\|_t^p \le \varepsilon^p$ for any finite set f; hence $\sum_{t \in J} \|\pi_t \psi^n - \psi_t\|_t^p \le \varepsilon^p$ if $n \ge N(\varepsilon)$ and $(\psi^n - \psi) \in \bigoplus_p \mathcal{N}_t$; since this set is a vector space, $(\psi - \psi^n) + \psi^n = \psi \in \bigoplus_p \mathcal{N}_t$. By the above inequality it also follows that $\psi^n \to \psi$ in $\bigoplus_p \mathcal{N}_t$. Therefore, $\bigoplus_p \mathcal{N}_t$ is complete. $\square$

EXERCISE **23.1.** Prove Proposition 23.1 for $p = \infty$. For all $1 \le p \le \infty, t \in J$, show that $\|\pi_t\| = 1$.

Definition 23.2. The expressions *direct sum* and *direct product* of normed spaces both refer to $\bigoplus_p \mathcal{N}_t$, and for any $1 \le p \le \infty$.

In case all $\mathcal{N}_t$ are Hilbert spaces, there is a natural inner product only if $p = 2$, and the term *direct sum of Hilbert spaces* will be reserved to this particular setting.

Proposition 23.3. *If* $\{\mathcal{H}_t\}_{t \in J}$ *is a collection of Hilbert spaces, then the direct sum of such spaces*

$$\bigoplus \mathcal{H}_t := \bigoplus_2 \mathcal{H}_t,$$

is a Hilbert space with the inner product

$$\langle \psi, \phi \rangle_\oplus := \sum_{t \in J} \langle \pi_t \psi, \pi_t \phi \rangle, \qquad \psi, \phi \in \bigoplus \mathcal{H}_t.$$

Proof. Due to Proposition 23.1, it is enough to verify that $\langle \psi, \phi \rangle_\oplus$ is well defined and is an inner product (which generates the right norm). It is left to the reader that this expression actually defines an inner product.

If $\psi, \phi \in \bigoplus \mathcal{H}_t$, by the inequality

$$\left| \sum_t \langle \pi_t \psi, \pi_t \phi \rangle \right| \leq \sum_t |\langle \pi_t \psi, \pi_t \phi \rangle| \leq \sum_t \|\pi_t \psi\| \|\phi_t \psi\|$$

$$\leq \frac{1}{2} \sum \left(\|\pi_t \psi\|^2 + \|\phi_t \psi\|^2 \right) < \infty,$$

it is found that $\langle \psi, \phi \rangle_\oplus$ converges absolutely and so it is well defined. □

EXERCISE **23.2.** Check that if $\mathcal{N}_j = \mathbb{F}$, for all $j \in J$, then $\bigoplus_p \mathcal{N}_t = l^p(J)$.

REMARK **23.4.** By construction, in $\bigoplus \mathcal{H}_t$ one has that "$\mathcal{H}_{t_1} \perp \mathcal{H}_{t_2}$" if $t_1 \neq t_2$, in the sense that if $\psi^1, \psi^2 \in \bigoplus \mathcal{H}_t$ and $\pi_t \psi^1 = 0$ for $t \neq t_1$, and $\pi_t \psi^2 = 0$, for $t \neq t_2$, then $\langle \psi^1, \psi^2 \rangle_\oplus = 0$.

Based on the above remark, one defines the *inner direct sum*, denoted by $\bigoplus^{\text{int}} \mathcal{H}_t$, of a family of closed subspaces $\{\mathcal{H}_t\}_{t \in J}$ of a Hilbert space $\mathcal{H}$, only if they are pairwise orthogonal, and one has $\bigoplus_{t \in J}^{\text{int}} \mathcal{H}_t = \overline{\text{Lin}}(\{\mathcal{H}_t\}_{t \in J})$. A particular case of inner direct sum is the decomposition $\mathcal{H} = E \oplus E^\perp$, with E being a closed subspace of the Hilbert space $\mathcal{H}$, described in Theorem 18.7.

Example 23.5. Let $\{(\Omega_t, \mathcal{A}_t, \mu_t)\}_{t \in J}$ be a family of measure spaces, Ω the disjoint union of $\{\Omega_t\}_{t \in J}$, $\mathcal{A}$ the σ-algebra $\{A \subset \Omega : A \cap \Omega_t \in \mathcal{A}_t, \forall t \in J\}$, and μ the measure $\mu(A) = \sum_t \mu_t(A \cap \Omega_t)$, for $A \in \mathcal{A}$. Thus, the mapping $V : \bigoplus L^2_{\mu_t}(\Omega_t) \to L^2_\mu(\Omega)$ defined as

$$(V\psi)(s) = (\pi_t \psi)(s), \qquad \text{if } s \in \Omega_t,$$

is a unitary mapping. Indeed, since V is linear and

$$\langle V\psi, V\psi \rangle = \int_\Omega |V\psi(s)|^2 \, d\mu(s) = \sum_t \int_{\Omega_t} |(\pi_t \psi)(s)|^2 \, d\mu_t(s)$$

$$= \sum_t \|\pi_t \psi\|_t^2 = \|\|\psi\|\|_2^2,$$

it follows that V is isometric. It only remains to show that V is surjective on $L^2_\mu(\Omega)$. If $u \in L^2_\mu(\Omega)$, put $\psi_t = u|_{\Omega_t}$, so that

$$\|u\|^2 = \int_\Omega |u(s)|^2 d\mu(s) = \sum_t \int_{\Omega_t} |\psi_t(s)|^2 d\mu_t(s),$$

that is, $\psi = \{\psi_t\}_{t \in J} \in \bigoplus L^2_{\mu_t}(\Omega_t)$ and $V\psi = u$, so V is surjective, and therefore, a unitary operator. ●

23.2 QUOTIENT SPACE

Let E be a closed subspace of a Banach space $\mathcal{B}$; the notation $\mathcal{B}/E$ indicates the collection of subsets of $\mathcal{B}$ of the form

$$\xi + E := \{\xi + \eta : \eta \in E\},$$

which will also be denoted by $[\xi]_E$, or simply $[\xi]$, when it is clear which is the subspace E under consideration. Each of such sets $[\xi]$ is referred to as a *coset* of E. Note that if $\xi \in E$, then $[\xi]_E = E$; in particular one has $[0]_E = E$.

There is a natural linear structure in $\mathcal{B}/E$, with the operations defined, for all $\xi, \eta \in \mathcal{B}$, $\alpha \in \mathbb{F}$, by

$$\begin{aligned} [\xi] + [\eta] &= \xi + E + \eta + E = \xi + \eta + E = [\xi + \eta], \\ \alpha[\xi] &= \alpha(\xi + E) = \alpha\xi + E = [\alpha\xi], \end{aligned}$$

(with $\alpha E = E$ and $\beta[E] = [E]$ if $\beta = 0$, since $[0]_E = E$) and $\mathcal{B}/E$ becomes a vector space called the *quotient space* of $\mathcal{B}$ modulus E.

Usually the elements of E have a common property and one wishes to "discard" it by identifying these elements, such as in the Lebesgue integral, in which functions are identified up to sets of measure zero.

REMARK 23.6. The cosets are equivalence classes. Note the identification of vectors ξ and η, that is, they belong to the same coset, if and only if $(\xi - \eta) \in E$. An illustrative view of the quotient space is obtained by taking E as a given straight line (containing the origin) in $\mathbb{R}^2$; hence, in $\mathbb{R}^2/E$, each coset $[\xi]$ is the straight line in $\mathbb{R}^2$, parallel to E, that contains ξ.

Example 23.7. If $\|\|\cdot\|\|$ is a seminorm on a vector space X and $E = \{\xi \in X : \|\|\xi\|\| = 0\}$, then E is a subspace of X and $\|[\xi]_E\| := \|\|\xi\|\|$ is a norm on X/E (check this!). This is a typical use of quotient spaces, in which vectors ξ, η are identified if $\|\|\xi - \eta\|\| = 0$. •

Proposition 23.8. *Let E be a closed subspace of the Banach space $\mathcal{B}$. Then:*

(i) The mapping $\|\cdot\|_Q : \mathcal{B}/E \to \mathbb{R}$,

$$\|[\xi]\|_Q := d(\xi, E) = \inf_{\eta \in E} \|\xi - \eta\|,$$

is a norm on $\mathcal{B}/E$, called the quotient norm.

(ii) For every $\xi \in \mathcal{B}$ one has $\|[\xi]\|_Q \leq \|\xi\|$ and, therefore, the mapping $(\mathcal{B}, \|\cdot\|) \to (\mathcal{B}/E, \|\cdot\|_Q)$, $\xi \mapsto [\xi]$, is continuous.

(iii) $(\mathcal{B}/E, \|\cdot\|_Q)$ is a Banach space.

Proof. (i) The verification of this item is simple. Note that the closeness of E is important; otherwise $\|\cdot\|_Q$ may be only a seminorm.

(ii) $\|[\xi]\|_Q = \inf_{\eta \in E} \|\xi - \eta\| \leq \|\xi - 0\| = \|\xi\|$. From this it also follows that the mapping $\xi \mapsto [\xi]$ is continuous.

(iii) Let $([\xi_n])_{n=1}^{\infty}$ be a Cauchy sequence in $\mathcal{B}/E$. It will be shown that there exists $\xi_0 \in \mathcal{B}$ so that $[\xi_n] \to [\xi_0]$. Pick a subsequence $([\xi_{n_j}])$ of $([\xi_n])$ for which, for all j, $\|[\xi_{n_j}] - [\xi_{n_{j+1}}]\|_Q < 2^{-j-1}$ (it exists, verify!). Pick the sequence $(\eta_j) \subset E$ so that, for all j,

$$\|(\xi_{n_j} - \eta_j) - (\xi_{n_{j+1}} - \eta_{j+1})\| < \|[\xi_{n_j}] - [\xi_{n_{j+1}}]\|_Q + \frac{1}{2^{j+1}} < \frac{1}{2^j}.$$

Since $\sum_{j=1}^{\infty} 2^{-j-1} < \infty$, it is found that $(\xi_{n_j} - \eta_j)$ is a Cauchy sequence in $\mathcal{B}$, and so it converges to some $\xi_0 \in \mathcal{B}$.

Since the mapping $\xi \mapsto [\xi]$ is continuous, it follows that (by using that $\eta_j \in E$)

$$[\xi_{n_j}] = [\xi_{n_j} - \eta_j] \to [\xi_0],$$

and since $([\xi_{n_j}])$ is a subsequence of the Cauchy sequence $([\xi_n])$, then $[\xi_n] \to [\xi_0]$ and so $\mathcal{B}/E$ is complete. □

EXERCISE 23.3. How are the elements of the sequence $(\eta_j) \subset E$ chosen in the proof of Proposition 23.8?

Example 23.9. Let $E = \{\psi \in C[-1,1] : \psi(0) = 0\}$. Then $C[-1,1]/E$ is identified with $\mathbb{F}$ (i.e., they are isomorphic as normed spaces), since each coset $[\phi]$ is identified with $\phi(0)$. •

As an application of the concept of a quotient space, the Open Mapping Theorem 8.3 will be obtained from the Closed Graph Theorem 9.9; since the converse was discussed in Chapter 9, then these important results are, in fact, equivalent.

Lemma 23.10. *If E is a closed subspace of $\mathcal{B}$, then the mapping $(\mathcal{B}, \|\cdot\|) \to (\mathcal{B}/E, \|\cdot\|_Q)$, $\xi \mapsto [\xi]$, is open.*

Proof. It is enough to show that the image (under $[\cdot]$) of any open ball $B = B_{\mathcal{B}}(\xi_0; r)$ is open in $\mathcal{B}/E$; indeed, it will be shown that $[B] = B_{\mathcal{B}/E}([\xi_0]; r)$ (recall that every open ball in a metric space is an open set); clearly $[B] = \{[\xi] : \xi \in \mathcal{B}, \|\xi - \xi_0\| < r\}$.

If $\xi \in B$, then from $\|\xi - \xi_0\| < r$ and the continuity of the mapping $[\cdot]$, one has

$$\|[\xi] - [\xi_0]\|_Q = \|[\xi - \xi_0]\|_Q \le \|\xi - \xi_0\| < r,$$

and so $[B] \subset B_{\mathcal{B}/E}([\xi_0]; r)$. Now, if $[\xi] \in B_{\mathcal{B}/E}([\xi_0]; r)$ one gets $\|[\xi] - [\xi_0]\|_Q = \|[\xi] - [\xi_0]\|_Q < r$ and so there exists $\eta \in E$ so that $\|(\xi - \xi_0) - \eta\| < r$, that is,

$$([\xi] - [\xi_0]) = [\xi - \xi_0] = [\xi - \xi_0 - \eta] \in [B_{\mathcal{B}}(0; r)];$$

therefore, $[\xi] \in [B]$ and finally $B_{\mathcal{B}/E}([\xi_0]; r) \subset [B]$. □

By Exercise 9.13, the Inverse Mapping Theorem 8.5 is a consequence of the Closed Graph Theorem. Thus, in order to reach the proposed goal, one may assume that Theorem 8.5 holds.

Let $T \in B(\mathcal{B}_1, \mathcal{B}_2)$ with rng $T = \mathcal{B}_2$; the goal is to show that T is an open mapping. $N(T)$ is a closed subspace and $T_Q : \mathcal{B}_1/N(T) \to \mathcal{B}_2$, defined by the relation $T \cdot = T_Q \circ [\cdot]_{N(T)}$ (i.e., $T_Q(\xi + N(T)) = T\xi$, $\xi \in \mathcal{B}_1$), is linear and bijective, for if

$T_Q[\xi] = 0$, then $T\xi = 0$, $\xi \in \mathrm{N}(T)$ and so $[\xi] = [0]$. Since for all $\eta \in \mathrm{N}(T)$ one has $T_Q[\xi] = T\xi = T(\xi - \eta)$, then $\|T_Q[\xi]\| \leq \|T\| \, \|\xi - \eta\|$, and

$$\|T_Q[\xi]\| \leq \|T\| \inf_{\eta \in \mathrm{N}(T)} \|\xi - \eta\| = \|T\| \, \|[\xi]\|_Q,$$

so that T_Q is bounded; consequently a homeomorphism by Theorem 8.5. This and Lemma 23.10 show that T is the composition of a homeomorphism T_Q and an open mapping $[\cdot]_{\mathrm{N}(T)}$, hence T is also an open mapping.

Notes

Some authors do not consider the terms *direct sum* and *direct product* synonyms; for example, some define direct product of normed spaces and call its completion the direct sum; others restrict the nomenclature to some specific norm $\||\cdot\||_p$.

The sum (as sets) of two closed subspaces of a Banach space is not necessarily closed; see Exercise 23.11.

Direct sum and quotient spaces have many applications, for instance in Spectral Theory, commutative W^*-algebras, representations of C^*-algebras, Quantum Field Theory, etc. There is also the notion of *direct integral* of Hilbert spaces (introduced by J. von Neumann), which generalizes the direct sum discussed here. It is also possible to define the *direct sum of operators*. Another important construction, that is not considered here, is the *tensor product* of spaces, which consists, in a rather naive way, of a generalization of the product of functions defined on different sets. In [BlExHa] it is possible to get in touch with such constructions and find an extensive list of references.

There are proofs of the Closed Graph Theorem that make no use of the Open Mapping Theorem; see, for instance, [Trenon].

Additional Exercises

EXERCISE **23.4.** Show that if some $\mathcal{N}_t$ is not complete, then $\bigoplus_p \mathcal{N}_t$ is not a Banach space, for $1 \leq p \leq \infty$.

EXERCISE **23.5.** If J is a finite set, show that all norms $\||\cdot\||_p$ are equivalent in the direct sum.

EXERCISE **23.6.** Show that $\bigoplus_p \mathcal{N}_t$, $1 \leq p < \infty$, is separable if only if J is countable and each $\mathcal{N}_t$ is separable. What about the case $p = \infty$?

EXERCISE **23.7.** If E is a closed subspace of $\mathcal{B}$, use the Open Mapping Theorem to show that the mapping $(\mathcal{B}, \|\cdot\|) \to (\mathcal{B}/E, \|\cdot\|_Q)$, $\xi \mapsto [\xi]$, is open.

EXERCISE **23.8.** Use the following outline to show that if E is a closed subspace of $\mathcal{B}$ and E^0 its annihilator, defined in Exercise 12.9, then $E^0 = (\mathcal{B}/E)^*$, with (as always) equality meaning that these spaces are isomorphic.

(a) Verify that if $f \in (\mathcal{B}/E)^*$, then $f \circ [\ ]_E \in E^0$ (with $(f \circ [\ ]_E)(\xi) = f([\xi]_E)$), and that the mapping $A : (\mathcal{B}/E)^* \to E^0$, $A(f) := f \circ [\ ]_E$ is a linear isometry.

(b) If $h \in E^0$, show that $f : \mathcal{B}/E \to \mathbb{F}$, $f([\xi]_E) := h(\xi)$ is a well-defined and linear mapping. Show then that for all $\eta \in E$, $|f([\xi]_E)| = |h(\xi + \eta)| \leq \|h\| \, \|\xi + \eta\|$, and hence, $\|f\| \leq \|h\|$. Conclude that $f \in (\mathcal{B}/E)^*$, $A(f) = h$, and that A is an isomorphism.

EXERCISE **23.9.** If E is a closed subspace of $\mathcal{H}$, show that $\mathcal{H}/E$ is isomorphic to $E^\perp$. Explicitly, show that the mapping $[\ ]_E : E^\perp \to \mathcal{H}/E$ is an isometric isomorphism. Conclude that $\mathcal{H}/E$ is a Hilbert space and that $(\mathcal{H}/E)^*$ is isomorphic to $E^\perp$. Compare with Exercise 23.8.

EXERCISE **23.10.** Let K be a finite subset of $\mathbb{N}$, $1 \le p \le \infty$, and

$$E_K = \{\xi \in l^p(\mathbb{N}) : \xi_k = 0, \, \forall k \in K\}.$$

Show that E_K is closed and that $l^p(\mathbb{N})/E_K$ can be identified with $l^p(\mathbb{N})$ (that is, isomorphic as normed spaces). What may happen if K is infinite?

EXERCISE **23.11.** It is interesting to note that the sum $E_1 + E_2 = \{\xi + \eta : \xi \in E_1, \eta \in E_2\}$ of two closed subspaces E_1, E_2 of a Banach space is not necessarily closed. Define E_1 as the vectors $\xi = (\xi_1, \xi_2, \cdots) \in l^2(\mathbb{N})$ so that $\xi_{2j-1} = 0$, for all j, and E_2 as the subspace of ξ with $\xi_{2j} = j\xi_{2j-1}$, for all j.

(a) Show that E_1, E_2 are closed subspaces.
(b) Check that any sequence in $E_1 + E_2$ can be uniquely written in the form

$$\xi = (0, \xi_2 - \xi_1, 0, \xi_4 - 2\xi_3, 0, \xi_6 - 3\xi_5, \cdots) + (\xi_1, \xi_1, \xi_3, 2\xi_3, \xi_5, 3\xi_5, \cdots).$$

(c) Verify that $E_1 + E_2$ contains all sequences with only a finite number of nonzero terms; conclude that $E_1 + E_2$ is dense in $l^2(\mathbb{N})$.
(d) Verify that $\xi = (1, 0, 1/2, 0, 1/3, 0, 1/4, 0, \cdots)$ belongs to $l^2(\mathbb{N})$ but not to $E_1 + E_2$; conclude that such sum is not closed in $l^2(\mathbb{N})$.

CHAPTER 24

Compact Operators

This chapter covers the basic properties of compact operators, an important class of bounded operators on normed spaces. When these operators act on infinite-dimensional spaces, they share some similarities with operators on finite-dimensional spaces, making them easier to handle in some cases. However, compact operators have shown significant potential in various applications. Notable examples of compact operators include integral operators and finite-rank operators. A key property of compact operators is that they map weakly convergent sequences to strongly convergent ones. As is typical, some results require complete normed spaces, i.e., Banach spaces.

Compact operators exhibit certain similarities with operators defined on finite-dimensional spaces, which leads to notable simplifications in the underlying theory. Despite this, compact operators have wide-ranging applications, frequently appearing as integral operators (historically one of the most prominent examples of compact operators).

Before proceeding, it is helpful to review a few definitions and results from the theory of metric spaces. A subset A in the metric space (X, d) is said to be *relatively compact*, or *precompact*, if its closure $\overline{A}$ is compact. A set A is called *totally bounded* if, for every $\varepsilon > 0$, it can be covered by finitely many open balls of radius ε in X of radii ε; hence, every totally bounded set is necessarily bounded. It is left as an exercise to verify that a precompact set is totally bounded; hence, it is bounded.

EXERCISE 24.1. If a subset A of (X, d) is totally bounded, verify that, for every $\varepsilon > 0$, A is contained in the union of a finite number of open balls of radii ε centered at points of A. Use this to conclude that a totally bounded set is separable (with the induced metric topology).

Lemma 24.1. *Let A be a subset of a complete metric space. If A is totally bounded, then it is precompact.*

Proof. Since A is a totally bounded, its closure is also totally bounded (Given a collection of open balls that covers a set in a metric space, the family of balls with the same centers and twice the radii forms a cover of the closure of the set). Since A is in a complete metric space, to check that its closure is compact it is enough to check that every sequence $(\xi_n) \subset \overline{A}$, in its closure, has a Cauchy subsequence. Since A is totally bounded, there exists a subsequence $(\xi_{1,n})$ of (ξ_n) contained in certain open

DOI: 10.1201/9781003656166-24

ball of radius 1. Similarly, there exists a subsequence $(\xi_{2,n})$ of $(\xi_{1,n})$ contained in an open ball of radius $1/2$; by continuing, one gets subsequences $(\xi_{k,n})_{n\geq 1}$ of $(\xi_{k-1,n})_{n\geq 1}$ in some open ball of radius $1/k$, for all $k \in \mathbb{N}$. In order to complete the proof, it is enough to note that $(\xi_{k,k})_{k\geq 1}$ is a Cauchy subsequence of the original one. □

Definition 24.2. A linear operator $T : \mathcal{N}_1 \to \mathcal{N}_2$ between normed spaces is said to be *compact*, sometimes also called *completely continuous*, if the range $T(A)$, of any bounded set $A \subset \mathcal{N}_1$, is precompact in $\mathcal{N}_2$. The collection of such compact operators will be denoted by $B_0(\mathcal{N}_1, \mathcal{N}_2)$ (or simply $B_0(\mathcal{N})$, in case $\mathcal{N}_1 = \mathcal{N}_2 = \mathcal{N}$).

EXERCISE **24.2.** Show that a linear operator $T : \mathcal{N}_1 \to \mathcal{N}_2$ is compact if for every bounded sequence $(\xi_n) \subset \mathcal{N}_1$, the image set $(T\xi_n)$ has a convergent subsequence in $\mathcal{N}_2$.

EXERCISE **24.3.** Is the identity operator compact?

Proposition 24.3. *Let T, S be elements of $B(\mathcal{N}_1, \mathcal{N}_2)$. Then:*

(i) The set of compact operators $B_0(\mathcal{N}_1, \mathcal{N}_2)$ is a vector subspace of the continuous operators $B(\mathcal{N}_1, \mathcal{N}_2)$.

(ii) If T or S is compact, then both TS and ST (if the compositions are well defined) are also compact operators.

Proof. (i) The verification that $B_0(\mathcal{N}_1, \mathcal{N}_2)$ is a vector space is left as an exercise. Let $T \in B_0(\mathcal{N}_1, \mathcal{N}_2)$ be compact. Then $T(S(0; 1))$ is precompact, so it is bounded. Hence, T is continuous.

(ii) If $E \subset \mathcal{N}_1$ is a bounded set, it follows that the image $S(E)$ is also bounded and $T(S(E))$ is precompact. Therefore, TS is a compact operator.

For each bounded set E, the image under T of any sequence $(\xi_n) \subset E$ has a convergent subsequence $(T\xi_{n_j})$, since T is compact. Since S is continuous, $(ST\xi_{n_j})$ is also convergent. Therefore, $ST(E)$ is precompact and ST is compact as well. □

REMARK **24.4.** A general mapping between metric spaces is said to be *compact* if the image under this mapping of bounded sets is precompact. The Dirichlet function $h : \mathbb{R} \to \mathbb{R}$, $h(t) = 1$ if $t \in \mathbb{Q}$ and $h(t) = 0$ otherwise, is compact, but not continuous at any point of its domain. Compare with Proposition 24.3.

Finite-rank operators form an important and illustrative class of compact operators, as discussed ahead.

Definition 24.5. An operator $T \in B(\mathcal{N}_1, \mathcal{N}_2)$ is called a *finite rank* operator if $\dim \operatorname{rng} T < \infty$. $B_f(\mathcal{N}_1, \mathcal{N}_2)$ will denote the vector space of such finite rank operators, with the companion notation $B_f(\mathcal{N})$ when $\mathcal{N}_1 = \mathcal{N}_2 = \mathcal{N}$.

Proposition 24.6. *Any finite rank operator is compact.*

Proof. Let $T \in B_f(\mathcal{N}_1, \mathcal{N}_2)$ and $E \subset \mathcal{N}_1$ a bounded set. $T(E)$ is bounded, because T is also a bounded operator, so its closure $\overline{T(E)}$ is a closed and bounded set. By using that $\dim \operatorname{rng} T < \infty$, it follows that $\overline{T(E)}$ is a compact set, so T is compact. □

Corollary 24.7. *If $\mathcal{N}$ is a normed space, then $\mathcal{N}^* = \mathrm{B}_0(\mathcal{N}, \mathbb{F}) = \mathrm{B}_f(\mathcal{N}, \mathbb{F})$.*

A useful property of compact operators, for both theoretical arguments and applications, appears in the

Proposition 24.8. *Let $T \in \mathrm{B}_0(\mathcal{N}_1, \mathcal{N}_2)$. If $\xi_n \rightharpoonup \xi$ in $\mathcal{N}_1$, then $T\xi_n \to T\xi$ in $\mathcal{N}_2$, i.e., a compact operator takes weakly convergent sequences to strongly convergent ones.*

Proof. Suppose $\xi_n \rightharpoonup \xi$ in $\mathcal{N}_1$; by Proposition 14.3, (ξ_n) is a bounded set. If $g \in \mathcal{N}_2^*$,

$$g(T\xi_n) = (T^{\mathrm{a}} g)(\xi_n) \to (T^{\mathrm{a}} g)(\xi) = g(T\xi),$$

showing that $T\xi_n \rightharpoonup T\xi$. If $T\xi_n$ does not converge strongly to $T\xi$, there exists $\varepsilon > 0$ and a subsequence $(T\xi_{n_j})$ with $\|T\xi_{n_j} - T\xi\| \geq \varepsilon$. Since T is a compact operator, $(T\xi_{n_j})$ has the strongly convergent subsequence, and since $T\xi_n \rightharpoonup T\xi$, it necessarily converges to $T\xi$. The contradiction with the above inequality proves the proposition. $\square$

Example 24.9. The identity operator on $l^1(\mathbb{N})$ is not compact, however any bounded linear operator on $l^1(\mathbb{N})$ takes weakly convergent sequences to strongly convergent ones, since in this case weak and strong convergences are equivalent (see Example 14.11). Thus, the property stated in Proposition 24.6 cannot be used to characterize compact operators. See, however, Proposition 25.6, which deals with reflexive spaces. ●

Theorem 24.10. $\mathrm{B}_0(\mathcal{N}, \mathcal{B})$ *is a closed subspace of* $\mathrm{B}(\mathcal{N}, \mathcal{B})$. *Therefore,* $\mathrm{B}_0(\mathcal{N}, \mathcal{B})$ *is a Banach space.*

Proof. Let $(T_n) \subset \mathrm{B}_0(\mathcal{N}, \mathcal{B})$, with $T_n \to T$ in $\mathrm{B}(\mathcal{N}, \mathcal{B})$. It will be shown that for all $r > 0$ the set $TB(0; r)$ is totally bounded, and therefore, precompact by Lemma 24.1. This implies that T is also a compact operator.

Fix $r > 0$. Given $\varepsilon > 0$, there is n such that $\|T_n - T\| < \varepsilon/r$. Since T_n is compact, the set $T_n B(0; r)$ is totally bounded and so, it is in the union of certain balls

$$B(T_n \xi_1; \varepsilon), B(T_n \xi_2; \varepsilon), \cdots, B(T_n \xi_m; \varepsilon),$$

with $\xi_j \in B(0; r)$, for all $1 \leq j \leq m$. Hence, if $\xi \in B(0; r)$, there is one of these ξ_j such that $T_n \xi \in B(T_n \xi_j; \varepsilon)$. From this one has

$$\begin{aligned}
\|T\xi - T\xi_j\| &\leq \|T\xi - T_n\xi\| + \|T_n\xi - T_n\xi_j\| + \|T_n\xi_j - T\xi_j\| \\
&< \|T - T_n\| \, \|\xi\| + \varepsilon + \|T_n - T\| \, \|\xi_j\| \\
&< \frac{\varepsilon}{r} r + \varepsilon + \frac{\varepsilon}{r} r = 3\varepsilon,
\end{aligned}$$

showing that $TB(0; r) \subset \bigcup_{j=1}^m B(T\xi_j; 3\varepsilon)$. Therefore $TB(0; r)$ is totally bounded for all $r > 0$. $\square$

Corollary 24.11. *If $(T_n) \subset \mathrm{B}_f(\mathcal{N}, \mathcal{B})$ and $T_n \to T$ in $\mathrm{B}(\mathcal{N}, \mathcal{B})$, then the operator T is compact.*

Proof. Combine Proposition 24.6 and Theorem 24.10 (see Example 24.15 and Proposition 25.6 for related results). □

Example 24.12. Let $(e_n)_{n\geq 1}$ be an orthonormal basis of the separable Hilbert space $\mathcal{H}$; recall that every element $\xi \in \mathcal{H}$ can be written in the form $\xi = \sum_{n=1}^{\infty}\langle e_n, \xi\rangle e_n$. Given a bounded sequence $(a_n)_{n\geq 1} \subset \mathbb{F}$ and the linear operator $T \in \mathrm{B}(\mathcal{H})$ with $Te_n = a_n e_n$, one has

$$T\Big(\sum_{n=1}^{\infty}\langle e_n, \xi\rangle e_n\Big) = \sum_{n=1}^{\infty}\langle e_n, \xi\rangle a_n e_n,$$

and so T is compact if and only if $\lim_{n\to\infty} a_n = 0$. Indeed, if $a_n \to 0$, by defining

$$T_N\Big(\sum_{n=1}^{\infty}\langle e_n, \xi\rangle a_n e_n\Big) = \sum_{n=1}^{N}\langle e_n, \xi\rangle a_n e_n,$$

which has finite rank for all N and, by Pythagoras and Parseval,

$$\|T_N\xi - T\xi\|^2 = \Big\|\sum_{n>N}\langle e_n, \xi\rangle a_n e_n\Big\|^2 \leq \Big(\sup_{n>N}|a_n|^2\Big)\|\xi\|^2,$$

showing that $T_N \to T$ and so concluding that T is compact.

If a_n does not vanish, there exists $\varepsilon > 0$ such that $J = \{j \in \mathbb{N} : |a_j| \geq \varepsilon\}$ is infinite; thus, for $j, k \in J$, $j \neq k$,

$$\|Te_j - Te_k\|^2 = |a_j|^2 + |a_k|^2 \geq 2\varepsilon^2$$

and so the sequence $(Te_j)_{j\in J}$ does not have a Cauchy subsequence, although $(e_j)_{j\in J}$ is bounded, concluding that T is not compact. ●

EXERCISE **24.4.** Show that the operator $T : l^p(\mathbb{N}) \hookleftarrow$, $1 \leq p < \infty$, $T(\xi_j) = (\xi_j/j)$ is compact.

Example 24.13. Let $(e_j)_{j=1}^{\infty}$ be an orthonormal basis of the separable Hilbert space $\mathcal{H}$. If $P_N : \mathcal{H} \hookleftarrow$ is the operator

$$P_N\xi = P_N\Big(\sum_{j=1}^{\infty}\langle e_j, \xi\rangle e_j\Big) = \sum_{j=1}^{N}\langle e_j, \xi\rangle e_j,$$

then each P_N is compact, $P_N \xrightarrow{\mathrm{s}} \mathbf{1}$, but $\mathbf{1}$ is not a compact operator. Hence, the norm convergence may not be replaced by strong convergence in Theorem 24.10. ●

Example 24.14. One of the main examples of compact operators is given by the integral operators $T_K : \mathrm{C}[a, b] \hookleftarrow$,

$$(T_K\psi)(t) = \int_a^b K(t, s)\,\psi(s)\,\mathrm{d}s, \qquad \psi \in \mathrm{C}[a, b],$$

with $K : [a, b] \times [a, b] \to \mathbb{F}$ continuous. T_K is named as *integral operator with kernel K*. See also Examples 25.4 and 25.5. ●

Proof. Write $Q = [a, b] \times [a, b]$ and $M = \max_{(t,s) \in Q} |K(t,s)| < \infty$. Then

$$\|T_K \psi\|_\infty \leq M(b-a)\|\psi\|_\infty,$$

i.e., $\|T_K\| \leq M(b-a)$ and, hence, for $\psi \in B(0; R)$ one has $\|T_K \psi\|_\infty \leq M(b-a)R$, and so T_K is uniformly bounded in this ball ($\forall R > 0$ fixed). The idea is to apply Ascoli's Theorem to $T_K B(0; R)$, being then necessary to show that this set is also equicontinuous.

Since Q is compact, K is uniformly continuous on Q, and so for any $\varepsilon > 0$ there exists $\delta > 0$ so that $|K(t,s) - K(r,s)| < \varepsilon$ if $|t-r| < \delta$ (δ is independent of s). Thus, if $\psi \in B(0; R)$ and $|t - r| < \delta$,

$$
\begin{aligned}
|(T_K\psi)(t) - (T_K\psi)(r)| &\leq \int_a^b |K(t,s) - K(r,s)|\,|\psi(s)|\,\mathrm{d}s \\
&\leq \varepsilon(b-a)\|\psi\|_\infty \leq \varepsilon(b-a)R,
\end{aligned}
$$

showing that the set $T_K B(0; R)$ is equicontinuous. Therefore, by Ascoli's Theorem, $T_K B(0; R)$ is precompact; since this holds for all $R > 0$, the integral operator $T_K : C[a, b] \hookleftarrow$ is compact. □

Example 24.15. This example shows that the assumption that $\mathcal{B}$ is complete cannot be omitted in Corollary 24.11 (nor in Theorem 24.10; see also Exercise 24.12). Let $\mathcal{B}_1 = C^1[0, 1]$ with the norm $\|\|\psi\|\| = \|\psi\|_\infty + \|\psi'\|_\infty$, $\mathcal{N}_1 = C^1[0, 1]$ with the norm $\|\psi\|_\infty$, and $\mathrm{I} \in B(\mathcal{B}_1, \mathcal{N}_1)$ given by $\mathrm{I}(\psi) = \psi$; here, both $\mathcal{B}_1$ and $\mathcal{N}_1$ are assumed to be real spaces. Note that $\|\mathrm{I}\| = 1$, but this operator is not compact; indeed, consider a bounded sequence (ψ_n) in $\mathcal{B}_1$ uniformly converging to, say, $\psi(t) = |t - \frac{1}{2}|$, which is not in $\mathcal{N}_1$; hence the closure of $\mathrm{I}(\psi_n)$ has no convergent subsequence in $\mathcal{N}_1$.

Define the sequence $(\mathrm{Ber}_n) \subset B_f(\mathcal{B}_1, \mathcal{N}_1)$ as the Bernstein polynomials

$$(\mathrm{Ber}_n\,\psi)\,(t) = \sum_{k=0}^n C_{n,k}\psi\left(\frac{k}{n}\right) t^k (1-t)^{n-k}, \quad \psi \in \mathcal{B}_1, \quad C_{n,k} = \binom{n}{k}.$$

The aim is to show that $\mathrm{Ber}_n \to \mathrm{I}$, which will complete the example since I is not a compact operator.

Differentiating Newton's binomial $\sum_{k=0}^n C_{n,k}\, t^k\,(1-t)^{n-k} = 1$, multiplying then by $t(1-t)$, one gets $\sum_{k=0}^n C_{n,k}\, t^k\,(1-t)^{n-k}(k-nt) = 0$. Differentiate the last relation and use the binomial again to obtain $\sum_{k=0}^n C_{n,k}\, t^{k-1}\,(1-t)^{n-k-1}(k-nt)^2 = n$; finally, by multiplying such expression by $t(1-t)$ one gets the searched relation (keep it!)

$$\sum_{k=0}^n C_{n,k}\, t^k\,(1-t)^{n-k}\left(t - \frac{k}{n}\right)^2 = \frac{t(1-t)}{n}.$$

Now, if $\|\|\psi\|\| \leq 1$, it follows that (by using again the binomial)

$$(\mathrm{I} - \mathrm{Ber}_n)(\psi)(t) = \sum_{k=0}^n C_{n,k}\left(\psi(t) - \psi\left(\frac{k}{n}\right)\right) t^k\,(1-t)^{n-k},$$

and since $|\psi(t) - \psi(s)| \leq |t - s|$, for all $t, s \in [0, 1]$, one concludes that

$$\|(\mathrm{I} - \mathrm{Ber}_n)\psi\| \leq \sum_{k=0}^n C_{n,k}\left|t - \frac{k}{n}\right| t^k\,(1-t)^{n-k}.$$

Given $\varepsilon > 0$, for each $t \in [0,1]$ divide this sum in $\sum'$ and $\sum''$, with the first one restricted to the values of k so that $|t - \frac{k}{n}| < \varepsilon$, and the second one to its complement. Thus, $\sum' \leq \varepsilon$ and from the above relation it follows that

$$\varepsilon^2 \sum_{|t-\frac{k}{n}|\geq\varepsilon} C_{n,k}\, t^k\,(1-t)^{n-k} \leq \frac{t(1-t)}{n} \leq \frac{1}{4n}.$$

By recalling that $|t - \frac{k}{n}| \leq 1$, one obtains

$$\sum'' \leq \sum_{|t-\frac{k}{n}|\geq\varepsilon} C_{n,k}\, t^k\,(1-t)^{n-k} \leq \frac{1}{4n\varepsilon^2},$$

and for $n > 1/(4\varepsilon^3)$ one gets, for all $t \in [0,1]$, $\sum'' \leq \varepsilon$. Therefore,

$$\|(I - \mathrm{Ber}_n)\psi\| \leq 2\varepsilon, \quad \text{if} \quad n > \frac{1}{4\varepsilon^3},$$

for all ψ with $\|\|\psi\|\| = 1$, which shows that $\mathrm{Ber}_n \to I$ in $B(\mathcal{B}_1, \mathcal{N}_1)$. ∙

Notes

In his studies of linear equations, Hilbert used bilinear forms, mainly on l^2, with emphasis to a class he called *completely continuous*; the original meaning of completely continuous referred to operators that take weakly convergent sequences to strongly convergent ones. This notion was adapted to l^p spaces, with $p \neq 2$, by F. Riesz who, in 1918, noticed that the important property was that such operators mapped bounded sets into precompact ones, hence the term *compact operator*. The adaptation of this concept to operators on normed spaces in general was immediate. That work by Riesz of 1918 discussed much of this theory, including results that will be presented in the chapter concerning spectrum of compact operators.

Bernstein's polynomials, as well as technical details in Example 24.15, are used in a proof of the famous Weierstrass' Theorem on uniform approximations of continuous functions by polynomials [Simm].

Additional Exercises

EXERCISE **24.5.** Show that $T \in B(\mathcal{N}, \mathcal{B})$ is compact if and only if $T(A)$ is totally bounded for all bounded $A \subset \mathcal{N}$.

EXERCISE **24.6.** (a) Show that if $T \in B_0(\mathcal{N}_1, \mathcal{N}_2)$, then $T(\mathcal{N}_1)$ is separable.
   (b) Let $T : \mathcal{N}_1 \to \mathcal{N}_2$ be linear. Show that it is compact if and only if $TB(0;1)$ is precompact in $\mathcal{N}_2$.

EXERCISE **24.7.** Let $T : \mathrm{dom}\, T \subset \mathcal{N} \to \mathcal{B}$ be a compact operator. If $\mathrm{dom}\, T$ is dense in $\mathcal{N}$, show that its extension $\overline{T} : \mathcal{N} \to \mathcal{B}$ is also compact (see Theorem 4.7).

EXERCISE **24.8.** Assume that $T \in B(\mathcal{B})$ satisfies at least one of the following relations (see Exercise 4.15):
(a) $T^n + a_{n-1}T^{n-1} + \cdots + a_1 T + a_0 \mathbf{1} = 0$, $a_0 \neq 0$;
(b) $\cos T = 0$;
(c) $\exp(aT) + \mathbf{1} = 0$, with $a \neq 0$.
Show that T is compact if and only if $\dim \mathcal{B} < \infty$.

EXERCISE **24.9.** Under what conditions $B_0(\mathcal{N}) = B(\mathcal{N})$?

EXERCISE **24.10.** Let $a, b : [a_0, b_0] \to [a_0, b_0]$ and $K : [a_0, b_0] \times [a_0, b_0] \to \mathbb{F}$ be continuous functions. Show that the operator $T_K : \mathrm{C}[a_0, b_0] \to \mathrm{C}[a_0, b_0]$,

$$(T_K \psi)(t) = \int_{a(t)}^{b(t)} K(t, s) \psi(s) \, \mathrm{d}s, \qquad \psi \in \mathrm{C}[a_0, b_0],$$

is compact.

EXERCISE **24.11.** Let $T \in \mathrm{B}_0(\mathcal{N})$ be bijective, with $\dim \mathcal{N} = \infty$. Show that T^{-1} is not bounded.

EXERCISE **24.12.** Complement Theorem 24.10: show that $\mathrm{B}_0(\mathcal{N}_1, \mathcal{N}_2)$ is complete if and only if $\mathcal{N}_2$ is Banach. Conclude that $\mathrm{B}(\mathcal{N}_1, \mathcal{N}_2)$ is complete if and only if $\mathrm{B}_0(\mathcal{N}_1, \mathcal{N}_2)$ is complete.

EXERCISE **24.13.** If $0 \neq \phi \in \mathrm{L}^2[a, b]$ is bounded, show that the multiplication operator by ϕ,

$$\mathcal{M}_\phi : \mathrm{L}^2[a, b] \hookleftarrow, \quad (\mathcal{M}_\phi \psi)(t) = \phi(t)\psi(t), \qquad \psi \in \mathrm{L}^2[a, b],$$

is not a compact operator.

EXERCISE **24.14.** (a) If $T \in \mathrm{B}_0(\mathcal{B})$ and E is a closed vector subspace of $\mathcal{B}$, invariant under T, show that the restriction $T|_E$ is compact.
(b) If $0 \neq \phi \in \mathrm{C}[a, b]$, show that the multiplication operator

$$\mathcal{M}_\phi : \mathrm{C}[a, b] \hookleftarrow, \quad (\mathcal{M}_\phi \psi)(t) = \phi(t)\psi(t), \qquad \psi \in \mathrm{C}[a, b],$$

is not a compact operator.

EXERCISE **24.15.** Here is Exercise 8.11 with a more appropriate language. If $T \in \mathrm{B}_0(\mathcal{B}_1, \mathcal{B}_2)$, show that $\mathrm{rng}\, T$ does not contain a closed subspace of $\mathcal{B}_2$ of infinite dimension. Conclude that if T is onto, then $\dim \mathcal{B}_2 < \infty$.

CHAPTER 25

Compact Operators on Hilbert Spaces

Some fundamental properties of compact operators on Hilbert spaces are discussed. Some of these properties justify the interpretation that compact operators exhibit similarities with operators in finite-dimensional spaces. It is shown that a bounded linear operator between Hilbert spaces is compact if and only if it can be uniformly approximated by finite-rank operators. Another important result presented here is that if the domain is a reflexive space (which includes Hilbert spaces), then an operator is compact if and only if it maps weakly convergent subsequences to strongly convergent ones. Finally, it is proven that a bounded linear operator between normed spaces is compact if and only if its adjoint is also compact.

In a Hilbert space, the closure (with the usual norm of $B(\mathcal{H})$) of the vector space of finite rank operators coincides with the set of compact operators; to show this the following technical result will be useful.

Lemma 25.1. *Let $T \in B_0(\mathcal{H}_1, \mathcal{H}_2)$; then* rng T *and* $N(T)^\perp$ *are separable vector subspaces of $\mathcal{H}_2$ and $\mathcal{H}_1$, respectively.*

Proof. rng T is separable by Exercise 24.6. Let $\{e_j\}_{j \in J}$ be an orthonormal basis of $N(T)^\perp$. If J is finite the result should be clear.

Suppose that J is not finite; the goal is to show that J is enumerable (see Exercise 21.2). Every sequence $(e_{j_j})_{j=1}^\infty$ of pairwise distinct elements of $\{e_j\}_{j \in J}$ weakly converges to zero (Exercise 21.5), and by Proposition 24.8, $T e_{j_j} \to 0$, as $j \to \infty$. Thus, for each $n \in \mathbb{N}$ there exists only a finite number of $j \in J$ with $\|Te_j\| \geq 1/n$. Hence, J is enumerable, for

$$ J = \bigcup_{n=1}^\infty \left\{ j : \|Te_j\| \geq \frac{1}{n} \right\}. $$

Recall that $Te_j \neq 0$, for all $j \in J$, since $e_j \in N(T)^\perp$. $\qquad \square$

Theorem 25.2. *An operator $T \in B(\mathcal{H}_1, \mathcal{H}_2)$ is compact if and only if there is a sequence of finite rank operators (T_n) that converges to T in $B(\mathcal{H}_1, \mathcal{H}_2)$.*

DOI: 10.1201/9781003656166-25

Proof. If T is the limit of finite rank operators, then T is compact by Corollary 24.11. Let $T \in B_0(\mathcal{H}_1, \mathcal{H}_2)$ and P the orthogonal projection onto $N(T)^\perp$, so that $T = TP$. If $\dim N(T)^\perp < \infty$, the result is clear; suppose then that $\dim N(T)^\perp = \infty$ and pick an orthonormal basis $(e_j)_{j=1}^\infty$ of $N(T)^\perp$, which is enumerable by Lemma 25.1. Denote by P_n the orthogonal projection onto $\mathrm{Lin}(\{e_1, \cdots, e_n\})$. Thus, the operator $T_n = TP_n$ has finite rank. It will be shown that $T_n \to T$, concluding then that T is compact.

For each n there exists $\xi_n \in \mathcal{H}_1$, $\|\xi_n\| = 1$, so that

$$\frac{1}{2}\|T - T_n\| \le \|(T - T_n)\xi_n\| = \|T(P - P_n)\xi_n\|.$$

Since $(P_n - P) \xrightarrow{s} 0$ and for all $\eta \in \mathcal{H}_1$ one has

$$|\langle \eta, (P - P_n)\xi_n\rangle| = |\langle (P - P_n)\eta, \xi_n\rangle| \le \|(P - P_n)\eta\|,$$

and so $(P - P_n)\xi_n \xrightarrow{w} 0$. Since T is a compact operator, by Proposition 24.8, it follows that $T(P - P_n)\xi_n \to 0$, and by the inequality above, it is found that $\|T - T_n\| \to 0$. □

EXERCISE 25.1. Let $T \in B(\mathcal{H})$, with $\mathcal{H}$ separable. Show that there is a sequence (T_n) of finite rank operators which converges strongly to T, that is, $T_n \xrightarrow{s} T$.

Corollary 25.3. *Let $T \in B(\mathcal{H}_1, \mathcal{H}_2)$. Then T is compact if and only if its Hilbert adjoint T^* is compact.*

Proof. T is compact if and only if there exists a sequence $(T_n) \subset B_f(\mathcal{H}_1, \mathcal{H}_2)$ so that $T_n \to T$. Since T_n^* has also finite rank (Exercise 25.4) and $\|T^* - T_n^*\| = \|(T - T_n)^*\| = \|T - T_n\|$, one concludes that T is compact if and only if T^* is compact. □

The next examples are related to Example 24.14.

Example 25.4. Let $Q = [a, b] \times [a, b]$ and $K : Q \to \mathbb{F}$ be continuous. Then the integral operator $T_K : L^2[a, b] \hookleftarrow$ given by

$$(T_K\psi)(t) = \int_a^b K(t, s)\psi(s)\,ds, \qquad \psi \in L^2[a, b],$$

is compact. ●

Proof. For each $t \in [a, b]$, the function $s \mapsto K(t, s)$ is an element of $L^2[a, b]$. Let $\psi \in B(0; R) \subset L^2[a, b]$ and $M = \max_{(t,s)\in Q} |K(t, s)|$. For all $t \in [a, b]$ one has

$$\begin{aligned} |(T_K\psi)(t)| &\le \int_a^b |K(t, s)||\psi(s)|\,ds \\ &\le \left(\int_a^b |K(t, s)|^2\,ds\right)^{\frac{1}{2}} \|\psi\|_2 \le M\sqrt{b - a}\,R, \end{aligned}$$

thus $\|T_K\psi\|_\infty \le M(b - a)^{1/2}R$ and $T_K B(0; R)$ is a bounded set in $C[a, b]$. This set is also equicontinuous, since by an argument similar to Example 24.14, for $\psi \in B(0; R)$ one has

$$\begin{aligned} |(T_K\psi)(t) - (T_K\psi)(r)| &\le \|K(t, \cdot) - K(r, \cdot)\|_2 \|\psi\|_2 \\ &\le \varepsilon\sqrt{b - a}\,R, \end{aligned}$$

if $|t - r| < \delta$. Hence, by Ascoli's Theorem, it is found that $T_K B(0; R)$ is precompact in $(C[a, b], \| \cdot \|_\infty)$. Since $\|\phi\|_2 \leq \sqrt{b - a}\|\phi\|_\infty$, for all continuous ϕ (in particular for $\phi = T_K \psi$), then $T_K B(0; R)$ is precompact in $L^2[a, b]$. ☐

EXERCISE **25.2.** Show that a precompact set (resp. compact) in $(C[a, b], \| \cdot \|_\infty)$ is precompact (resp. compact) in $L^2[a, b]$. This occurs because the identity mapping

$$\mathbf{1} : (C[a, b], \| \cdot \|_\infty) \to L^2[a, b]$$

is continuous.

Example 25.5. Let $K \in L^2(Q)$, with $Q = [a, b] \times [a, b]$. Then the integral operator $T_K : L^2[a, b] \hookleftarrow$ given by $(T_K \psi)(t) = \int_a^b K(t, s)\psi(s)\, ds$, for $\psi \in L^2[a, b]$, is compact.
●

Proof. Since the set of continuous functions on Q is dense in $L^2(Q)$, there exists a sequence $K_n : Q \to \mathbb{F}$ of continuous functions so that $\|K - K_n\|_{L^2(Q)} \to 0$. Thus, by defining $T_n : L^2[a, b] \hookleftarrow$,

$$(T_n \psi)(t) = \int_a^b K_n(t, s)\psi(s)\, ds, \qquad \psi \in L^2[a, b],$$

and using estimates similar to those in preceding examples, one obtains

$$\|T_n \psi - T_K \psi\|_2 \leq \|K_n - K\|_{L^2(Q)}\|\psi\|_2,$$

and $\|T_n - T_K\| \leq \|K_n - K\|_{L^2(Q)}$, which vanishes as $n \to \infty$. By Example 25.4, each T_n is a compact operator and so T_K is compact (Theorem 24.10). ☐

For reflexive spaces, and Hilbert spaces in particular, it is possible to give more information than Proposition 24.8. See Example 24.9 for the case of a nonreflexive Banach space, for which the following characterization of compact operators does not hold.

Proposition 25.6. *Let $\mathcal{B}$ be a reflexive Banach space and $T \in B(\mathcal{B}, \mathcal{N})$. Then T is compact if and only if $(T\xi_n)$ is convergent in $\mathcal{N}$ for all weakly convergent sequences (ξ_n) in $\mathcal{B}$. Note that this characterization holds if $\mathcal{B}$ is a Hilbert space.*

Proof. If $\dim \mathcal{B} < \infty$ the proof is quite simple. Suppose that $\dim \mathcal{B} = \infty$. Taking into account the hypotheses and Proposition 24.8, it is enough to show that for every bounded sequence (ξ_n) in $\mathcal{B}$ the sequence $(T\xi_n)$ has a convergent subsequence in $\mathcal{N}$. By Theorem 16.5, (ξ_n) has a weakly convergent subsequence (ξ_{n_j}); by hypothesis, $(T\xi_{n_j})$ is convergent. Thus, the image of every bounded sequence admits a convergent subsequence, and so, T is a compact operator. ☐

Due to Corollary 25.3, it is natural to ask about the relation between a compact operator T (between normed spaces) and its adjoint T^a. The next result will clarify this question.

Theorem 25.7. *Let $\mathcal{E}$ and $\mathcal{N}$ be normed spaces. Then $T \in \mathrm{B}(\mathcal{E}, \mathcal{N})$ is compact if and only if its adjoint T^{a} is compact.*

Proof. Recall that $T^{\mathrm{a}} \in \mathrm{B}(\mathcal{N}^*, \mathcal{E}^*)$. Suppose that $T \in \mathrm{B}(\mathcal{E}, \mathcal{N})$ is compact. For each $r > 0$, denote $B_r = \overline{B}_{\mathcal{E}}(0; r)$ and $B_r^* = \overline{B}_{\mathcal{N}^*}(0; r)$. To conclude that T^{a} is compact, it will be shown that $(T^{\mathrm{a}} g_n)$ has a convergent subsequence in $\mathcal{E}^*$ for any sequence (g_n) in B_r^* (and for any fixed $r > 0$).

Since T is compact, $\overline{TB_1} \subset \mathcal{N}$ is a compact set. Since for all $\eta, \zeta \in \mathcal{N}$

$$|g_n(\eta) - g_n(\zeta)| \le \|g_n\| \, \|\eta - \zeta\| \le r\|\eta - \zeta\|,$$

then (g_n) is equicontinuous on $\mathcal{N}$. For $\eta \in TB_1$ one has $|g_n(\eta)| \le \|g_n\| \, \|\eta\| \le r\|T\|$, for all n, and by continuity this also holds for all $\eta \in \overline{TB_1}$, and so $\{g_n(\overline{TB_1})\}$ is uniformly bounded. By Ascoli's Theorem, (g_n) is a precompact subset of $\mathrm{C}(\overline{TB_1})$, and so it has a Cauchy subsequence in this space. Recall that $g_n(T) = T^{\mathrm{a}}(g_n)$; then

$$
\begin{aligned}
\|T^{\mathrm{a}} g_n - T^{\mathrm{a}} g_m\|_{\mathcal{E}^*} &= \sup_{\xi \in B_1} |g_n(T\xi) - g_m(T\xi)| \\
&= \|g_n - g_m\|_{\mathrm{C}(TB_1)} \le \|g_n - g_m\|_{\mathrm{C}(\overline{TB_1})},
\end{aligned}
$$

and $(T^{\mathrm{a}} g_n)$ has a Cauchy subsequence in $\mathcal{E}^*$, which is convergent since $\mathcal{E}^*$ is complete. Therefore, T^{a} is compact.

Suppose now that T^{a} is compact, and denote by $\eta \mapsto \check{\eta}$ the canonical mapping of $\mathcal{E}$ into $\mathcal{E}^{**}$. By the arguments above, T^{aa} is compact, and if $B_r^{**} = \overline{B}_{\mathcal{E}^{**}}(0; r)$, then $T^{\mathrm{aa}} B_r^{**} \subset \mathcal{N}^{**}$ is a precompact set. Since $\mathcal{N} \subset \mathcal{N}^{**}$, by considering the canonical mapping $\xi \mapsto \hat{\xi}$, $\xi \in \mathcal{N}$, and using that $\widehat{T\eta} = T^{\mathrm{aa}} \check{\eta}$, for all $\eta \in \mathcal{E}$, one gets

$$\widehat{TB_r} \subset T^{\mathrm{aa}} B_r^{**},$$

and therefore, $\widehat{TB_r}$ is a precompact subset. Since $\hat{\ }$ is an isometric mapping (Proposition 12.7), one concludes that TB_r is precompact in $\mathcal{N}$ for all $r > 0$, and so T is a compact operator. $\qquad\square$

EXERCISE 25.3. Let $\mathcal{E}$ and $\mathcal{N}$ be normed spaces and $T \in \mathrm{B}(\mathcal{E}, \mathcal{N})$. Show that $\widehat{T\eta} = T^{\mathrm{aa}} \check{\eta}$, for all $\eta \in \mathcal{E}$ (notation as in the proof of Theorem 25.7).

Notes

In 1930, J. Schauder published the first version of Theorem 25.7. There is an alternative proof that uses Alaoglu's Theorem (see [Conway], page 178).

As stated in Exercise 25.7, the presence of a Schauder basis in a Banach space assures that every compact operator can be approximated in norm by finite rank operators. The example by Per Enflo, showing that there are separable (and reflexive) Banach spaces that do not admit Schauder bases, consisted in showing that, in a particular case, the set of finite rank operators is not dense in the set of compact operators. See the Notes in Chapter 3.

Additional Exercises

EXERCISE 25.4. Let $T \in \mathrm{B}(\mathcal{H}_1, \mathcal{H}_2)$. Show that T has finite rank if and only if its Hilbert adjoint T^* has finite rank, and in this case one has $\dim \mathrm{rng}\, T = \dim \mathrm{rng}\, T^*$.

EXERCISE **25.5.** Give necessary and sufficient conditions for an orthogonal projection operator on a Hilbert space be compact.

EXERCISE **25.6.** Use Theorem 25.7 to prove Corollary 25.3.

EXERCISE **25.7.** Show that if the Banach space $\mathcal{B}$ has a Schauder basis, then the conclusions of Theorem 25.2 characterize the set $B_0(\mathcal{B})$.

EXERCISE **25.8.** If X is a compact metric space, show that $B_f(C(X))$ is dense in $B_0(C(X))$. This shows that the characterization of compact operators on Hilbert spaces, given in Theorem 25.2, can hold in some nonreflexive Banach spaces.

EXERCISE **25.9.** Use Theorem 25.7 and Exercise 13.8 to show that if $\mathcal{B}$ is reflexive, then $B(c_0(\mathbb{N}), \mathcal{B}) = B_0(c_0(\mathbb{N}), \mathcal{B})$.

EXERCISE **25.10.** Let $T \in B(\mathcal{H})$ and (T_n) a sequence in $B(\mathcal{H})$. Show that:

(i) If $T_n^* \xrightarrow{s} T^*$, then $ST_n \to ST$ in $B(\mathcal{H})$, for all compact operators $S \in B_0(\mathcal{H})$.

(ii) If $T_n \xrightarrow{s} T$, then $T_n S \to TS$ in $B(\mathcal{H})$, for all compact operators $S \in B_0(\mathcal{H})$.

(iii) By considering the shift operators on $l^2(\mathbb{N})$, check that it is possible to have a strongly convergent sequence of bounded linear operators, while the sequence of their respective adjoint does not converge.

EXERCISE **25.11.** Let $\mathcal{H}$ be a separable infinite-dimensional Hilbert space.

(a) If (ξ_j) is an orthonormal basis of $\mathcal{H}$, show that the operator $T\xi = \sum_{j=1}^{\infty} a_j \langle \xi_j, \xi \rangle \xi_j$, with a_j taking the values 0 and 1, is bounded. Show that the distance, in $B(\mathcal{H})$, between two of such operators (distinct, of course) is always equal to 1. Conclude that $B(\mathcal{H})$ is not separable.

(b) Show that there exists a sequence (Q_n), of finite rank operators in $B(\mathcal{H})$, with $Q_n \xrightarrow{s} 1$, and so for every compact operator $T \in B_0(\mathcal{H})$ one has $Q_n T Q_n \to T$. Conclude that $B_0(\mathcal{H})$ is a separable subset of $B(\mathcal{H})$.

EXERCISE **25.12.** Let $\mathcal{H} = l^2(\mathbb{N})$. For each $t \in (0, 1)$ define $T_t : \mathcal{H} \hookleftarrow$ by

$$T_t(\xi_1, \xi_2, \xi_3, \cdots) = (t\xi_1, t^2\xi_2, t^3\xi_3, \cdots).$$

Show that each T_t is compact, $\mathrm{w} \cdot \lim_{t \uparrow 1} T_t = \mathbf{1}$, and for each compact operator $S \in B_0(\mathcal{H})$ one has $\lim_{t \uparrow 1} ST_t = S$ in $B(\mathcal{H})$.

EXERCISE **25.13.** If $T \in B(\mathcal{H}_1, \mathcal{H}_2)$ and T^*T is compact, show that T is compact. Use this to present an alternative proof of the following fact: if $T \in B(\mathcal{H}_1, \mathcal{H}_2)$ is compact, then T^* is compact.

EXERCISE **25.14.** If $\mathcal{B}$ is reflexive, show that $B(\mathcal{B}, l^1(\mathbb{N})) = B_0(\mathcal{B}, l^1(\mathbb{N}))$.

CHAPTER 26

Hilbert-Schmidt Operators

A particularly important category of compact operators on Hilbert spaces is formed by the Hilbert-Schmidt operators, which will be explored in this chapter. In some cases, the most direct way to establish that an operator is compact is to verify that it is Hilbert-Schmidt. A key property of Hilbert-Schmidt operators is that they form a Hilbert space themselves, which is quite rare among operator classes. Another feature is that many important examples of compact operators, such as integral operators with square-integrable kernels, belong to this class. In some situations, this condition characterizes Hilbert-Schmidt operators completely.

Definition 26.1. A linear operator $T \in B(\mathcal{H}_1, \mathcal{H}_2)$ is said to be *Hilbert-Schmidt* if there exists an orthonormal basis $\{e_j\}_{j \in J}$ of $\mathcal{H}_1$ so that

$$\|T\|_{\mathrm{HS}} := \left(\sum_{j \in J} \|Te_j\|^2 \right)^{\frac{1}{2}} < \infty.$$

The collection of such Hilbert-Schmidt operators will be denoted by $\mathrm{HS}(\mathcal{H}_1, \mathcal{H}_2)$ or, as usual, by $\mathrm{HS}(\mathcal{H})$ in case $\mathcal{H}_1 = \mathcal{H}_2 = \mathcal{H}$.

Proposition 26.2. *If $T \in B(\mathcal{H}_1, \mathcal{H}_2)$, then:*

(i) $\|T\|_{\mathrm{HS}}$ is independent of the choice of orthonormal basis.

(ii) $T \in \mathrm{HS}(\mathcal{H}_1, \mathcal{H}_2)$ if and only if its adjoint $T^* \in \mathrm{HS}(\mathcal{H}_2, \mathcal{H}_1)$, and in this case $\|T\|_{\mathrm{HS}} = \|T^*\|_{\mathrm{HS}}$.

Proof. Pick $\{e_j\}_{j \in J}$ and $\{f_k\}_{k \in K}$ orthonormal bases of $\mathcal{H}_1$ and $\mathcal{H}_2$, respectively; then, Parseval Identity (Theorem 21.12) implies that

$$\sum_{j \in J} \|Te_j\|^2 = \sum_{\substack{j \in J \\ k \in K}} |\langle Te_j, f_k \rangle|^2 = \sum_{\substack{j \in J \\ k \in K}} |\langle e_j, T^* f_k \rangle|^2 = \sum_{k \in K} \|T^* f_k\|^2.$$

Since this holds for any orthonormal bases, $\|T\|_{\mathrm{HS}} = \|T^*\|_{\mathrm{HS}}$, and so such values independent of the orthonormal bases considered. $\square$

Corollary 26.3. *Suppose that S and T are bounded operators between Hilbert spaces. In case one of them is Hilbert-Schmidt, then the composition (if defined) TS is also Hilbert-Schmidt.*

DOI: 10.1201/9781003656166-26

Proof. If S is Hilbert-Schmidt, then for any orthonormal basis $\{e_j\}_{j \in J}$ of its domain

$$\|TS\|_{\mathrm{HS}}^2 = \sum_{j \in J} \|TSe_j\|^2 \leq \|T\|^2 \sum_{j \in J} \|Se_j\|^2 = \|T\|^2 \|S\|_{\mathrm{HS}}^2,$$

and it follows that TS is Hilbert-Schmidt.

If the operator T is Hilbert-Schmidt, then, by Proposition 26.2 and the initial part of this corollary, it follows that $S^* T^*$ is also Hilbert-Schmidt. Noting that $TS = (S^* T^*)^*$, one concludes that TS is Hilbert-Schmidt as well. □

Theorem 26.4. *The set* $\mathrm{HS}(\mathcal{H}_1, \mathcal{H}_2)$ *forms a vector subspace of* $\mathrm{B}(\mathcal{H}_1, \mathcal{H}_2)$ *and becomes a Hilbert space when equipped with the so-called Hilbert-Schmidt norm* $\| \cdot \|_{\mathrm{HS}}$, *which is induced by the Hilbert-Schmidt inner product*

$$\langle T, S \rangle_{\mathrm{HS}} := \sum_{j \in J} \langle Te_j, Se_j \rangle, \qquad T, S \in \mathrm{HS}(\mathcal{H}_1, \mathcal{H}_2),$$

with $\{e_j\}_{j \in J}$ *denoting any orthonormal basis of* $\mathcal{H}_1$. *Moreover, the norm inequality* $\|T\| \leq \|T\|_{\mathrm{HS}}$ *holds for all such operators.*

Proof. Suppose that $T, S \in \mathrm{HS}(\mathcal{H}_1, \mathcal{H}_2)$; then for every orthonormal basis $\{e_j\}_{j \in J}$ of $\mathcal{H}_1$ and every $t \in \mathbb{F}$, by Cauchy-Schwarz applied to the inner product $\sum_{j \in J} \|Te_j\| \|Se_j\|$ in l^2, one obtains

$$
\begin{aligned}
\|T &+ tS\|_{\mathrm{HS}}^2 \\
&\leq \sum_{j \in J} \|Te_j\|^2 + |t|^2 \sum_{j \in J} \|Se_j\|^2 + 2|t| \sum_{j \in J} \|Te_j\| \|Se_j\| \\
&\leq (\|T\|_{\mathrm{HS}} + |t| \|S\|_{\mathrm{HS}})^2,
\end{aligned}
$$

and so $\mathrm{HS}(\mathcal{H}_1, \mathcal{H}_2)$ is a vector space. Such inequality also shows that $\| \cdot \|_{\mathrm{HS}}$ is a norm.

Next it is checked that $\langle T, S \rangle_{\mathrm{HS}}$ is well defined and does not depend on the considered orthonormal basis. By Cauchy-Schwarz one gets

$$
\begin{aligned}
\sum_{j \in J} |\langle Te_j, Se_j \rangle| &\leq \sum_{j \in J} \|Te_j\| \|Se_j\| \\
&\leq \left(\sum_{j \in J} \|Te_j\|^2 \right)^{\frac{1}{2}} \left(\sum_{j \in J} \|Se_j\|^2 \right)^{\frac{1}{2}} \\
&= \|T\|_{\mathrm{HS}} \|S\|_{\mathrm{HS}},
\end{aligned}
$$

so $|\langle T, S \rangle_{\mathrm{HS}}| \leq \|T\|_{\mathrm{HS}} \|S\|_{\mathrm{HS}}$ and the series that defines $\langle T, S \rangle_{\mathrm{HS}}$ is absolutely convergent. It is left to the reader to check that the properties of inner product holds. The polarization identity applied to $\langle T, S \rangle_{\mathrm{HS}}$ (or similar to the proof of Proposition 26.2) implies that

$$\sum_j \langle Te_j, Se_j \rangle = \sum_k \langle S^* f_k, T^* f_k \rangle,$$

for every orthonormal basis $\{f_k\}$ of $\mathcal{H}_2$; hence $\langle T, S \rangle_{\mathrm{HS}}$ is independent of the orthonormal basis and therefore, well defined.

If $\xi \in \mathcal{H}_1$, $\|\xi\| = 1$, take a particular form of an orthonormal basis of $\mathcal{H}_1$, that is, $\{\xi, \eta_l\}_{l \in M}$. Hence, $\|T\xi\|^2 \leq \sum_l \|T\eta_l\|^2 + \|T\xi\|^2 = \|T\|_{\text{HS}}^2$, and so concluding that $\|T\| \leq \|T\|_{\text{HS}}$.

Now, it suffices to prove that the space $\text{HS}(\mathcal{H}_1, \mathcal{H}_2)$ is complete; for any Cauchy sequence $(T_n) \subset \text{HS}(\mathcal{H}_1, \mathcal{H}_2)$, by the inequality $\|\cdot\|_{\text{B}(\mathcal{H}_1, \mathcal{H}_2)} \leq \|\cdot\|_{\text{HS}}$ one has that (T_n) is a Cauchy sequence in $\text{B}(\mathcal{H}_1, \mathcal{H}_2)$ and therefore, it converges to some $T \in \text{B}(\mathcal{H}_1, \mathcal{H}_2)$. The task now is to show that $T \in \text{HS}(\mathcal{H}_1, \mathcal{H}_2)$ and also that $T_n \to T$ in this space.

Given $\varepsilon > 0$, there is $N(\varepsilon)$ so that $\|T_n - T_m\|_{\text{HS}}^2 < \varepsilon$ for any $n, m \geq N(\varepsilon)$. Consider an orthonormal basis $\{e_j\}_{j \in J}$ of $\mathcal{H}_1$. If $F \subset J$ is finite, then

$$\sum_{j \in F} \|T_n e_j - T_m e_j\|^2 \leq \|T_n - T_m\|_{\text{HS}}^2 < \varepsilon.$$

By taking $m \to \infty$, it follows that $\sum_{j \in F} \|(T_n - T)e_j\|^2 \leq \varepsilon$, for every finite subsets F. Hence, $\|T_n - T\|_{\text{HS}}^2 = \sum_{j \in J} \|(T_n - T)e_j\|^2 \leq \varepsilon$, and so $(T - T_n) \in \text{HS}(\mathcal{H}_1, \mathcal{H}_2)$ and $(T_n - T) \to 0$ in this space. Since $\text{HS}(\mathcal{H}_1, \mathcal{H}_2)$ is a vector space, one finds that $T = (T - T_n) + T_n$ belongs to $\text{HS}(\mathcal{H}_1, \mathcal{H}_2)$, and so this space is Hilbert. $\quad\square$

Now, the compactness of Hilbert-Schmidt operators can be established using the developed tools.

Theorem 26.5. $\text{HS}(\mathcal{H}_1, \mathcal{H}_2) \subset \text{B}_0(\mathcal{H}_1, \mathcal{H}_2)$.

Proof. Select $T \in \text{HS}(\mathcal{H}_1, \mathcal{H}_2)$ and a sequence $(\xi_n) \subset \mathcal{H}_1$, so that $\xi_n \rightharpoonup \xi$. By Proposition 25.6, in order to show that T is compact it is enough to check that $T\xi_n \to T\xi$. One should observe that, by linearity, it is enough to consider the case $\xi_n \rightharpoonup 0$.

Let $\{e_j\}_{j \in J}$ be an orthonormal basis of $\mathcal{H}_2$. For every n, it is known that the set $J_n = \{j \in J : \langle e_j, T\xi_n \rangle \neq 0\}$ is countable; thus, $\cup_n J_n$ is also countable, and for notational simplicity, it will be denoted by the natural numbers (in case it is finite, the argument ahead is adapted with no special difficult). Hence,

$$\|T\xi_n\|^2 = \sum_{j=1}^{\infty} |\langle e_j, T\xi_n \rangle|^2 \leq \sum_{j=1}^{N} |\langle T^* e_j, \xi_n \rangle|^2 + M \sum_{j=N+1}^{\infty} \|T^* e_j\|^2,$$

with $M = \sup_{n \in \mathbb{N}} \|\xi_n\|^2$ (M is finite since every weakly convergent sequence is bounded; see Proposition 14.3).

Given $\varepsilon > 0$, select N so that

$$\sum_{j=N+1}^{\infty} \|T^* e_j\|^2 < \varepsilon/M;$$

such N does exist because $T^* \in \text{HS}(\mathcal{H}_2, \mathcal{H}_1)$. Since $\xi_n \rightharpoonup 0$, there is K so that $\sum_{j=1}^{N} |\langle T^* e_j, \xi_n \rangle|^2 < \varepsilon$ if $n \geq K$. Therefore, if $n \geq K$ one has $\|T\xi_n\|^2 < 2\varepsilon$, and it follows that $T\xi_n \to 0$. $\quad\square$

EXERCISE 26.1. Let $T : l^2(\mathbb{N}) \hookleftarrow$ be the linear operator with action $(T\xi)_n = \sum_{j=1}^{\infty} a_{nj} \xi_j$, $n \in \mathbb{N}$, with $(a_{nj})_{n,j \in \mathbb{N}}$ an infinite matrix satisfying $\sum_{n,j \in \mathbb{N}} |a_{nj}|^2 < \infty$. Verify that T is a Hilbert-Schmidt operator and find its Hilbert-Schmidt norm.

As a preparation for an important example, take account of the next lemma.

Lemma 26.6. *Consider two separable spaces* $\mathcal{H}_1 = L^2_\mu(\Omega)$ *and* $\mathcal{H}_2 = L^2_\nu(\Lambda)$, *with* μ, ν *being* σ-*finite measures, and* $\mathcal{H}_3 = L^2_{\mu\times\nu}(\Omega\times\Lambda)$. *If* (ψ_n) *and* (ϕ_j) *are (countable) orthonormal bases of* $\mathcal{H}_1$ *and* $\mathcal{H}_2$, *respectively, then* $(\overline{\psi_n}\phi_j)$ *is an orthonormal basis of* $\mathcal{H}_3$; *hence, the latter space is also separable.*

Proof. By Fubini, $(\overline{\psi_n}\phi_j)$ is an orthonormal set in $\mathcal{H}_3$. By using Proposition 21.7 to prove this lemma, it suffices to check that if $f \in \mathcal{H}_3$ satisfies $\langle f, \overline{\psi_n}\phi_j\rangle_{\mathcal{H}_3} = 0$, for all n, j, then it is the null function $f = 0$.

For each $s \in \Lambda$, consider the sector function $f^s : \Omega \to \mathbb{F}$, $f^s(t) = f(t, s)$, which is an element of $\mathcal{H}_1$ for s in a set of total measure ν; for each n, write $F_n(s) = \langle \overline{f^s}, \psi_n\rangle_{\mathcal{H}_1}$ (it is measurable since ν is σ−finite), so that $\langle f, \overline{\psi_n}\phi_j\rangle_{\mathcal{H}_3} = \langle F_n, \phi_j\rangle_{\mathcal{H}_2}$. By Cauchy-Schwartz Inequality, ν−a.e. one has $|F_n(s)| \le \|f^s\|_{\mathcal{H}_1}$, and so $F_n \in \mathcal{H}_2$ for all n, since

$$\|F_n\|^2_{\mathcal{H}_2} \le \int_\Lambda \|f^s\|^2_{\mathcal{H}_1} \, \mathrm{d}\nu(s) = \|f\|^2_{\mathcal{H}_3}.$$

Hence, one reduces the condition $\langle f, \overline{\psi_n}\phi_j\rangle_{\mathcal{H}_3} = 0$ to $\langle F_n, \phi_j\rangle_{\mathcal{H}_2} = 0$, for every n, j. Since (ϕ_j) is a basis of $\mathcal{H}_2$, for all n, it follows that $F_n(s) = 0$ ν−a.e. and since (ψ_n) is a basis of $\mathcal{H}_1$, it is found that $f^s = 0$ (in $\mathcal{H}_1$) ν−a.e. Consequently, the intended result $\|f\|^2_{\mathcal{H}_3} = \int_\Lambda \langle f^s, f^s\rangle_{\mathcal{H}_1} \, \mathrm{d}\nu(s) = 0$ is obtained. . $\qquad\square$

Example 26.7. Let $\mathcal{H}_1$, $\mathcal{H}_2$, and $\mathcal{H}_3$ be spaces as in Lemma 26.6. Then, an operator $T \in \mathrm{HS}(\mathcal{H}_1, \mathcal{H}_2)$ if and only if there exits $K \in \mathcal{H}_3$ such that

$$(T\psi)(t) = (T_K\psi)(t) := \int_\Omega K(t, s)\psi(s) \, \mathrm{d}\mu(s), \qquad \psi \in \mathcal{H}_1.$$

Moreover, $\|T\|_{\mathrm{HS}} = \|K\|_{\mathcal{H}_3}$. •

Proof. Pick orthonormal bases (ψ_n) and (ϕ_j) of $\mathcal{H}_1$ and $\mathcal{H}_2$, respectively. By Lemma 26.6, one has that $(\overline{\psi_n}\phi_j)$ is an orthonormal basis of $\mathcal{H}_3$. Now, suppose that $T = T_K$; thus

$$\sum_n \|T_K\psi_n\|^2_{\mathcal{H}_2} = \sum_{n,j} |\langle T_K\psi_n, \phi_j\rangle_{\mathcal{H}_2}|^2 = \sum_{n,j} |\langle K, \overline{\psi_n}\phi_j\rangle_{\mathcal{H}_3}|^2 = \|K\|^2_{\mathcal{H}_3},$$

so that $T_K \in \mathrm{HS}(\mathcal{H}_1, \mathcal{H}_2)$ and $\|T_K\|_{\mathrm{HS}} = \|K\|_{\mathcal{H}_3}$.

Let $T \in \mathrm{HS}(\mathcal{H}_1, \mathcal{H}_2)$. It follows that

$$\sum_{n,j} |\langle \phi_j, T\psi_n\rangle_{\mathcal{H}_2}|^2 = \sum_n \|T\psi_n\|^2 = \|T\|^2_{\mathrm{HS}} < \infty,$$

so one concludes that the function $K_0(t, s) := \sum_{n,j}\langle \phi_j, T\psi_n\rangle_{\mathcal{H}_2}\overline{\psi_n(s)}\phi_j(t)$ is well defined in the space $\mathcal{H}_3$; one also finds that $\|K_0\|_{\mathcal{H}_3} = \|T\|_{\mathrm{HS}}$. The task now is to check that $T = T_{K_0}$.

If $\psi \in \mathcal{H}_1$ and $\phi \in \mathcal{H}_2$, by using that T is bounded and that the inner product is continuous, it is found that

$$
\begin{aligned}
\langle \phi, T_{K_0}\psi \rangle_{\mathcal{H}_2} &= \int_\Lambda \mathrm{d}\nu(t) \left(\overline{\phi(t)} \int_\Omega K_0(t,s)\psi(s)\, \mathrm{d}\mu(s) \right) \\
&= \langle \phi\overline{\psi}, K_0 \rangle_{\mathcal{H}_3} = \sum_{n,j} \langle \phi_j, T\psi_n \rangle_{\mathcal{H}_2} \langle \phi\overline{\psi}, \phi_j\overline{\psi_n} \rangle_{\mathcal{H}_3} \\
&= \sum_{n,j} \langle \phi_j, T\psi_n \rangle_{\mathcal{H}_2} \langle \phi, \phi_j \rangle_{\mathcal{H}_2} \langle \psi_n, \psi \rangle_{\mathcal{H}_1} \\
&= \left\langle \sum_j \langle \phi_j, \phi \rangle_{\mathcal{H}_2} \phi_j, \sum_n \langle \psi_n, \psi \rangle_{\mathcal{H}_1} T\psi_n \right\rangle_{\mathcal{H}_2} \\
&= \left\langle \phi, \sum_n \langle \psi_n, \psi \rangle_{\mathcal{H}_1} T\psi_n \right\rangle_{\mathcal{H}_2} = \left\langle \phi, T \sum_n \langle \psi_n, \psi \rangle_{\mathcal{H}_1} \psi_n \right\rangle_{\mathcal{H}_2} \\
&= \langle \phi, T\psi \rangle_{\mathcal{H}_2}.
\end{aligned}
$$

Therefore, $T = T_{K_0}$. □

Notes

The term *Hilbert-Schmidt operator* is after D. Hilbert and his student E. Schmidt; they have investigated symmetric compact bilinear forms in the beginning of the XXth century.

There is a family of compact operators in $B(\mathcal{H})$ for each $1 \le p < \infty$, so that certain norm $\|T\|_p < \infty$ (this norm is based on that of l^p); the Hilbert-Schmidt operators are obtained with the choice $p = 2$. The case $p = 1$ is important in Mathematical Physics, particularly in Statistical Mechanics and Scattering Theory, and such operators are called *trace class* ($\|T\|_1$ is a generalization of the trace of the absolute values of the entries of a matrix). It is possible to show that an operator is trace class if and only if it is the product of two Hilbert-Schmidt operators. A reference to this subject is the book [Schatt].

Additional Exercises

EXERCISE **26.2.** Give another proof of Theorem 26.5: with the notation used in that proof, let P_N be the orthogonal projection onto $\mathrm{Lin}(\{e_1, \cdots, e_N\})$ and define $T_N = P_N T$. Show that T_N has finite rank and that $T_N \to T$ in the norm $\| \cdot \|_{HS}$.

EXERCISE **26.3.** (a) Let $(e_j)_{j=1}^\infty$ be an orthonormal basis of the separable Hilbert space $\mathcal{H}$, and T a linear operator defined on a subset of $\mathcal{H}$ so that $Te_j = e_1$, for all j. Check that T has finite rank but is not Hilbert-Schmidt.

(b) Show that $HS(\mathcal{H}_1, \mathcal{H}_2)$ is the closure of the set of finite rank operators in $HS(\mathcal{H}_1, \mathcal{H}_2)$ with the norm $\| \cdot \|_{HS}$.

EXERCISE **26.4.** Comment the statement: If Ω is a compact metric space and μ a Borelian $\sigma-$finite measure over Ω, then to each operator $T \in B(L_\mu^2(\Omega))$ it is possible to associate a function $K \in L_{\mu\times\mu}^2(\Omega \times \Omega)$ so that $T = T_K$, with

$$
(T_K\psi)(t) = \int_\Omega K(t,s)\psi(s)\, \mathrm{d}\mu(s), \quad \psi \in L_\mu^2(\Omega).
$$

EXERCISE **26.5.** Let $T, S \in B(\mathcal{H})$.

(a) If T and S are invertible, show that $(T - S)$ is Hilbert-Schmidt if and only if $(T^{-1} - S^{-1})$ is Hilbert-Schmidt.

(b) If $(T - S)$ is Hilbert-Schmidt, show that $(T^2 - S^2)$ is Hilbert-Schmidt. What can be said about $(T^n - S^n)$?

EXERCISE **26.6.** Fix $\eta \in \mathcal{H}$ with $\|\eta\| = 1$. Let $T_\eta : \mathcal{H} \to \mathcal{H}$ be defined as $T_\eta \xi = \langle \eta, \xi \rangle \eta$, $\xi \in \mathcal{H}$. Show that T_η is linear and find its Hilbert-Schmidt norm $\|T\|_{\mathrm{HS}}$.

EXERCISE **26.7.** Let $\mathcal{H}$ be separable and $T \in \mathrm{B}(\mathcal{H})$ an operator whose eigenvectors form an orthonormal basis (ξ_j) of $\mathcal{H}$, that is, for all j, $T\xi_j = \lambda_j \xi_j$, $\lambda_j \in \mathbb{F}$. Present conditions for $T \in \mathrm{HS}(\mathcal{H})$. Verify that on infinite-dimensional Hilbert spaces there always are compact operators that are not Hilbert-Schmidt.

EXERCISE **26.8.** Are there sequences $(T_n) \subset \mathrm{HS}(\mathcal{H})$ that converge in $\mathrm{B}(\mathcal{H})$ but do not converge in $\mathrm{HS}(\mathcal{H})$?

EXERCISE **26.9.** Show that in order to conclude that $T = T_{K_0}$ in Example 26.7, it is enough to check that $\langle \phi_j, T\psi_n \rangle = \langle \phi_j, T_{K_0}\psi_n \rangle$, for all j, n.

CHAPTER 27

The Spectrum

This chapter introduces the spectrum of bounded linear operators on complex Banach spaces. While the general theory of spectrum is discussed here, the particular cases of self-adjoint operators and compact operators will be explored in more detail in later chapters. The spectrum of a linear operator is defined as the set of (complex) numbers that describe how the operator fails to be invertible in some sense. It generalizes the concept of eigenvalues from finite-dimensional linear algebra to infinite-dimensional spaces, where operators can exhibit more intricate behaviors. Furthermore, it is shown that the spectrum of an operator acting on a Banach space is always nonempty. The chapter also covers the technical resolvent identities.

The spectrum is a generalization of the set of eigenvalues of linear operators. Spectral theory is closely connected to the solvability and uniqueness of solutions to linear equations in Banach spaces, boundary value problems, linear approximations of nonlinear problems, and stability analysis. Moreover, it plays a fundamental role in the mathematical framework of Quantum Mechanics. Unless explicitly stated otherwise, all Banach spaces considered from this point onward will be assumed to be complex.

As is customary, it is convenient to begin by identifying what does not belong to the spectrum.

Definition 27.1. Let $\mathcal{B} \neq \{0\}$ be a complex Banach space. The *resolvent set* of a linear operator $T : \text{dom } T \subset \mathcal{B} \to \mathcal{B}$, denoted by $\rho(T)$, is the set of points $\lambda \in \mathbb{C}$ for which the *resolvent operator* of T at λ, given by

$$R_\lambda(T) : \mathcal{B} \to \text{dom } T, \qquad R_\lambda(T) := (T - \lambda\mathbf{1})^{-1},$$

exists and is bounded, i.e., $R_\lambda(T) \in \text{B}(\mathcal{B})$.

Definition 27.2. The *spectrum* of the linear operator T is the complement of its resolvent set, that is, $\sigma(T) = \mathbb{C}\backslash\rho(T)$.

REMARK **27.3.** (a) Any eigenvalue λ of T (i.e., there is an eigenvector $\xi \neq 0$ with $T\xi = \lambda\xi$) belongs to the spectrum of T, for $(T - \lambda\mathbf{1})$ is not invertible in this case.

(b) If $T \in \text{B}(\mathcal{B})$ and the operator $(T - \lambda\mathbf{1})$ is bijective with range $\mathcal{B}$, then, by Open Mapping, $R_\lambda(T) \in \text{B}(\mathcal{B})$ and $\lambda \in \rho(T)$.

DOI: 10.1201/9781003656166-27

(c) Usually one simplifies the notation by writing $T_\lambda = T - \lambda\mathbf{1}$, and if it is clear which operator T is being considered, $R_\lambda = R_\lambda(T)$.

The definition of the spectrum is not limited to real numbers, as this broader perspective ensures that the spectrum remains nonempty for continuous operators (as will be seen later in this chapter). For instance, when $\dim \mathcal{B} < \infty$, the spectrum coincides with the set of eigenvalues. However, consider the operator representing a rotation by a right angle on $\mathbb{R}^2$ (see Example 20.6); it has no real eigenvalues (verify this!).

EXERCISE 27.1. Let $T : \mathcal{B} \hookleftarrow$ be linear with $\dim \mathcal{B} < \infty$. Show that $\sigma(T)$ is the set of eigenvalues of T, and by the Fundamental Theorem of Algebra, conclude that $\sigma(T) \neq \emptyset$ in this case.

Proposition 27.4. *Let $T :$ dom $T \subset \mathcal{B} \to \mathcal{B}$ be linear. Then the eigenvectors $\{\xi_j\}_{j \in J}$ of T, corresponding to pairwise distinct eigenvalues $\{\lambda_j\}_{j \in J}$, form a linearly independent subset of* dom T.

Proof. If $\alpha_1, \alpha_2, \cdots, \alpha_m$ are scalars, by applying the operator $T_{\lambda_{j_2}} T_{\lambda_{j_3}} \cdots T_{\lambda_{j_m}}$ to $\alpha_1 \xi_{\lambda_{j_1}} + \alpha_2 \xi_{\lambda_{j_2}} + \cdots + \alpha_m \xi_{\lambda_{j_m}} = 0$ one gets

$$\alpha_1 (\lambda_{j_2} - \lambda_{j_1})(\lambda_{j_3} - \lambda_{j_1}) \cdots (\lambda_{j_m} - \lambda_{j_1}) \xi_{\lambda_{j_1}} = 0,$$

and so $\alpha_1 = 0$, since the eigenvalues are pairwise different. Similarly one shows that $\alpha_j = 0$, for all $2 \leq j \leq m$, and so $\{\xi_j\}_{j \in J}$ is linearly independent. $\square$

Proposition 27.5. *Let $T \in \mathrm{B}(\mathcal{B})$ and $\lambda, \mu \in \rho(T)$. Then any operator $S \in \mathrm{B}(\mathcal{B})$ that commutes with T, i.e., $TS = ST$, also commutes with $R_\lambda(T)$, and one has the first resolvent identity*

$$R_\lambda(T) - R_\mu(T) = (\lambda - \mu) R_\lambda(T) R_\mu(T).$$

Furthermore, $R_\lambda(T)$ commutes with $R_\mu(T)$ (see Exercise 27.8 for the second resolvent identity).

Proof. If $\lambda \in \rho(T)$ and $ST = TS$, then $ST_\lambda = T_\lambda S$. Since $\mathbf{1} = R_\lambda T_\lambda = T_\lambda R_\lambda$, it follows that

$$R_\lambda S = R_\lambda S T_\lambda R_\lambda = R_\lambda T_\lambda S R_\lambda = S R_\lambda,$$

i.e., R_λ commutes with S. Since R_μ commutes with T, by using the above argument it is possible to conclude that R_μ commutes with R_λ, for all $\mu, \lambda \in \rho(T)$.

Consider now

$$R_\lambda - R_\mu = R_\lambda T_\mu R_\mu - R_\lambda T_\lambda R_\mu = R_\lambda (T_\mu - T_\lambda) R_\mu = (\lambda - \mu) R_\lambda R_\mu,$$

which shows the validity of the first resolvent identity. $\square$

EXERCISE 27.2. Use the first resolvent identity to show that, for $T \in \mathrm{B}(\mathcal{B})$, $R_\lambda(T)$ commutes with $R_\mu(T)$, for all $\lambda, \mu \in \rho(T)$.

EXERCISE **27.3.** If for $T \in B(\mathcal{B})$, there exists $\lambda_0 \in \rho(T)$ such that $R_{\lambda_0}(T)$ is compact, show that $R_\lambda(T)$ is compact for all $\lambda \in \rho(T)$ (these are the so-called *operators with compact resolvent*).

Theorem 27.6. *If $T \in B(\mathcal{B})$ and $\lambda_0 \in \rho(T)$, then for any λ in the disk $|\lambda - \lambda_0| < 1/\|R_{\lambda_0}(T)\|$ in the complex plane, one has $R_\lambda(T) \in B(\mathcal{B})$ and*

$$R_\lambda(T) = \sum_{j=0}^{\infty}(\lambda - \lambda_0)^j R_{\lambda_0}(T)^{j+1},$$

with convergent series in $B(\mathcal{B})$.

Proof. The first remark is that $R_{\lambda_0}(T) \neq 0$, since it is the inverse of an operator. From

$$T - \lambda\mathbf{1} = T - (\lambda_0 + (\lambda - \lambda_0))\mathbf{1} = T_{\lambda_0}\left[\mathbf{1} + (\lambda_0 - \lambda)R_{\lambda_0}\right],$$

it would follow formally that

$$R_\lambda = \left(\sum_{j=0}^{\infty}(\lambda - \lambda_0)^j R_{\lambda_0}^j\right)R_{\lambda_0}.$$

Now it is necessary to give grounds for this expression and to prove that it actually acts as $(T - \lambda\mathbf{1})^{-1}$ in $B(\mathcal{B})$. For $|\lambda - \lambda_0| < 1/\|R_{\lambda_0}(T)\|$, this series is absolutely convergent in $B(\mathcal{B})$ and defines an operator satisfying

$$\left(\sum_{j=0}^{N}(\lambda - \lambda_0)^j R_{\lambda_0}^{j+1}\right)(T - \lambda\mathbf{1})$$

$$= \sum_{j=0}^{N}(\lambda - \lambda_0)^j R_{\lambda_0}^{j+1}(T - (\lambda_0 + (\lambda - \lambda_0))\mathbf{1})$$

$$= \sum_{j=0}^{N}(\lambda - \lambda_0)^j R_{\lambda_0}^j - \sum_{j=0}^{N}(\lambda - \lambda_0)^{j+1}R_{\lambda_0}^{j+1}$$

$$= \mathbf{1} - [(\lambda - \lambda_0)R_{\lambda_0}]^{N+1}.$$

Since $\lim_{N\to\infty}\left[(\lambda - \lambda_0)R_{\lambda_0}\right]^N = 0$ in $B(\mathcal{B})$, because $|\lambda - \lambda_0| < 1/\|R_{\lambda_0}(T)\|$, it follows that

$$\left(\sum_{j=0}^{\infty}(\lambda - \lambda_0)^j R_{\lambda_0}^{j+1}\right)(T - \lambda\mathbf{1}) = \mathbf{1}.$$

In the same way one obtains that

$$(T - \lambda\mathbf{1})\left(\sum_{j=0}^{\infty}(\lambda - \lambda_0)^j R_{\lambda_0}^{j+1}\right) = \mathbf{1},$$

completing the proof. □

EXERCISE **27.4.** Verify that if a, with $|a| \leq \|R_{\lambda_0}(T)\|$, is an eigenvalue of $R_{\lambda_0}(T)$, then for all $|\lambda - \lambda_0|\|R_{\lambda_0}(T)\| < 1$, the scalar

$$\sum_{j=0}^{\infty} (\lambda - \lambda_0)^j a^{j+1}$$

is an eigenvalue of $R_\lambda(T)$.

Corollary 27.7. *If $T \in B(\mathcal{B})$, then $\rho(T)$ is an open subset and $\sigma(T)$ is a closed subset of $\mathbb{C}$. Furthermore, if $|\lambda| > \|T\|$, then $\lambda \in \rho(T)$ and $\|R_\lambda(T)\| \to 0$ as $|\lambda| \to \infty$.*

Proof. One sees that $\rho(T)$ is open directly from Theorem 27.6, hence $\sigma(T)$ is closed. Following the proof of the above theorem (write $T - \lambda \mathbf{1} = -\lambda(1 - T/\lambda)$), one concludes that the representation of $R_\lambda(T)$ by the series, called *Neumann's series of T*,

$$R_\lambda(T) = -\frac{1}{\lambda} \sum_{j=0}^{\infty} \left(\frac{T}{\lambda}\right)^j$$

is convergent in norm if $|\lambda| > \|T\|$ and, in this case, one has

$$\|R_\lambda(T)\| \leq \frac{1}{|\lambda|} \sum_{j \geq 0} \left(\frac{\|T\|}{\lambda}\right)^j = \frac{1}{(|\lambda| - \|T\|)}.$$

It then follows that the spectrum $\sigma(T) \subset \{\lambda \in \mathbb{C} : |\lambda| \leq \|T\|\}$ and also that

$$\lim_{|\lambda| \to \infty} \|R_\lambda(T)\| = 0.$$

$\square$

EXERCISE **27.5.** If $T \in B(\mathcal{B})$, conclude that $\rho(T) \neq \emptyset$ and that it is an unbounded subset of $\mathbb{C}$.

Corollary 27.8. *Let $T \in B(\mathcal{B})$. The mapping $\rho(T) \to B(\mathcal{B})$ given by $\lambda \mapsto R_\lambda(T)$ is continuous and uniformly holomorphic, that is, it has a derivative in $B(\mathcal{B})$ defined by the limit*

$$\frac{dR_\lambda(T)}{d\lambda} := \lim_{h \to 0} \frac{R_{\lambda+h}(T) - R_\lambda(T)}{h} = R_\lambda(T)^2,$$

for all λ in an open ball centered at each point $\lambda_0 \in \rho(T)$.

Proof. Pick $\lambda_0 \in \rho(T)$ and $\lambda \in \mathbb{C}$ with $|\lambda - \lambda_0| < 1/\|R_{\lambda_0}\|$. By Theorem 27.6,

$$\|R_\lambda(T) - R_{\lambda_0}(T)\| \leq \sum_{j=1}^{\infty} |\lambda - \lambda_0|^j \|R_{\lambda_0}(T)\|^{j+1}$$

$$\leq |\lambda - \lambda_0| \|R_{\lambda_0}(T)\|^2 \sum_{j=0}^{\infty} |\lambda - \lambda_0|^j \|R_{\lambda_0}(T)\|^j$$

$$= \frac{|\lambda - \lambda_0| \|R_{\lambda_0}(T)\|^2}{1 - |\lambda - \lambda_0| \|R_{\lambda_0}(T)\|} \longrightarrow 0 \text{ as } \lambda \to \lambda_0,$$

which shows that the mapping $\lambda \mapsto R_\lambda(T)$, $\lambda \in \rho(T)$, is continuous.

Apply the first resolvent identity to obtain $(R_{\lambda+h} - R_\lambda)/h = R_{\lambda+h}R_\lambda$; finally, taking $h \to 0$ combined with the continuity shown above, one obtains the existence of the derivative and $\mathrm{d}R_\lambda(T)/\mathrm{d}\lambda = R_\lambda(T)^2$ follows. □

Corollary 27.9. *If $T \in \mathrm{B}(\mathcal{B})$, then $\sigma(T) \neq \emptyset$.*

Proof. Let f be an element of the dual space $\mathrm{B}(\mathcal{B})^*$ of $\mathrm{B}(\mathcal{B})$, and consider $F : \rho(T) \to \mathbb{C}$ by $F(\lambda) = f(R_\lambda(T))$. By Corollary 27.8, it follows that, for every $\lambda \in \rho(T)$,

$$\frac{\mathrm{d}F(\lambda)}{\mathrm{d}\lambda} = \lim_{h \to 0} \frac{F(\lambda + h) - F(\lambda)}{h} = f\left(R_\lambda(T)^2\right),$$

which is continuous; therefore, F is holomorphic in $\rho(T)$. Now, use the inequality $|F(\lambda)| \leq \|f\| \|R_\lambda(T)\|$ and Corollary 27.7 to get $\lim_{|\lambda| \to \infty} F(\lambda) = 0$.

If $\sigma(T) = \emptyset$, then $\rho(T) = \mathbb{C}$; by continuity, the function F is bounded in any ball in $\mathbb{C}$, and taking into account that it converges to zero as $|\lambda| \to \infty$, it follows that $F : \mathbb{C} \to \mathbb{C}$ is an entire and bounded function, hence constant by Liouville's Theorem. Since $\lim_{|\lambda| \to \infty} F(\lambda) = 0$, one gets $F(\lambda) = f(R_\lambda(T)) = 0$ for all $\lambda \in \mathbb{C}$, $f \in \mathrm{B}(\mathcal{B})^*$. By the corollaries of Hahn-Banach's Theorem (Theorem 12.1), one has $R_\lambda(T) = 0$, for all $\lambda \in \mathbb{C}$, but this may not occur, since $R_\lambda(T)$ is the inverse of some operator. The resulting contradiction establishes that $\sigma(T) \neq \emptyset$. □

Corollary 27.10. *If $T \in \mathrm{B}(\mathcal{B})$, then $\|R_\lambda(T)\| \geq 1/d(\lambda, \sigma(T))$ for all $\lambda \in \rho(T)$ (with $d(\lambda, \sigma(T)) := \inf_{\mu \in \sigma(T)} |\mu - \lambda|$).*

EXERCISE **27.6.** Prove Corollary 27.10.

Definition 27.11. The *spectral radius* of $T \in \mathrm{B}(\mathcal{B})$ is $r_\sigma(T) := \sup_{\lambda \in \sigma(T)} |\lambda|$.

The proof of the next result will be discussed in Chapter 28, as well as some of its consequences.

Theorem 27.12. *If $T \in \mathrm{B}(\mathcal{B})$, then $r_\sigma(T) = \lim_{n \to \infty} \|T^n\|^{1/n} \leq \|T\|$.*

Example 27.13. Let $S_l : l^\infty(\mathbb{N}) \hookleftarrow$ be the (left) shift operator

$$S_l(\xi_1, \xi_2, \xi_3, \cdots) = (\xi_2, \xi_3, \xi_4, \cdots),$$

which has norm $\|S_l\| = 1$. Hence, one has $\sigma(S_l) \subset \overline{B}(0; 1)$. Moreover, if $|\lambda| \leq 1$ it is an eigenvalue of S_l, since $S_l \xi^\lambda = \lambda \xi^\lambda$ has the solution $\xi^\lambda = (1, \lambda, \lambda^2, \lambda^3, \cdots)$ in $l^\infty(\mathbb{N})$. Hence $\sigma(S_l) = \overline{B}(0; 1)$, $r_\sigma(S_l) = 1$, and each point of its spectrum is an eigenvalue.
●

Example 27.14. The Volterra operator, $T : C[0,1] \hookleftarrow$, $(T\psi)(t) = \int_0^t \psi(s)\,\mathrm{d}s$, illustrates many issues related to spectral theory. First, it has no eigenvalues. Indeed, by the eigenvalue equation $(T\psi)(t) = \lambda\psi(t) = \int_0^t \psi(s)\,\mathrm{d}s$, it follows that $\lambda\psi'(t) = \psi(t)$ (ψ is differentiable since it is the integral of a continuous function). If $\lambda = 0$, then $\psi = 0$ and zero may not be an eigenvalue; if $\lambda \neq 0$, the solutions to this differential equation are $\psi(t) = C\exp(t/\lambda)$, and since $\psi(0) = 0$ it follows that the constant $C = 0$, and so $\psi = 0$ and no $\lambda \in \mathbb{C}$ is an eigenvalue of T.

The inequality $|(T\psi)(t)| \le t\|\psi\|_\infty$ implies, by induction, that

$$|(T^2\psi)(t)| \le \int_0^t s\|\psi\|_\infty\,\mathrm{d}s = \frac{t^2}{2}\|\psi\|_\infty, \qquad |(T^n\psi)(t)| \le \frac{t^n}{n!}\|\psi\|_\infty.$$

Hence, $\|T^n\| \le 1/n!$ and, by Theorem 27.12, $r_\sigma(T) \le \lim_{n\to\infty}(1/n!)^{1/n} = 0$. So $r_\sigma(T) < \|T\|$, $\sigma(T) = \{0\}$ (since it is nonempty) and T has no eigenvalues. •

Notes

Although it is not a recent reference, in [Steen] there is a broad view of the interesting history of Spectral Theory.

The spectral theory presented here has arisen, probably, with works of Fourier and his method of separation of variables, and with the efforts of generalizing Sturm-Liouville theory in the XIX century. It is possible "to notice" the continuous spectrum in such cases, either in the limit of the period going to infinity in Fourier expansions, for which the sum is replaced by an integral, or by following the eigenvalues of some Sturm-Liouville problems that "fill out intervals" in the limit of boundary conditions at infinity. Hilbert noticed clearly the existence of the continuous spectrum (see Chapter 28) in the beginning of XXth century. Hilbert has also popularized the term "spectrum" in Mathematical Analysis; it is interesting to observe that the set of energy levels of an atom (or molecule) is also called spectrum, which can be theoretically studied as the spectrum of some differential Schrödinger or Dirac operators (a curious linguistic coincidence!).

Important contributions to the initial development of spectral theory were given by F. Riesz, E. Hellinger, and J. von Neumann. For instance, Riesz was the first to propose the definition of spectrum as presented here; von Neumann has adapted the definitions of resolvent operator, spectrum, Hilbert adjoint, and others, to unbounded normal operators in Hilbert spaces (whose abstract definition he had also introduced). The spectral radius formula is due to I. Gelfand, who has shown it in the context of Banach algebras, around 1940. The first resolvent identity is due to Hilbert.

Additional Exercises

EXERCISE **27.7.** If $T, S \in B(\mathcal{B})$ and S commutes with $R_\lambda(T)$ for some $\lambda \in \rho(T)$, show that S commutes with T and with $R_\mu(T)$, for all $\mu \in \rho(T)$.

EXERCISE **27.8.** For $T, S \in B(\mathcal{B})$ and $\lambda \in \rho(T) \cap \rho(S)$, verify the *second resolvent identity*

$$R_\lambda(T) - R_\lambda(S) = R_\lambda(T)(S - T)R_\lambda(S) = R_\lambda(S)(S - T)R_\lambda(T).$$

EXERCISE **27.9.** Let $A, T \in B(\mathcal{H})$ and $S = T + A^*A$ be a "perturbed operator." Use the second resolvent identity to show that for any $\lambda \in \rho(T) \cap \rho(S)$ one has $AR_\lambda(S)A^* = \mathbf{1} - (\mathbf{1} + AR_\lambda(T)A^*)^{-1}$.

EXERCISE **27.10.** Show that the spectral radius formula (Theorem 27.12) for a bounded operator gives the same result if equivalent norms are considered in the Banach space.

EXERCISE **27.11.** Consider the operator represented by the matrix $T = \begin{pmatrix} 1 & A \\ 0 & 1 \end{pmatrix}$ acting on $\mathbb{C}^2$, and with nonzero A. Verify that its spectrum $\sigma(T) = \{1\}$, hence $r_\sigma(T) = 1 < \|T\| = (1 + |A|^2)^{1/2}$. Check that $\lim_{n\to\infty}\|T^n\|^{1/n} = 1$. Is this operator normal?

EXERCISE **27.12.** If $T : C[0, 1] \hookleftarrow$ is the Volterra operator in Example 27.14, use Neumann series to show that, for $\lambda \neq 0$,

$$(R_\lambda(T)\psi)(t) = \frac{\psi(t)}{\lambda} + \frac{1}{\lambda^2} \int_0^t e^{(t-s)/\lambda} \psi(s) \, ds, \qquad \psi \in C[0, 1].$$

EXERCISE **27.13.** If $K : [a, b] \times [a, b] \to \mathbb{R}$ is continuous and $T_K : C[a, b] \hookleftarrow$,

$$(T_K\psi)(t) = \int_a^t K(t, s)\psi(s) \, ds, \qquad \psi \in C[a, b],$$

show that $r_\sigma(T_K) = 0$. This T_K is also called Volterra operator.

EXERCISE **27.14.** Show that if $ST = TS$, $T, S \in B(\mathcal{B})$, then $r_\sigma(TS) \leq r_\sigma(T)r_\sigma(S)$, and by using the Newton's binomial, show that $r_\sigma(T + S) \leq r_\sigma(T) + r_\sigma(S)$.

EXERCISE **27.15.** If $T \in B(\mathcal{B})$ is invertible, with $T^{-1} \in B(\mathcal{B})$, show that $\sigma(T^{-1}) = \{\lambda^{-1} : \lambda \in \sigma(T)\}$.

EXERCISE **27.16.** If $T \in B(\mathcal{B})$, show that $\lim_{|\lambda| \to \infty} \lambda R_\lambda(T) = -1$.

EXERCISE **27.17.** By mirroring Example 27.13, present an alternative proof that $l^\infty(\mathbb{N})$ is not separable (see Example 3.7).

EXERCISE **27.18.** For $T \in B(\mathcal{B})$, define $V(t) = e^{tT}$, $t \in \mathbb{R}$, as in Exercise 4.11. Show that: (a) The mapping $t \mapsto V(t) \in B(\mathcal{B})$ is continuous with $V(0) = 1$ and $V(t + s) = V(t)V(s)$. (b) If $S \in B(\mathcal{B})$ commutes with T, then it also commutes with $V(t)$, for all t. (c) This mapping is uniformly holomorphic and $dV(t)/dt = TV(t)$

EXERCISE **27.19.** In $C[0, \infty)$, restricted to bounded functions, infer that the eigenvalues λ of the operator $(T\psi)(t) = \int_0^{t+\frac{\pi}{2}} \psi(s) \, ds$ are related to the corresponding eigenvectors ψ_λ by (suppose $\psi_\lambda(0) \neq 0$)

$$\lambda = \frac{1}{\psi_\lambda(0)} \int_0^{\frac{\pi}{2}} \psi_\lambda(s) \, ds.$$

Verify that $\lambda_k = (-1)^k (2k - 1)^{-1}$, $k \in \mathbb{N}$, are eigenvalues of T and that the associated eigenvectors satisfy the above relation, and that zero is an eigenvalue of infinite multiplicity (i.e., dim $N(T) = \infty$). This should be compared with the (similar?!) Volterra operator $(V\psi)(t) = \int_0^t \psi(s) \, ds$, that has no eigenvalues ($\psi \in C[0, \infty)$).

EXERCISE **27.20.** Use the following steps to show that $\lim_{n\to\infty} \|T^n\|^{1/n}$ exists. Denote by $r = \inf_n \|T^n\|^{1/n}$. Given $\varepsilon > 0$, pick m such that $\|T^m\|^{1/m} \leq r + \varepsilon$. Write out each positive integer $n = p_n m + q_n$, with $0 \leq q_n < m$, show that $\|T^n\|^{1/n} \leq (r + \varepsilon)^{mp_n/n} \|T\|^{q_n/n}$, and conclude that

$$\lim_{n\to\infty} (r + \varepsilon)^{mp_n/n} \|T\|^{q_n/n} = r + \varepsilon.$$

Therefore, there exists $N(\varepsilon)$ with $(r+\varepsilon)^{mp_n/n}\|T\|^{q_n/n} < (r+2\varepsilon)$ if $n \geq N(\varepsilon)$. Thus, $r \leq \|T^n\|^{1/n} < r + 2\varepsilon$ if $n \geq N(\varepsilon)$; conclude then that $\lim_{n\to\infty} \|T^n\|^{1/n}$ exists and is equal to r.

CHAPTER 28

Spectral Classification

In this chapter, Theorem 27.12 is proven, and some of its consequences are discussed. In addition to eigenvalues, infinite-dimensional spaces have two other important components of the spectrum: the continuous spectrum and the residual spectrum. This traditional spectral classification is provided here. It is also shown that the spectrum of unitary operators on a Hilbert space is a subset of the unit circle in the complex plane. Finally, it is shown that a bounded operator in a Banach space has the same spectrum as its adjoint.

Proof. (Theorem 27.12) Due to Corollary 27.7, $r_\sigma(T) \leq \|T\|$. In order to conclude Theorem 27.12, results from the Theory of Holomorphic Functions will be used combined with Theory combined with "any weakly convergent sequence is bounded," and the remark: if $\lambda \in \mathbb{C}$ and $\lambda_1, \lambda_2, \cdots, \lambda_n$ are its nth roots in $\mathbb{C}$, then

$$T^n - \lambda \mathbf{1} = T_{\lambda_1} T_{\lambda_2} \cdots T_{\lambda_n}.$$

Hence $\lambda \in \sigma(T^n)$ if and only if $\lambda_j \in \sigma(T)$ for some $1 \leq j \leq n$. So, $\sigma(T^n) = \sigma(T)^n$, where

$$\sigma(T)^n := \{\lambda^n : \lambda \in \sigma(T)\}.$$

From this one obtains that, for every $n \in \mathbb{N}$, $r_\sigma(T) = r_\sigma(T^n)^{1/n} \leq \|T^n\|^{1/n}$.

For every f in the dual of $B(\mathcal{B})$, consider $F : \rho(T) \to \mathbb{C}$ by $F(\lambda) = f(R_\lambda(T))$, which is a holomorphic function (as in the proof of Corollary 27.9). If $|\lambda| > \|T\|$, by the Neumann series

$$F(\lambda) = -\frac{1}{\lambda} \sum_{n=0}^{\infty} \frac{1}{\lambda^n} f(T^n),$$

and the uniqueness of Laurent expansion, it follows that the above series is convergent for every $\lambda \in \mathbb{C}$ in the region $|\lambda| > r_\sigma(T)$ (or Taylor expansion if the variable $s = 1/\lambda$, with $F(0) = 0$, is considered).

Given $\varepsilon > 0$, for $r_\sigma(T) < \alpha < r_\sigma(T) + \varepsilon$ and all $f \in B(\mathcal{B})^*$, the series $\sum_{n=0}^{\infty} f(T^n/\alpha^n)$ converge. Thus, the sequence T^n/α^n converges weakly to zero in $B(\mathcal{B})$, and hence, it is bounded and there exists $C = C(\alpha) > 0$ so that

$$\|T^n/\alpha^n\| \leq C \implies \|T^n\|^{1/n} \leq \alpha\, C^{1/n}, \qquad \forall n \in \mathbb{N}.$$

DOI: 10.1201/9781003656166-28

Since $\lim_{n\to\infty} C^{1/n} = 1$, there is $N(\varepsilon) > 0$ such that

$$\|T^n\|^{\frac{1}{n}} < r_\sigma(T) + \varepsilon, \qquad \forall n \geq N(\varepsilon).$$

This, combined with $r_\sigma(T) \leq \|T^n\|^{1/n}$, which was verified above, concludes that $\lim_{n\to\infty} \|T\|^{1/n}$ exists and equals $r_\sigma(T)$. $\qquad\square$

EXERCISE **28.1.** If any pair of the operators in the set $\{T_1, \cdots, T_n\} \subset \mathrm{B}(\mathcal{B})$ is commuting, prove that the product $T_1 T_2 \cdots T_n$ is invertible with bounded inverse if and only if each T_j is invertible in $\mathrm{B}(\mathcal{B})$.

Corollary 28.1. *For $T \in \mathrm{B}(\mathcal{B})$, then $\sigma(T^n) = \sigma(T)^n$ and $r_\sigma(T^n) = r_\sigma(T)^n$.*

Corollary 28.2. *If $T \in \mathrm{B}(\mathcal{H})$ is normal, then $r_\sigma(T) = \|T\|$ (hence this equality holds for self-adjoint and unitary operators, in particular).*

Proof. If T is normal, then $\|T^{2^n}\| = \|T\|^{2^n}$ (see Proposition 20.7) for all $n \in \mathbb{N}$; thus

$$r_\sigma(T) = \lim_{n\to\infty} \|T^{2^n}\|^{1/2^n} = \|T\|.$$

Recall that, if a limit exists, one may use any subsequence to evaluate it. $\qquad\square$

If $T \in \mathrm{B}(\mathcal{B})$, then both $\rho(T)$ and $\sigma(T)$ are nonempty. By removing the continuity hypothesis of the operator, such properties may not hold. The following examples illustrate such phenomena. It is necessary to define the spectrum in the case of unbounded linear operators. The complex number λ belongs to the resolvent set $\rho(T)$ of the linear operator (not necessarily bounded) $T : \operatorname{dom} T \subset \mathcal{B} \to \mathcal{B}$, if there exists a bounded linear operator $R_\lambda : \mathcal{B} \to \operatorname{dom} T$ in such a way that $R_\lambda(T - \lambda\mathbf{1}) = \mathbf{1}$, the identity operator on $\operatorname{dom} T$, and $(T - \lambda\mathbf{1})R_\lambda = \mathbf{1}$, the identity on $\mathcal{B}$. The spectrum of this operator is the set $\sigma(T) = \mathbb{C}\backslash\rho(T)$.

Example 28.3. Let $D : \operatorname{dom} D = \mathrm{C}^1[0,1] \subset \mathrm{C}[0,1] \to \mathrm{C}[0,1]$ and $(D\psi)(t) = \psi'(t)$, which is a closed and unbounded operator. If $\lambda \in \mathbb{C}$, the function $\psi_\lambda(t) = e^{\lambda t} \in \operatorname{dom} D$ and $D\psi_\lambda = \lambda\psi_\lambda$, showing that $\sigma(D) = \mathbb{C}$ and it makes up exclusively of eigenvalues. Therefore $\rho(D) = \emptyset$. $\quad\bullet$

Example 28.4. Let $\operatorname{dom} d = \{\psi \in (\mathrm{C}^1[0,1], \|\cdot\|_\infty) : \psi(0) = 0\}$, $d : \operatorname{dom} d \to \mathrm{C}[0,1]$, $(d\psi)(t) = \psi'(t)$, which is a closed and unbounded operator. If $\lambda \in \mathbb{C}$, the operator $W_\lambda : \mathrm{C}[0,1] \to \operatorname{dom} d$, $(W_\lambda\phi)(t) = e^{\lambda t} \int_0^t e^{-\lambda s}\phi(s)\,ds$, $\phi \in \mathrm{C}[0,1]$ is bounded and satisfies $(d - \lambda\mathbf{1})W_\lambda = \mathbf{1}$ (identity on $\mathrm{C}[0,1]$) and $W_\lambda(d - \lambda\mathbf{1}) = \mathbf{1}$ (identity in $\operatorname{dom} d$). Therefore W_λ is the resolvent of the operator d at λ and so $\rho(d) = \mathbb{C}$, showing that $\sigma(d) = \emptyset$ (the action of the resolvent W_λ was obtained by considering the solution to the differential equation $\psi' - \lambda\psi = \phi$ with $\psi(0) = 0$). $\quad\bullet$

EXERCISE **28.2.** Fill out the missing details in Example 28.4.

The *point spectrum* of $T \in \mathrm{B}(\mathcal{B})$ is the set of its eigenvalues, denoted by $\sigma_\mathrm{p}(T)$; in the case of $\dim \mathcal{B} < \infty$ one has $\sigma(T) = \sigma_\mathrm{p}(T)$, but if $\dim \mathcal{B} = \infty$ the operator T_λ can be injective with $\operatorname{rng} T_\lambda \neq \mathcal{B}$. If T_λ^{-1} does exist, one distinguishes two cases:

the *continuous spectrum* of T, denoted by $\sigma_c(T)$, is the set of $\lambda \in \mathbb{C}$ so that rng T_λ is dense in $\mathcal{B}$ (but not equal to $\mathcal{B}$; if rng $T_\lambda = \mathcal{B}$, then $\lambda \in \rho(T)$ by the Open Mapping Theorem) with unbounded resolvent $R_\lambda(T)$, whereas the set of $\lambda \in \mathbb{C}$ such that $\overline{\text{rng } T_\lambda} \neq \mathcal{B}$ is called the *residual spectrum* of T, denoted by $\sigma_r(T)$. With such definitions,

$$\sigma(T) = \sigma_p(T) \cup \sigma_c(T) \cup \sigma_r(T),$$

with disjoint union (check this!). This spectral classification may also be adapted to unbounded operators. The reader should be aware that in more specific studies of self-adjoint operators (see, for instance, [BlExHa] and [de Oliv]), there is a conflict with the above nomenclature, and the expression "continuous spectrum" has a different conceptual meaning; the residual spectrum of self-adjoint operators is always empty (see Corollary 29.4).

It is interesting to quickly analyze some implications related to the different spectral types. The term *resolvent* is motivated by the equation (for given $\eta \in \mathcal{B}$)

$$T\xi - \lambda\xi = \eta;$$

if $\lambda \in \rho(T)$, then $R_\lambda(T)\eta$ is the unique solution to this problem (for all η) and depends continuously on η. If $\lambda \in \sigma_p(T)$ and ξ is a solution to that equation, then $\xi + \xi_\lambda$, for all $0 \neq \xi_\lambda \in N(T_\lambda)$, are also solutions and the uniqueness does not hold. If $\lambda \in \sigma_c(T)$, the corresponding solution is obtained by means of a discontinuous operator, and finally, if $\lambda \in \sigma_r(T)$ that equation has no solution for η in an open set of $\mathcal{B}$ (and when those solutions exist, they do not depend continuously on η).

Example 28.5. The operator in Example 27.13 has *pure point spectrum*, i.e., $\sigma(T) = \sigma_p(T)$. ●

Example 28.6. For the Volterra operator, Example 27.14, one has $\{0\} = \sigma_r(T) = \sigma(T)$, that is, *pure residual spectrum* (use that $\psi(0) = 0$ and check this!). ●

Example 28.7. Let $\mathcal{M}_\phi$ on $L^2[0,1]$, with $\phi(t) = t$. Then $\mathcal{M}_\phi$ has no eigenvalues, since from $\mathcal{M}_\phi\psi = \lambda\psi$ it follows that $(t - \lambda)\psi(t) = 0$, or $\psi(t) = 0$ for all $t \neq \lambda$, i.e., $\psi = 0$ in $L^2[0,1]$.

The residual spectrum of $\mathcal{M}_\phi$ is empty. Indeed, if $\psi \in L^2[0,1]$, let

$$\psi_n(t) = \chi_{[\frac{1}{n},1]}(t)\frac{\psi(t)}{t} \in L^2[0,1];$$

(recall that χ_A denotes the characteristic function of the set A) so,

$$(\mathcal{M}_\phi\psi_n)(t) - \psi(t) \to 0, \quad \text{a.e.},$$

and since $|(\mathcal{M}_\phi\psi_n)(t) - \psi(t)|^2 \leq |\psi(t)|^2 \in L^1[0,1]$, by Dominated Convergence it is found $\|(\mathcal{M}_\phi\psi_n) - \psi\|_2 \to 0$, showing that rng $\mathcal{M}_\phi$ is dense in $L^2[0,1]$, and so $0 \notin \sigma_r(T)$. In a similar way, it is founded that for all λ, rng $(\mathcal{M}_\phi - \lambda\mathbf{1}) = $ rng $\mathcal{M}_{\phi-\lambda}$ is dense in $L^2[0,1]$ and so $\sigma_r(\mathcal{M}_\phi) = \emptyset$. Therefore, this operator has *pure continuous spectrum*, i.e., $\sigma(\mathcal{M}_\phi) = \sigma_c(\mathcal{M}_\phi)$. ●

EXERCISE **28.3.** Show that in Example 28.7 one has $\sigma(\mathcal{M}_\phi) = [0, 1]$.

It is not difficult to verify that any eigenvalue of a unitary operator U in $B(\mathcal{H})$ has unity absolute value; indeed, if $U\xi_\lambda = \lambda\xi_\lambda$, then $\langle\xi_\lambda, \xi_\lambda\rangle = \langle U\xi_\lambda, U\xi_\lambda\rangle = |\lambda|^2\langle\xi_\lambda, \xi_\lambda\rangle$, so that $|\lambda| = 1$. This property extends to its entire spectrum:

Proposition 28.8. *If $U : \mathcal{H} \to \mathcal{H}$ is a unitary operator, then $\sigma(U)$ is a subset of $\{\lambda \in \mathbb{C} : |\lambda| = 1\}$.*

Proof. Since $\|U\| = 1$, then by Corollary 27.7, it follows that, if $|\lambda| > 1$, then λ belongs to the resolvent set of U. Since $U^{-1} = R_0(U)$ is unitary and $UU^{-1} = \mathbf{1} = U^{-1}U$, then $0 \in \rho(U)$, and if $|\lambda| = |\lambda - 0| < 1/\|U^{-1}\| = 1$, by Theorem 27.6 λ belongs to the resolvent set of U. Therefore, if $\lambda \in \sigma(U)$, necessarily $|\lambda| = 1$. $\square$

The next result does not hold if the Banach adjoint is replaced by the Hilbert adjoint; for instance, the operator $T = i\mathbf{1}$ on $\mathcal{H}$ has spectrum $\{i\}$, whereas $\sigma(T^*) = \{-i\}$; see Exercise 28.10.

Theorem 28.9. *If $T \in B(\mathcal{B})$, then $\sigma(T^a) = \sigma(T)$.*

Proof. It will be shown that $\rho(T) = \rho(T^a)$. If $\lambda \in \rho(T)$, then $R_\lambda(T) \in B(\mathcal{B})$, and since $(T - \lambda\mathbf{1})R_\lambda(T) = \mathbf{1} = R_\lambda(T)(T - \lambda\mathbf{1})$,

$$R_\lambda(T)^a(T^a - \lambda\mathbf{1}) = \mathbf{1} = (T^a - \lambda\mathbf{1})R_\lambda(T)^a, \quad \text{on} \quad \mathcal{B}^*.$$

This shows that $\lambda \in \rho(T^a)$. Therefore, $\rho(T) \subset \rho(T^a)$.

Suppose now that $\lambda \in \rho(T^a)$, so that $R_\lambda(T^a) \in B(\mathcal{B}^*)$, and

$$R_\lambda(T^a)(T^a - \lambda\mathbf{1}) = \mathbf{1} = (T^a - \lambda\mathbf{1})R_\lambda(T^a), \quad \text{on} \quad \mathcal{B}^*.$$

By taking adjoints and denoting $S = R_\lambda(T^a)^a \in B(\mathcal{B}^{**})$, it is found that

$$(T^{aa} - \lambda\mathbf{1})S = \mathbf{1} = S(T^{aa} - \lambda\mathbf{1}), \quad \text{on} \quad \mathcal{B}^{**}.$$

Recall that the canonical application $\hat{\ }: \mathcal{B} \to \mathcal{B}^{**}$ is an isometry and satisfies $\widehat{T\xi} = T^{aa}\hat{\xi}$, for all $\xi \in \mathcal{B}$ (Exercise 25.3); then

$$S(T^{aa} - \lambda\mathbf{1})\hat{\xi} = S\widehat{T_\lambda\xi} = \hat{\xi}, \qquad \forall\xi \in \mathcal{B},$$

and $\|\xi\| = \|\hat{\xi}\| = \|S\widehat{T_\lambda\xi}\| \leq \|S\|\,\|T_\lambda\xi\|$. Hence $N(T_\lambda) = \{0\}$ and $R_\lambda(T)$ is well defined. This condition also guarantees that rng T_λ is closed. Indeed, let $\eta \in \overline{\text{rng } T_\lambda}$ and $(\xi_j) \subset \mathcal{B}$ with $T_\lambda\xi_j \to \eta$. The inequality above implies that (ξ_j) is a Cauchy sequence in $\mathcal{B}$, and therefore, it converges to some $\xi \in \mathcal{B}$. Since T_λ is continuous, then $T_\lambda\xi = \eta$ and rng T_λ is closed.

In order to conclude that $\lambda \in \rho(T)$ it suffices to show that the resolvent operator $R_\lambda(T) \in B(\mathcal{B})$, which is reduced to show that rng $T_\lambda = \mathcal{B}$, by the Open Mapping Theorem (Corollary 8.5).

Let $f \in \mathcal{B}^*$ so that the restriction $f|_{\text{rng } T_\lambda} = 0$. Thus, $0 = f(T_\lambda\xi) = (T_\lambda^a f)(\xi)$, for all $\xi \in \mathcal{B}$, and so $f \in N(T^a - \lambda\mathbf{1})$, and $f = 0$ since $\lambda \in \rho(T^a)$. By Corollary 12.3, rng T_λ is dense in $\mathcal{B}$; since rng T_λ is closed, rng $T_\lambda = \mathcal{B}$. Summing up, $\rho(T^a) \subset \rho(T)$. This completes the proof of the theorem. $\square$

Notes

There are other spectral classifications, in general, adequate for certain classes of operators and the kind of problem considered; the classification presented here has the advantage of being applicable to any linear operator. Some authors use the term "point spectrum" to designate the closure of the set of eigenvalues, and the reader should be aware of this. Just to cite some terms not considered here: essential spectrum, absolutely continuous spectrum, singular continuous spectrum.

Currently, there is a great interest in the Hausdorff dimension of the spectra of linear operators and generalizations. A recommendation about Hausdorff measures is [Rogers], and for its relation with the dynamics of Schrödinger operators consult [Last] and [de Oliv].

The subspace X generated by the eigenvectors of a bounded linear operator T is invariant under T, i.e., if $\xi \in X$, then $T\xi \in X$. The Volterra operator, Example 27.14, although it has no eigenvalues, it has invariant subspaces different from $\{0\}$ and the whole space (i.e., nontrivial spaces). Indeed, for each $s \in (0,1)$, $E_s = \{\psi \in C[0,1] : \psi(t) = 0, 0 \le t \le s\}$ is a proper invariant subspace under this operator. This kind of result holds in a more general setting. In the work [ArSmi], the authors showed that any linear compact operator acting on a complex Banach space has nontrivial invariant subspaces; in 1973, V. I. Lomonosov generalized this result and simplified the proof (there are also extensions due to C. Pearcy and A. L. Shields, and P. Rosenthal, both in 1974). A book about this subject is [RadRos].

Additional Exercises

EXERCISE **28.4.** Let $V : \mathcal{B}_1 \to \mathcal{B}_2$ be an isomorphism (surjective isometric linear operator), $T \in$ B$(\mathcal{B}_1)$ and $S = VTV^{-1}$. Show that $S \in$ B$(\mathcal{B}_2)$ and $\sigma_j(T) = \sigma_j(S)$, for $j \in \{\mathrm{p}, \mathrm{r}, \mathrm{c}\}$.

EXERCISE **28.5.** (a) If $T \in$ B$(\mathcal{B})$ is idempotent, i.e., $T^2 = T$, show that $\sigma(T) \subset \{0, 1\}$. How is the spectrum of an idempotent operator T that is distinct from the identity and the null operators?

(b) Show that if $T^n = 0$ for some n, then $\sigma(T) = \{0\}$. Examine the spectrum of $T : l^p(\mathbb{N}) \hookleftarrow$, $1 \le p \le \infty$, $T(\xi_1, \xi_2, \cdots) = (0, \xi_1, \xi_2, \cdots, \xi_n, 0, 0, 0, \cdots)$.

EXERCISE **28.6.** Verify that if T is normal and $\sigma(T) = \{\lambda_0\}$, then $T = \lambda_0 \mathbf{1}$.

EXERCISE **28.7.** If $T \in$ B$(\mathcal{B})$, show that $R_\lambda(T^\mathrm{a}) = R_\lambda(T)^\mathrm{a}$, $\forall \lambda \in \rho(T)$. Show also that $\sigma_\mathrm{r}(T) \subset \sigma_\mathrm{p}(T^\mathrm{a})$ and $\sigma_\mathrm{p}(T) \subset \sigma_\mathrm{p}(T^\mathrm{a}) \cup \sigma_\mathrm{r}(T^\mathrm{a})$.

EXERCISE **28.8.** Combine Proposition 20.7 with Corollary 28.2 to show that if $T \in$ B$(\mathcal{H})$, then $\|T\| = \sqrt{r_\sigma(T^*T)}$.

EXERCISE **28.9.** By using the Exercises 19.2 and 27.15, show that $\sigma(U) \subset \{\lambda \in \mathbb{C} : |\lambda| = 1\}$ for each unitary operator U (Proposition 28.8).

EXERCISE **28.10.** If $T \in$ B$(\mathcal{H})$, take adjoints in the relation $T_\lambda R_\lambda(T) = \mathbf{1} = R_\lambda(T)T_\lambda$ to show that $\sigma(T^*)$ is the complex conjugate of $\sigma(T)$. Show the same result from $(T_\lambda^*)^{-1} = (T_\lambda^{-1})^*$.

EXERCISE **28.11.** Is it possible to use directly Theorem 28.9 to show that if $T \in$ B$(\mathcal{H})$, then $\sigma(T^*)$ is the complex conjugate of $\sigma(T)$?

EXERCISE **28.12.** Show that if $r_\sigma(T) > 1$, then the series $\sum_{j=0}^{\infty} T^j$ does not converge.

EXERCISE **28.13.** If for certain $T \in$ B$(\mathcal{B})$ one has $\lim_{n \to \infty} \|T^n\| = 0$, show that the spectral radius $r_\sigma(T) < 1$ and conclude that $\sum_{n=0}^{\infty} T^n = (\mathbf{1} - T)^{-1} \in$ B$(\mathcal{B})$.

EXERCISE **28.14.** Show that if $T \in$ B$(\mathcal{H})$ is normal, then $\|T^n\| = \|T\|^n$, for all $n \in \mathbb{N}$.

EXERCISE **28.15.** If $0 \ne \phi \in C[a, b]$, show that $\lambda \in \mathbb{C}$ is an eigenvalue of the multiplication operator $\mathcal{M}_\phi : C[a, b] \hookleftarrow$, $(\mathcal{M}_\phi \psi)(t) = \phi(t)\psi(t)$, $\psi \in C[a, b]$, if and only if the set $\phi^{-1}(\lambda)$ has nonempty interior.

EXERCISE **28.16.** If $T : C[0, 2\pi] \hookleftarrow$, $(T\psi)(t) = e^{it}\psi(t)$, find $\sigma_j(T)$, for $j \in \{p, r, c\}$.

EXERCISE **28.17.** Let $K \in \mathcal{H} = L^2[-1, 1]$, periodically extended to the set of real numbers, and $T_K : \mathcal{H} \hookleftarrow$ given by

$$(T_K\psi)(t) = \int_{-1}^{1} K(t - s)\psi(s)\, ds, \qquad \psi \in \mathcal{H}.$$

If K is an even function, determine the eigenvalues of T_K, show that

$$\sigma_{\mathrm{p}}(T_K) = \left\{ \int_{-1}^{1} K(s)\cos(\pi n s)\, ds : n \in \mathbb{Z} \right\},$$

and $\sigma(T_K) = \{0\} \cup \sigma_{\mathrm{p}}(T_K)$. Adapt this result to the case of an odd function K. Generalize to any $K \in \mathcal{H}$ and show that zero is always belongs to $\sigma(T_K)$.

CHAPTER 29

Spectra of Bounded Self-Adjoint Operators

Some important results on the spectral theory of bounded self-adjoint operators are discussed in this chapter. It will become evident that while many properties can be easily verified for eigenvalues, checking for the general spectrum often requires more sophisticated arguments. It is shown that the spectrum of a self-adjoint operator is always a subset of the real numbers, and it is also proven that the residual spectrum of a self-adjoint operator is empty. Additionally, the chapter explores how the spectrum of a self-adjoint operator is connected to its operator norm.

Many spectral properties are simple to check for eigenvalues, however more elaborate tools are needed to deal with the general case (as in Proposition 28.8). For instance, it is easy to see that each eigenvalue of a self-adjoint operator is a real number (Exercise 20.2; see ahead), but to show that all its spectrum is real the following result will be used. The Hilbert spaces in this chapter are complex unless otherwise indicated.

Theorem 29.1. *Let* $T \in \mathrm{B}(\mathcal{H})$ *be a self-adjoint operator. For each* $\lambda \in \mathbb{C}$*, the following assertions are equivalent:*

(i) $\lambda \in \rho(T)$*.*

(ii) $\mathrm{rng}\, T_\lambda = \mathcal{H}$*.*

(iii) There exists $C > 0$ *so that* $\|T_\lambda \xi\| \geq C\|\xi\|$*, for all* $\xi \in \mathcal{H}$ *(and in this case* $C \leq 1/\|R_\lambda(T)\|$*).*

Proof. If $T\xi = \lambda\xi$, then (since T is self-adjoint) $\lambda\langle \xi, \xi \rangle = \langle \xi, T\xi \rangle = \langle T\xi, \xi \rangle = \bar{\lambda}\langle \xi, \xi \rangle$ and $\lambda \in \mathbb{R}$, so that $\sigma_\mathrm{p}(T)$ is real. It is worth underlining that if *(iii)* holds, so if $T_\lambda \xi = 0$, then $\xi = 0$, that is, λ is not an eigenvalue of T.

$(i) \Rightarrow (ii)$ It follows straightly from the definition of resolvent set.

$(ii) \Rightarrow (i)$ Suppose that $\mathrm{rng}\, T_\lambda = \mathcal{H}$. Then

$$\{0\} = (\mathrm{rng}\, T_\lambda)^\perp = \mathrm{N}(T^* - \bar{\lambda}\mathbf{1}) = \mathrm{N}(T - \bar{\lambda}\mathbf{1})$$

and $\bar{\lambda} \notin \sigma_\mathrm{p}(T)$. Since $\sigma_\mathrm{p}(T) \subset \mathbb{R}$, then $\lambda \notin \sigma_\mathrm{p}(T)$ and $R_\lambda(T)$ is well defined with

DOI: 10.1201/9781003656166-29

domain $\mathcal{H}$. By the Open Mapping Theorem, it follows that $R_\lambda(T) \in \mathrm{B}(\mathcal{H})$ and $\lambda \in \rho(T)$.

$(i) \Rightarrow (iii)$ If $\lambda \in \rho(T)$, then for each $\xi \in \mathcal{H}$ one has

$$\|\xi\| = \|R_\lambda(T)T_\lambda\xi\| \leq \|R_\lambda(T)\| \, \|T_\lambda\xi\|,$$

and so (iii) holds with $C = 1/\|R_\lambda(T)\|$.

$(iii) \Rightarrow (i)$ and (ii) If (iii) holds then T_λ is injective and $R_\lambda(T)$ exists with domain rng T_λ. It will be shown that rng T_λ is closed and dense in $\mathcal{H}$, and so (i) and (ii) will follow simultaneously.

Let $\eta \in \overline{\mathrm{rng}\, T_\lambda}$ and $(\xi_j) \subset \mathcal{H}$ with $T_\lambda\xi_j \to \eta$. From (iii) one has that (ξ_j) is Cauchy in $\mathcal{H}$, and hence, converges to some $\xi \in \mathcal{H}$. Since T_λ is continuous it follows that $T_\lambda\xi = \eta$ and rng T_λ is closed in $\mathcal{H}$.

Let $\xi_0 \in (\mathrm{rng}\, T_\lambda)^\perp = \mathrm{N}(T^* - \bar{\lambda}\mathbf{1}) = \mathrm{N}(T - \bar{\lambda}\mathbf{1})$; thus $T\xi_0 = \bar{\lambda}\xi_0$, and if $\xi_0 \neq 0$ then $\bar{\lambda} \in \sigma_\mathrm{p}(T) \subset \mathbb{R}$; therefore $\lambda \in \mathbb{R}$. Suppose then that $\lambda \in \mathbb{R}$; from (iii) one finds $0 = \|T_\lambda\xi_0\| \geq C\|\xi_0\|$ and $\xi_0 = 0$; this contradiction shows that rng T_λ is dense in $\mathcal{H}$. Therefore (i) and (ii) hold.

If $\eta \in \mathcal{H}$ and (i) hold, then there exists $\xi \in \mathcal{H}$ with $\eta = T_\lambda\xi$. Thus,

$$\|R_\lambda(T)\eta\| = \|R_\lambda(T)T_\lambda\xi\| = \|\xi\| \leq \frac{1}{C}\|T_\lambda\xi\| = \frac{1}{C}\|\eta\|,$$

so that $\|R_\lambda(T)\| \leq 1/C$, and the additional remark in (iii) has also been checked. $\qquad\square$

Corollary 29.2. *If $T \in \mathrm{B}(\mathcal{H})$ is self-adjoint, then $\sigma(T)$ is a subset of the real numbers, it is contained in the interval $[-\|T\|, \|T\|]$, and $-\|T\|$ or $\|T\|$ (or both) belongs (belong) to the spectrum of T.*

Proof. If $\lambda = a + ib$, $a, b \in \mathbb{R}$, then for all $\xi \in \mathcal{H}$ one has

$$\|T_\lambda\xi\|^2 = \|T_a\xi - ib\xi\|^2 = \|T_a\xi\|^2 + b^2\|\xi\|^2 \geq b^2\|\xi\|^2.$$

Therefore, $\lambda \in \rho(T)$ if $\mathrm{Im}\,\lambda = b \neq 0$ and $\sigma(T)$ is real by Theorem 29.1. By combining this result with Corollary 28.2, and recalling that $\sigma(T)$ is closed, the other assertions follow. $\qquad\square$

From the proof of Corollary 29.2, together with Theorem 29.1, one has

Corollary 29.3. *If $\mathrm{Im}\,\lambda \neq 0$ and the operator $T \in \mathrm{B}(\mathcal{H})$ is self-adjoint, then $\|R_\lambda(T)\| \leq 1/|\mathrm{Im}\,\lambda|$.*

Corollary 29.4. *If $T \in \mathrm{B}(\mathcal{H})$ is self-adjoint, then $\sigma_\mathrm{r}(T) = \emptyset$.*

Proof. If $\sigma_\mathrm{r}(T) \neq \emptyset$, there exists $\lambda \in \sigma_\mathrm{r}(T) \subset \mathbb{R}$ so that rng T_λ is not dense in $\mathcal{H}$. Let $0 \neq \eta \in (\mathrm{rng}\, T_\lambda)^\perp = \mathrm{N}(T_\lambda)$; thus, $T_\lambda\eta = 0$ and so $\lambda \in \sigma_\mathrm{p}(T)$, which is a contradiction with $\lambda \in \sigma_\mathrm{r}(T)$. $\qquad\square$

EXERCISE **29.1.** Let $\{e_j\}_{j\in J}$ be an orthonormal basis of $\mathcal{H}$ and $\{\lambda_j\}_{j\in J}$ a bounded set of $\mathbb{R}$. Show that an operator $T \in \mathrm{B}(\mathcal{H})$ with $Te_j = \lambda_j e_j$, for all $j \in J$, is self-adjoint and $\sigma(T)$ is the closure of the set $\{\lambda_j\}_{j\in J}$.

For a self-adjoint $T \in B(\mathcal{H})$, $\langle \xi, T\xi \rangle$ is real for all $\xi \in \mathcal{H}$, and such values are directly related to the spectrum of T. Observe that if $\|\xi\| = 1$ one has $|\langle \xi, T\xi \rangle| \leq \|T\|$.

Corollary 29.5. *Let $T \in B(\mathcal{H})$ be self-adjoint,*

$$m = \inf_{\|\xi\|=1} \langle \xi, T\xi \rangle \quad \text{and} \quad M = \sup_{\|\xi\|=1} \langle \xi, T\xi \rangle.$$

Then $\sigma(T) \subset [m, M]$ and $|m|, |M| \leq \|T\|$.

Proof. It is enough to consider $\lambda \in \mathbb{R}$. It will be shown that $\lambda \in \rho(T)$ if $\lambda > M$. The case $\lambda < m$ can be handled in a similar way. If $\lambda > M$, for all $\xi \in \mathcal{H}$

$$\langle \xi, (\lambda \mathbf{1} - T)\xi \rangle \geq (\lambda - M) \|\xi\|^2.$$

On the other hand,

$$|\langle \xi, (\lambda \mathbf{1} - T)\xi \rangle| \leq \|\xi\| \, \|T_\lambda \xi\|,$$

that is, $\|T_\lambda \xi\| \geq (\lambda - M)\|\xi\|$, for all $\xi \in \mathcal{H}$, and so $\lambda \in \rho(T)$ by Theorem 29.1 with $C = (\lambda - M)$. □

EXERCISE **29.2.** Recall that an operator $T \in B(\mathcal{H})$ is positive if $\langle \xi, T\xi \rangle \geq 0$, for all $\xi \in \mathcal{H}$. If T is positive, show that the mapping

$$\xi, \eta \mapsto [\xi, \eta] := \langle \xi, T\eta \rangle, \qquad \xi, \eta \in \mathcal{H},$$

fulfills the properties that define an inner product on $\mathcal{H}$, except (maybe) $[\xi, \xi] = 0 \Rightarrow \xi = 0$. Conclude then the Cauchy-Schwarz inequality

$$|\langle \xi, T\eta \rangle| \leq \langle \xi, T\xi \rangle^{1/2} \, \langle \eta, T\eta \rangle^{1/2}.$$

See Exercise 17.7.

Proposition 29.6. *Both m and M, introduced in Corollary 29.5, belong to the spectrum of the self-adjoint operator T.*

Proof. It will be shown that $m \in \sigma(T)$; for M the argument is quite similar. The operator $T_m = T - m\mathbf{1}$ is positive and, by Exercise 29.2, for each $\xi \in \mathcal{H}$ one has

$$|\langle T_m \xi, \eta \rangle|^2 \leq \langle \xi, T_m \xi \rangle \, \langle \eta, T_m \eta \rangle.$$

By picking $\eta = T_m \xi$, one gets

$$\|T_m \xi\|^4 \leq \langle \xi, T_m \xi \rangle \, \langle T_m \xi, T_m^2 \xi \rangle \leq \langle \xi, T_m \xi \rangle \, \|T_m\|^3 \, \|\xi\|^2.$$

Thus,

$$\inf_{\|\xi\|=1} \|T_m \xi\|^4 \leq \|T_m\|^3 \inf_{\|\xi\|=1} (\langle \xi, T\xi \rangle - m) = 0,$$

and there exists no $C > 0$ as in Theorem 29.1(iii). Therefore $m \in \sigma(T)$. See also Exercise 29.4. □

Corollary 29.7. *By using the notation introduced in Corollary 29.5, if T is self-adjoint, then $\|T\| = \max\{|m|, |M|\} = \sup_{\|\xi\|=1} |\langle \xi, T\xi \rangle|$.*

Proof. The second equality is clear. Since $m, M \in \sigma(T)$, the first equality follows from the relations

$$\sigma(T) \subset [m, M] \subset [-\|T\|, \|T\|],$$

and either $-\|T\|$ or $\|T\|$ is in $\sigma(T)$, since $r_\sigma(T) = \|T\|$. □

REMARK **29.8.** By Corollary 29.7, if the operator $T \in B(\mathcal{H})$ is self-adjoint and $\langle T\xi, \xi \rangle = 0$, for all $\xi \in \mathcal{H}$, then $T = 0$, even in real Hilbert spaces. Compare with Proposition 19.11 and Example 19.12.

EXERCISE **29.3.** If $T \in B(\mathcal{H})$ is self-adjoint, then verify that: (a) $\|T\| = \sup_{\|\xi\|=\|\eta\|=1} |\langle T\xi, \eta \rangle|$; (b) The eigenvectors of T corresponding to different eigenvalues are orthogonal.

Notes

The standard spectral theory of self-adjoint operators, bounded or not, culminates with the so-called Spectral Theorem (there are similar results for normal operators). It is one of the most interesting set of results in Mathematics. Such results are not addressed in this text; nonetheless, a version for compact normal operators is discussed in Chapter 30. See the article [Steen] for historical information; the full Spectral Theorem is discussed in many texts, for instance in [Conway, ReedSi], and a guide to different proofs can be found in [de Oliv].

Additional Exercises

EXERCISE **29.4.** Consider a self-adjoint $T \in B(\mathcal{H})$.

(a) Show that $\lambda \in \sigma(T)$ if and only if there exists a sequence (ξ_n) in $\mathcal{H}$, with $\|\xi_n\| = 1$, for all n, and $T_\lambda \xi_n \to 0$.

(b) Show that $\lambda \in \sigma(T)$ if and only if $\inf_{\|\xi\|=1} \|T_\lambda \xi\| = 0$.

(c) The characterizations in (a) and (b) above do not hold for all operators (although they hold for normal operators). Verify this by considering the right shift S_r on $l^2(\mathbb{N})$, and showing that zero belongs to its spectrum, but since this operator is an isometry, (a) and (b) do not hold for $\lambda = 0$.

EXERCISE **29.5.** As a complement to Exercise 29.4, given $S \in B(\mathcal{H})$ (not necessarily self-adjoint), show that $\lambda \in \sigma(S)$ if there exists (ξ_n) in $\mathcal{H}$, $\|\xi_n\| = 1$, with $S_\lambda \xi_n \to 0$. If λ is a boundary point of $\sigma(S)$, show that the converse also holds.

EXERCISE **29.6.** Use Proposition 29.6 to show that the spectrum of a bounded self-adjoint operator is not empty.

EXERCISE **29.7.** Show that two unitarily equivalent self-adjoint operators have the same spectra. What does happen if such operators are not self-adjoint? (Note that it is a particular case of Exercise 28.4.)

EXERCISE **29.8.** If P_E is the orthogonal projection operator onto a proper closed subspace $E \subset \mathcal{H}$, determine m and M for P_E.

EXERCISE **29.9.** Show that if $T \in B(\mathcal{H})$ is self-adjoint (or normal) and $T^n = 0$ for some $n \in \mathbb{N}$, then $T = 0$.

EXERCISE **29.10.** Let $T, S \in B(\mathcal{H})$. Show that:

(a) If $T \geq S$ (that is, $(T - S)$ is positive) and $S \geq T$, then $T = S$.

(b) The spectrum of T^*T is real and contained in $[0, \infty)$.

(c) S is positive if and only if S is self-adjoint and $\sigma(S) \subset [0, \infty)$.

(d) If T is self-adjoint, then $(\mathbf{1} + \lambda T^2)^{-1} \in B(\mathcal{H})$ for all $\lambda \geq 0$.

(e) $(\mathbf{1} + \lambda T^*T)^{-1} \in B(\mathcal{H})$ for all $\lambda \geq 0$ (here T is not necessarily self-adjoint).

EXERCISE **29.11.** Prove the following version of Theorem 29.1 for bounded operators on Banach spaces: If $T \in B(\mathcal{B})$, then the following assertions are equivalent:

(i) $\lambda \in \rho(T)$

(ii) There exists $C > 0$ with $\|T_\lambda \xi\| \geq C\|\xi\|$, for all $\xi \in \mathcal{B}$, and rng T_λ is dense in $\mathcal{B}$ (and in this case one has $C \leq 1/\|R_\lambda(T)\|$).

EXERCISE **29.12.** Let $T \in B(\mathcal{H})$. Use Exercise 29.11 to show that if there exists $C > 0$ with

$$\|(T - \lambda\mathbf{1})\xi\| \geq C\|\xi\| \quad \text{and} \quad \|(T^* - \bar{\lambda}Id)\xi\| \geq C\|\xi\|, \qquad \forall \xi \in \mathcal{H},$$

then $\lambda \in \rho(T)$.

EXERCISE **29.13.** For this exercise some familiarity with the general theory of measure and integration is needed. Show that $\sigma(\mathcal{M}_\phi)$, $\mathcal{M}_\phi : L^2_\mu(\Omega) \hookleftarrow$, with $\phi \in L^\infty_\mu(\Omega)$, is given by $\lambda \in \mathbb{C}$ so that $\mu(\phi^{-1}B(\lambda; \varepsilon)) > 0$ for all $\varepsilon > 0$ ($B(\lambda; \varepsilon)$ denotes an open ball in $\mathbb{C}$).

CHAPTER 30

Spectra of Compact Operators

Some general properties of the spectra of compact linear operators on Banach spaces are discussed. It is shown that, with the possible exception of zero, each eigenvalue of a compact operator has finite multiplicity, and the set of such eigenvalues can only accumulate at zero. The chapter then considers the important case of normal compact operators, particularly self-adjoint ones, on Hilbert spaces. It is shown that the eigenvectors of a normal operator can be chosen to form an orthonormal basis for the Hilbert space, which is used to present a version of the Spectral Theorem for normal operators.

30.1 COMPACT OPERATORS

As already mentioned, the spectral theory of compact linear operators exhibits some features of the case of linear operators on finite-dimensional spaces. For instance, except possibly for the point zero, every eigenvalue of a compact operator has finite multiplicity. Nevertheless, it is important to note that there exist compact operators that possess no eigenvalues at all (see Exercise 30.7).

Proposition 30.1. *If $T \in B_0(\mathcal{B})$, then any $0 \neq \lambda \in \sigma_p(T)$ has finite multiplicity, namely, $\dim N(T_\lambda) < \infty$.*

Proof. Denote by B the closed ball centered at zero and radius one in the null space $N(T_\lambda)$. Once it is verified that B is compact, Theorem 2.2 implies that $\dim N(T_\lambda) < \infty$. The fact that T is compact implies that for any sequence $(\xi_n) \subset B$, the sequence $(T\xi_n = \lambda\xi_n)$ has a convergent subsequence $(T\xi_{n_j})$; write $(\xi_{n_j} = T\xi_{n_j}/\lambda)$ and see that this subsequence also converges to an element of B; hence B is a compact set. $\square$

EXERCISE **30.1.** An alternative proof of Proposition 30.1 is the following: If it is supposed that B is not compact, there exists a sequence $(\xi_n) \subset B$ with no convergent subsequence; use that T is compact in order to get a contradiction.

Proposition 30.2. *For all $\varepsilon > 0$, the number of eigenvalues λ with $|\lambda| \geq \varepsilon$, counting with multiplicities, of any compact operator $T \in B_0(\mathcal{B})$, is finite.*

DOI: 10.1201/9781003656166-30

Proof. Assume that there exists $\epsilon > 0$ such that the operator T has infinitely many eigenvalues $\{\lambda_j\}_{j\in\mathbb{N}}$ with $|\lambda_j| \geq \varepsilon$. By Proposition 30.1, these eigenvalues can be taken to be pairwise distinct. Let $\{\xi_j\}_j$ denote the corresponding eigenvectors. Recall that this set is linearly independent, as established in Proposition 27.4.

Consider the closed subspaces $E_0 = \{0\}$ and $E_n = \text{Lin}(\{\xi_1, \cdots, \xi_n\})$. By Riesz Lemma 2.1, there is a sequence $\{\eta_n\}$, $\eta_n \in E_n$, $\|\eta_n\| = 1$ and $\|\eta_n - \xi\| \geq 1/2$, for every $\xi \in E_{n-1}$. It will be shown that $\|T\eta_n - T\eta_m\| \geq \varepsilon/2$ for all pairs $n \neq m$, which would imply that it has no convergent subsequence, but this is not possible by the compactness of T.

If $m < n$, then $T\eta_n - T\eta_m = \lambda_n\eta_n + [(T - \lambda_n\mathbf{1})\eta_n - T\eta_m]$. Note that $T\eta_m \in E_m$ and put $\eta_n = \sum_{j=1}^{n} \alpha_j\xi_j$; then

$$(T - \lambda_n\mathbf{1})\eta_n = \left[\sum_{j=1}^{n-1} \alpha_j(\lambda_j - \lambda_n)\xi_j\right] \in E_{n-1},$$

and so $\zeta_m := -[(T - \lambda_n\mathbf{1})\eta_n - T\eta_m]/\lambda_n$ is an element of the subspace E_{n-1}. Hence,

$$\|T\eta_n - T\eta_m\| = |\lambda_n|\|\eta_n - \zeta_m\| \geq \frac{|\lambda_n|}{2} \geq \varepsilon/2,$$

and $\{T\eta_n\}$ has no convergent subsequence. □

Corollary 30.3. *If $T \in \text{B}_0(\mathcal{B})$, then:*

(i) The unique possible accumulation point of $\sigma_\text{p}(T)$ is zero.

(ii) $\sigma_\text{p}(T)$ is countable, and if $\lambda \neq 0$, then $\dim \text{N}(T_\lambda) < \infty$.

(iii) If $\sigma_\text{p}(T)$ is infinite, then the eigenvalues of T can be ordered in a sequence converging to zero.

(iv) If $\dim \mathcal{B} = \infty$, then zero belongs to the spectrum of T.

EXERCISE **30.2.** Present the details of the proof of Corollary 30.3.

Example 30.4. Consider the operator $T : l^2(\mathbb{N}) \hookleftarrow$,

$$T(\xi_1, \xi_2, \xi_3, \cdots) = (\xi_2/2, \xi_3/3, \xi_4/4, \cdots).$$

Then T is compact, $T(1, 0, 0, \cdots) = 0$ and so $0 \in \sigma_\text{p}(T)$. •

Example 30.5. Consider the operator $T : l^2(\mathbb{N}) \hookleftarrow$,

$$T(\xi_1, \xi_2, \xi_3, \cdots) = (0, \xi_1/1, \xi_2/2, \xi_3/3, \cdots).$$

T is compact and $0 \in \sigma_\text{r}(T)$ (see Corollary 30.3*(iv)*). This operator has no eigenvalues (Exercise 30.7). •

Example 30.6. Consider the operator $T : l^2(\mathbb{N}) \hookleftarrow$,

$$T(\xi_1, \xi_2, \xi_3, \cdots) = (\xi_1/1, \xi_2/2, \xi_3/3, \cdots).$$

T is compact and $0 \in \sigma_\text{c}(T)$. Indeed, zero is not an eigenvalue of T, however it is an element of its spectrum, since $\{1, 1/2, 1/3, \cdots\}$ is a subset of $\sigma_\text{p}(T)$ and the spectrum is closed. Since it is a self-adjoint operator $\sigma_\text{r}(T) = \emptyset$. Therefore zero belongs to its continuous spectrum. One may also directly deduce that the resolvent operator $R_0(T)$ exists and has a dense domain, although it is not bounded. •

30.2 NORMAL OPERATORS

Now the particular case of normal compact operators on a Hilbert space is considered. Many of the results presented ahead can be adapted for compact operators on Banach spaces, but with different arguments (not discussed here).

Lemma 30.7. *Let* $T \in B(\mathcal{H})$ *be a nonzero compact and self-adjoint operator. Then either* $-\|T\|$ *or* $\|T\|$ *is an eigenvalue of* T, *so* T *always has a nonzero eigenvalue.*

Proof. By self-adjointness and Corollary 29.2, $-\|T\|$ or $\|T\|$ belongs to the spectrum of T; by the compactness of T it will follow that one of them is an eigenvalue, which is equivalent to find $0 \neq \zeta \in \mathcal{H}$ with $(T^2 - \|T\|^2 \mathbf{1})\zeta = 0$. Denote $\|T\| = \lambda$.

Pick (ξ_n), $\|\xi_n\| = 1$, for all n, so that $\|T\xi_n\| \to \lambda$. By the compactness of T, there is a convergent subsequence of $(T\xi_n)$, also denoted by $(T\xi_n)$; since T is continuous, it follows that $(T^2\xi_n)$ also converges.

The inequalities

$$\begin{aligned} 0 \; &\leq \; \left\| T^2\xi_n - \|T\xi_n\|^2 \xi_n \right\|^2 = \|T^2\xi_n\|^2 - \|T\xi_n\|^4 \\ &\leq \; \lambda^2 \|T\xi_n\|^2 - \|T\xi_n\|^4 \longrightarrow 0 \quad \text{as} \quad n \to \infty, \end{aligned}$$

imply that the sequence $\eta_n = T^2\xi_n - \|T\xi_n\|^2\xi_n$ converges to zero and so

$$\xi_n = \left(T^2\xi_n - \eta_n \right) / \|T\xi_n\|^2$$

converges to some vector ζ with $\|\zeta\| = 1$. Now, since T is a continuous operator, $0 = T^2\zeta - \lambda^2\zeta = T_\lambda T_{-\lambda}\zeta$. Therefore, either $T_{-\lambda}\zeta = 0$ and $-\|T\|$ is an eigenvalue of T, or $T_{-\lambda}\zeta \neq 0$ and $\|T\|$ is an eigenvalue of T. □

Theorem 30.8 (Hilbert-Schmidt). *If* $T \in B_0(\mathcal{H})$ *is self-adjoint, then*

$$\mathcal{H} = \left[\bigoplus_{0 \neq \lambda \in \sigma_{\mathrm{p}}(T)} \mathrm{N}(T_\lambda) \right] \oplus \mathrm{N}(T).$$

Proof. One has $\mathrm{N}(T_\lambda) \perp \mathrm{N}(T_\mu)$ if $\lambda \neq \mu$, because T is self-adjoint, and so the direct sum above is well defined. Write $E = \bigoplus_{0 \neq \lambda \in \sigma_{\mathrm{p}}(T)} \mathrm{N}(T_\lambda)$; if $\eta \in E^\perp$, then for every $\xi_\lambda \in \mathrm{N}(T_\lambda)$ one finds $\langle T\eta, \xi_\lambda \rangle = \langle \eta, T\xi_\lambda \rangle = \lambda \langle \eta, \xi_\lambda \rangle = 0$, and so $T\eta \in \mathrm{N}(T_\lambda)^\perp$. Note that this occurs for all $\lambda \in \sigma_{\mathrm{p}}(T)$; then, it follows that $T\eta \in E^\perp$, that is, $E^\perp$ is invariant under T; moreover, $\mathcal{H} = E \oplus E^\perp$ and $\mathrm{N}(T) \subset E^\perp$.

To complete the proof, it will be shown that $E^\perp = \mathrm{N}(T)$. Since E is also invariant under T, then $S = T|_{E^\perp}$, the operator restriction of T to $E^\perp$, is well defined and also a self-adjoint compact operator. If $S \neq 0$, by Lemma 30.7, there exists a nonzero eigenvalue and an eigenvector $0 \neq \zeta$ of S. Hence, by construction, the vector $\zeta \in E$ and $\zeta \in E^\perp$, and automatically $\zeta = 0$. One then concludes that $S = 0$, that is, $E^\perp = \mathrm{N}(T)$. □

Corollary 30.9. *If* $T \in B_0(\mathcal{H})$ *is self-adjoint, then the Hilbert space* $\mathcal{H}$ *has an orthonormal basis of eigenvectors of* T.

Proof. If $\lambda \neq 0$ is an eigenvalue of T, denote by $d_\lambda = \dim \mathrm{N}(T_\lambda) < \infty$ and select an orthonormal basis $\{\xi_j^\lambda\}_{j=1}^{d_\lambda}$ of $\mathrm{N}(T_\lambda)$. Also select an orthonormal basis $\{\eta_j\}_{j \in J}$ of $\mathrm{N}(T)$. By Theorem 30.8, the set

$$\left[\bigcup_{0 \neq \lambda \in \sigma_\mathrm{p}(T)} \{\xi_j^\lambda\}_{j=1}^{d_\lambda} \right] \cup \{\eta_j\}_{j \in J}$$

is an orthonormal basis of $\mathcal{H}$. □

EXERCISE 30.3. Verify that the following operator is compact and find its spectrum: $T_K : \mathrm{L}^2[0,1] \hookleftarrow$, given by $(T_K \psi)(t) = \int_0^1 K(t,s)\psi(s)\,\mathrm{d}s$, with $K(t,s) = \min\{t,s\}$. (Hint: Note that T is the inverse of the operator $\psi \mapsto -\psi''$, with boundary conditions $\psi(0) = 0$ and $\psi'(1) = 0$, that is, $T\psi'' = -\psi$, and similarly $-(T\psi)'' = \psi$, $(T\psi)(0) = 0$, $(T\psi)'(1) = 0$; so $\psi_n(t) = \sin(\alpha_n t)$, with $\alpha_n = \pi(n - 1/2)$, $n \in \mathbb{N}$, are its eigenfunctions.)

In order to generalize this last result to normal compact operators, the next lemma will be useful. Recall that two operators R, S are commuting if $RS = SR$.

Lemma 30.10. *If $R, S \in \mathrm{B}_0(\mathcal{H})$ are self-adjoint and commuting, then an orthonormal basis of $\mathcal{H}$ consisting of vectors that are simultaneous eigenvectors of both R and S.*

Proof. If λ is an eigenvalue of S, $S\xi^\lambda = \lambda\xi^\lambda$, and $S(R\xi^\lambda) = R(S\xi^\lambda) = \lambda R\xi^\lambda$, so $\mathrm{N}(S_\lambda)$ is invariant under R (as well as its orthogonal complement). Note that the restriction operator $R|_{\mathrm{N}(S_\lambda)}$ is self-adjoint and compact, then there exists an orthonormal basis of $\mathrm{N}(S_\lambda)$, as in Corollary 30.9, composed of eigenvectors of R, and so, they are also eigenvectors of S. It is enough to take the union over such eigenvectors corresponding to all eigenvalues of S to obtain the result, again by Corollary 30.9. □

Corollary 30.11. *If $T \in \mathrm{B}_0(\mathcal{H})$ is normal, then $\mathcal{H}$ admits an orthonormal basis consisting of eigenvectors of T, and the decomposition of $\mathcal{H}$ described in Hilbert-Schmidt's Theorem 30.8 is valid.*

Proof. Since T^* is also compact by Corollary 25.3, it follows that $T = T_R + iT_I$, with T_R, T_I self-adjoint and compact, and since T is normal, T_R commutes with T_I (Exercise 20.4). Now, apply Lemma 30.10. Also observe that if $T\xi^\lambda = \lambda\xi^\lambda$, then $T_R\xi^\lambda = (\mathrm{Re}\,\lambda)\xi^\lambda$ and $T_I\xi^\lambda = (\mathrm{Im}\,\lambda)\xi^\lambda$, and eigenvectors corresponding to distinct eigenvalues are orthogonal. □

Theorem 30.12 (Spectral Theorem). *Let $0 \neq T$ be a normal compact linear operator on $\mathcal{H}$, let $\{\lambda_j\} \subset \mathbb{C}$ denote the nonzero eigenvalues of T, and let P_j be the orthogonal projection onto $\mathrm{N}(T_{\lambda_j})$, for each j. Then*

$$T = \sum_j \lambda_j P_j,$$

with convergent series in $\mathrm{B}(\mathcal{H})$.

Proof. Include the notation P_0 for the orthogonal projection onto $N(T)$, i.e., onto $N(T_\lambda)$ for $\lambda = 0$, and since $\mathbf{1} = P_0 + \sum_j P_j$ (see Corollary 30.11), for every $\xi \in \mathcal{H}$

$$T\xi = TP_0\xi + T\sum_j P_j\xi = \sum_j T(P_j\xi) = \sum_j \lambda_j P_j\xi.$$

The orthogonality $P_j P_k = 0$, if $j \neq k$, implies that

$$
\begin{aligned}
\left\|(T - \sum_{j=1}^n \lambda_j P_j)\xi\right\|^2 &= \sum_{j=n+1}^\infty |\lambda_j|^2 \|P_j\xi\|^2 \\
&\leq \left(\max_{j \geq n+1} |\lambda_j|^2\right) \sum_{j=n+1}^\infty \|P_j\xi\|^2 \\
&\leq \left(\max_{j \geq n+1} |\lambda_j|^2\right) \|\xi\|^2
\end{aligned}
$$

and without loss it was assumed that j is running over $\mathbb{N}$.

Recall that $\lambda_j \to 0$, that combined with $\|T - \sum_{j=1}^n \lambda_j P_j\|^2 \leq \max_{j \geq n+1} |\lambda_j|^2$, results in $T = \lim_{n\to\infty} \sum_{j=1}^n \lambda_j P_j$ in $B(\mathcal{H})$. □

EXERCISE **30.4.** Show that every operator $\sum_j \lambda_j P_j$, with P_j orthogonal projections onto pairwise orthogonal and finite-dimensional spaces, and $\lambda_j \to 0$, is compact and normal. Verify, furthermore, that this operator is self-adjoint if and only if $\{\lambda_j\} \subset \mathbb{R}$.

Corollary 30.13. *For each positive compact operator $T \in B_0(\mathcal{H})$, there exists a square root positive compact operator S that satisfies $S^2 = T$. Usually such operator is denoted by $T^{1/2}$ or $\sqrt{T}$.*

Proof. The point is that the positivity of T implies that it is self-adjoint with all eigenvalues $\lambda_j \geq 0$. The Spectral Theorem gives the representation $T = \sum_j \lambda_j P_j$, which can be employed to introduce the operator S by $S = \sum_j \sqrt{\lambda_j} P_j$, which is compact since $\lambda_j \to 0$ as $j \to \infty$, and S can be approximated by finite rank operators in $B(\mathcal{H})$ (indeed, by $\sum_{j=1}^n \sqrt{\lambda_j} P_j$). The property $S^2 = T$ is left to the readers. □

EXERCISE **30.5.** Find the square root of the operator in Exercise 30.3.

REMARK **30.14.** It is possible to show that S in Corollary 30.13 is the unique positive compact operator with such properties. See Exercise 30.24.

EXERCISE **30.6.** If $T \in B_0(\mathcal{H})$ is positive, check that the spectrum of $\sqrt{T}$ is the set of squared roots of the eigenvalues of T.

Notes

It was F. Riesz who noted that the important property in the Fredholm alternative (see Exercise 30.12) was the compactness of the linear operators. In a work of 1916 (written in Hungarian), Riesz followed Fredholm's view of operators, instead of the bilinear forms of Hilbert (as mentioned in other Notes), and developed most of the spectral theory of compact operators on Banach spaces.

Additional Exercises

EXERCISE 30.7. Verify that for the operator T defined in Example 30.5, one has $\sigma_p(T) = \sigma_c(T) = \emptyset$. Hence, there are compact operator with no eigenvalues; is this operator normal?

EXERCISE 30.8. Find the spectrum of the compact operator $T_K : C[0,1] \hookleftarrow$, given by $(T_K\psi)(t) = \int_0^1 K(t,s)\psi(s)\,ds$, with $K(t,s) = (t-s)$.

EXERCISE 30.9. Let $T \in B(\mathcal{H})$, with $\dim\mathcal{H} = \infty$. Show that if there exists $C > 0$ with $\|T\xi\| \geq C\|\xi\|$, for all $\xi \in \mathcal{H}$, then T is not compact.

EXERCISE 30.10. If $T \in B_0(\mathcal{H})$ is self-adjoint, show that $m = \inf_{\|\xi\|=1}\langle\xi, T\xi\rangle$ and $M = \sup_{\|\xi\|=1}\langle\xi, T\xi\rangle$ are the smallest and greatest eigenvalue of T, respectively.

EXERCISE 30.11. Prove Proposition 30.2 for Hilbert spaces and self-adjoint compact operators, without using the Riesz Lemma.

EXERCISE 30.12. [Fredholm Alternative] Let $T \in B_0(\mathcal{H})$ be a normal operator. Consider the equation $T\xi - \lambda\xi = \eta$, $\lambda \in \mathbb{C}$, $\eta \in \mathcal{H}$, and the corresponding homogeneous equation $T\xi - \lambda\xi = 0$. Show that for each $\lambda \neq 0$, one, and only one, of the following possibilities occurs (note that, in this case, uniqueness implies the existence of solution!):

(i) The homogeneous equation has only the trivial solution and the original equation has a unique solution for each $\eta \in \mathcal{H}$.

(ii) The homogeneous equation has $0 < \dim N(T_\lambda) < \infty$ linearly independent solutions, and the original equation either has infinitely many solutions or no solution at all.

EXERCISE 30.13. Find the eigenvalues and eigenvectors of the compact operator $T_K : C[0,1] \hookleftarrow$, given by $(T_K\psi)(t) = \int_0^1 K(t,s)\psi(s)\,ds$, with $K(t,s) = ts(1-ts)$.

EXERCISE 30.14. Let $0 \neq \varphi \in L^2[0,1]$ and $K(t,s) = \varphi(t)\overline{\varphi(s)}$, $t,s \in [0,1]$. Show that $\lambda = \|\varphi\|_2^2$ is the unique nonzero eigenvalue of the operator $T_K : L^2[0,1] \hookleftarrow$, $(T_K\psi)(t) = \int_0^1 K(t,s)\psi(s)\,ds$. Find the corresponding eigenfunction (note that it is usual the term *eigenfunction* to designate an eigenvector in a function vector space). Determine also the eigenfunctions corresponding to the zero eigenvalue.

EXERCISE 30.15. Fix $\eta \in \mathcal{H}$ with $\|\eta\| = 1$. Let $T_\eta : \mathcal{H} \to \mathcal{H}$ defined as $T_\eta\xi = \langle\eta, \xi\rangle\,\eta$, $\xi \in \mathcal{H}$ (Exercise 26.6). Determine the spectrum and the spectral radius of T_η.

EXERCISE 30.16. Let $T : l^2(\mathbb{Z}) \to l^2(\mathbb{Z})$ be given by $(T\xi)_n = (1/i)(\xi_{n+1} - \xi_{n-1})$, with $\xi = (\cdots, \xi_{-2}, \xi_{-1}, \xi_0, \xi_1, \xi_2, \cdots)$. Show that T is bounded and that its spectrum is real; then find its spectral radius.

EXERCISE 30.17. Find the eigenvalues and eigenvectors of the compact operator $T_K : L^2[0,1] \hookleftarrow$, $(T_K\psi)(t) = \int_0^1 K(t,s)\psi(s)\,ds$, with $K(t,s) = a\sin(t-s)$, $0 \neq a \in \mathbb{C}$.

EXERCISE 30.18. Let $U \in B(\mathcal{H})$ be unitary, so normal. Show that if it is compact then $\dim\mathcal{H} < \infty$. Thus, the Spectral Theorem 30.12 does not hold for unitary operators in infinite-dimensional spaces (see Exercise 30.4).

EXERCISE 30.19. Analyze the behavior of the spectrum of T_t in Exercise 25.12 for $t \uparrow 1$. Is this convergence of operators uniform?

EXERCISE 30.20. For $n \in \mathbb{N}$, denote by $S_{(n)}$ a nth root of the operator $T \in B(\mathcal{H})$, that is, $(S_{(n)})^n = T$ (check the existence of $S_{(n)}$ for positive compact operators). Show that T is invertible if and only if $S_{(n)}$ is invertible for some $n \geq 2$ (and so $S_{(n)}$ is invertible for all $n \geq 1$).

EXERCISE **30.21.** Show that a sequence (T_n) of positive operators on $B_0(\mathcal{H})$ converges strongly to T if and only if $\sqrt{T_n} \xrightarrow{s} \sqrt{T}$. Use this to conclude that for increasing sequences of operators (that is, $(T_n - T_{n-1}) \geq 0$) on $B_0(\mathcal{H})$, the strong convergence of such operators is equivalent to weak convergence.

EXERCISE **30.22.** Let $A, T \in B(\mathcal{H})$, with T self-adjoint so that for all $z \in \mathbb{C}$, with $\operatorname{Im} z \neq 0$, $AR_z(T)A^*$ is compact. Consider $S = T + A^*A$, i.e., a "perturbation of T." If the real number $\lambda \in \rho(T)$ and $(1 + AR_\lambda(T)A^*)$ has no inverse in $B(\mathcal{H})$, show that λ is an eigenvalue of S (see also Exercise 27.9).

EXERCISE **30.23.** Let $A, T \in B(\mathcal{H})$ be self-adjoint operators, with A compact and $TA = AT$. Show that $\sigma(T + A) \subset \sigma(T) \cup \sigma_p(T + A)$.

EXERCISE **30.24.** Show that there is a unique compact and positive operator S such that $S^2 = T$ in Corollary 30.13.

EXERCISE **30.25.** Let $T : l^2(\mathbb{N}) \to l^2(\mathbb{N})$, $T(\xi_1, \xi_2, \xi_3, \cdots) = (0, \xi_2, \xi_3, \cdots)$. Show that T is bounded, positive and self-adjoint, and that $\sqrt{T} = T$.

EXERCISE **30.26.** Show that the left shift operator $S_l : l^2(\mathbb{N}) \to l^2(\mathbb{N})$,

$$S_l(\xi_1, \xi_2, \xi_3, \cdots) = (\xi_2, \xi_3, \xi_4, \cdots),$$

has no square root.

Solutions to Selected Exercises

Sol. 1.6: Suppose that $\|\cdot\|_1$ and $\|\cdot\|_2$ generate the same topology in X; it will be shown that these norms are equivalent.

Write $X_1 = (X, \|\cdot\|_1)$ and $X_2 = (X, \|\cdot\|_2)$. Since $B_{X_1}(0;1)$ is also open in X_2, this ball contains a neighborhood of the origin and there exists $\varepsilon > 0$ with $B_{X_2}(0;\varepsilon) \subset B_{X_1}(0;1)$. For $0 \neq \xi \in X$, the vector $\varepsilon\xi/(2\|\xi\|_2)$ belongs to $B_{X_1}(0;1)$; thus

$$\left\|\frac{\varepsilon\xi}{2\|\xi\|_2}\right\|_1 < 1 \implies \|\xi\|_1 < \frac{2}{\varepsilon}\|\xi\|_2.$$

The other inequality is obtained in a similar way.

Sol. 1.7: (a) Every compact subset of a metric space is bounded and closed. Thus it is sufficient to show that, if the normed space is finite dimensional, then for any $r > 0$ the closed ball $\overline{B}(0;r)$ is compact.

Let $\{e_1, \cdots, e_n\}$ be a basis of $\mathcal{N}$; thus, any $\xi \in \mathcal{N}$ can be written as $\xi = \sum_{j=1}^n \xi_j e_j$. First, it will be shown that in this case of finite dimension, there exists $K > 0$ so that $\|\xi\| \geq K \sum_{j=1}^n |\xi_j|$, for all $\xi \in \mathcal{N}$, or equivalently, there exists $K > 0$ so that

$$\|\eta\| \geq K, \quad \forall \eta = \sum_{j=1}^n \eta_j e_j \text{ with } \sum_{j=1}^n |\eta_j| = 1;$$

the case of null vector is trivial.

If such a relation does not hold, there exists a sequence $\eta^m = \sum_{j=1}^n \eta_j^m e_j$, with $\sum_{j=1}^n |\eta_j^m| = 1$ and $\|\eta^m\| \to 0$ as $m \to \infty$. Since each sequence $(\eta_j^m)_{m=1}^\infty$ is bounded in $\mathbb{F}$, it has a convergent subsequence, and by using the usual process of taking successive subsequences, one finds $(m_r) \subset \mathbb{N}$ so that, for each $1 \leq j \leq n$, there exists $\eta_j^0 \in \mathbb{F}$, with $\eta_j^{m_r} \to \eta_j^0$ as $r \to \infty$. Setting $\eta^0 = \sum_{j=1}^n \eta_j^0 e_j$, one gets a contradiction with the linearly independence of $\{e_j\}$, since $\sum_{j=1}^n |\eta_j^0| \neq 0$, whereas $\|\eta^0\| = \lim_{r \to \infty} \|\eta^{m_r}\| = 0$.

To conclude the compactness of $\overline{B}(0;r)$, consider a sequence $\eta^m = \sum_{j=1}^n \eta_j^m e_j$ in this ball, and by using the relation above it is found that $\sum_{j=1}^n |\eta_j^m| \leq r/K$, for all m, and then it is possible to follow similar arguments to find a convergent subsequence of (η^m) in $\overline{B}(0;r)$. Therefore this closed ball is compact.

Sol. 1.17: For $D(t,s)$ consider the sequence $t_n = n$. Since $D(t_n, t_m) = D(n,m) \to 0$ as $n, m \to \infty$, then (t_n) is Cauchy in $(\mathbb{R}, D)$; on the other hand, for any $t \in \mathbb{R}$ the limit

$$\lim_{n\to\infty} D(t_n, t) = \lim_{n\to\infty} \left| \frac{n}{1+n} - \frac{t}{1+|t|} \right|$$

may not vanish, since one would get $\frac{t}{1+|t|} = 1$, which has no solution in $\mathbb{R}$. Therefore, $(\mathbb{R}, D)$ is not complete.

Sol. 2.11: If $\dim \mathcal{N} < \infty$, the answer is affirmative, since $S(0;1)$ is compact and ψ is continuous, it then follows that $\psi(S(0;1))$ is compact in $\mathbb{R}$, hence a bounded set.

If $\dim \mathcal{N} = \infty$, the answer is negative. Indeed, consider the sequence $(\xi_n)_{n=1}^\infty$ in the sphere $S(0;1)$, constructed from Riesz Lemma 2.1, so that $\|\xi_n - \xi_k\| \geq 1/2$, for all $n \neq k$.

For each n let $f_n : \mathcal{N} \to \mathbb{R}$ be a continuous function taking one in $\overline{B}(\xi_n; 1/8)$ and zero in the complement of $B(\xi_n; 1/5)$; by Uryshon Lemma, such functions exist, and note that for $n \neq k$ the support of f_n is disjoint from the support of f_k. Thus, the function $\psi : \mathcal{N} \to \mathbb{R}$ defined by

$$\psi(\xi) = \sum_{n=1}^\infty n\, f_n(\xi), \qquad \xi \in \mathcal{N},$$

is continuous and $\psi(S(0;1))$ is not bounded.

Sol. 3.1: It will be shown a more general result, i.e., that any subset of a separable metric space is also separable. Let (X, d) be separable and $E \subset X$. If (x_n) is dense in X, for each $j \in \mathbb{N}$ one has $X = \cup_n B(x_n; 1/j)$. Pick $y_{n,j} \in E_{n,j} := E \cap B(x_n; 1/j)$, $n, j = 1, 2, 3, \cdots$, and let Y be the union of such $y_{n,j}$. Claim: $Y \subset E$ is dense in E, hence E is separable. Indeed, if $x \in E$, for any j there is n so that $x \in E_{n,j}$; thus, $d(x, y_{n,j}) < 2/j$ (by triangle inequality).

Sol. 4.12: (b) A key element is that rng f is one dimensional. If $f \in \mathcal{N}^*$, then $\mathrm{N}(f)$ is closed by continuity.

Now, assume that for certain linear functional f on $\mathcal{N}$ the kernel $\mathrm{N}(f)$ is closed and $f \neq 0$ (the case of the null functional is trivial). Then there exists $0 \neq \xi \in \mathcal{N}$ with $f(\xi) \neq 0$. If f is not continuous, there exist $\|\xi_n\| = 1$ and $(a_n) \subset \mathbb{C}$ with $|a_n| \to \infty$, so that $f(\xi_n) = a_n f(\xi)$, since $f(\xi)$ generates rng f.

Hence, $\eta_n = (\xi_n - a_n \xi) \in \mathrm{N}(f)$ and since $\xi_n/a_n \to 0$, from the relation $\eta_n/a_n = \xi_n/a_n - \xi$, one gets $\eta_n/a_n \to -\xi$, and being the kernel closed, ξ should belong to $\mathrm{N}(f)$. This contradiction shows that f is bounded.

Sol. 4.14: Write $S = T - (T - S) = T(1 - T^{-1}(T - S))$. Thus, if $\|T^{-1}(T - S)\| < 1$, then S is invertible (see Exercise 4.13). The condition $\|T^{-1}\| \, \|(T - S)\| < 1$, in the exercise, implies that $\|T^{-1}(T - S)\| < 1$.

Sol. 4.20:   Showing that $\mathcal{N}$ is complete is simple. If $0 < r < 1$, then

$$f_r(\psi) \le \int_0^1 \frac{\|\psi\|_\infty}{t^r}\, dt = \frac{\|\psi\|_\infty}{1-r}, \qquad \forall \psi \in \mathcal{N},$$

and so $f_r \in \mathcal{N}^*$ and $\|f_r\| \le 1/(1-r)$. The sequence $\psi_n(t)$ ahead (for n large) shows that $\|f_r\| = 1/(1-r)$.

For $1 \le r < 2$, consider $\psi_n(t) = nt$ if $t \le 1/n$ and equal to 1 if $t \ge 1/n$. Then $\|\psi_n\|_\infty = 1$ and

$$f_r(\psi_n) = \frac{n^{r-1}}{2-r} + \begin{cases} \left(n^{r-1} - 1\right)/(r-1) & \text{if } r > 1 \\ \ln n & \text{if } r = 1 \end{cases},$$

so that $\lim_{n \to \infty} f_r(\psi_n) = \infty$ for $1 \le r < 2$; hence f_r is not bounded in these cases. For $r \ge 2$ the functions ψ_n do not belong to the domain of f_r, so such functionals are not in the dual space.

Sol. 5.19:   Let $\psi_0(t) = 1$ and $\phi_0(t) = 0$. Since $d(\psi_0, \phi_0) = 1$ and $d(T\psi_0, T\phi_0) = 3a/2$, it follows that T is not a contraction if $a \ge 2/3$. For any $\psi, \phi \in C[0,1]$, one has

$$\begin{aligned}
|T\psi(t) - T\phi(t)| &\le \frac{3a\pi}{4} \int_0^t |\psi(s) - \phi(s)|\, \cos(\pi s/2)\, ds \\
&\le d(\psi, \phi)\frac{3a\pi}{4} \int_0^t \cos(\pi s/2)\, ds = d(\psi, \phi)\frac{3a}{2}\sin(\pi t/2).
\end{aligned}$$

Hence,

$$\begin{aligned}
\left|T^2\psi(t) - T^2\phi(t)\right| &\le \frac{3a\pi}{4} \int_0^t |T\psi(s) - T\phi(s)|\, \cos(\pi s/2)\, ds \\
&\le \frac{3a\pi}{4} d(\psi, \phi)\frac{3a}{2} \int_0^t \sin(\pi s/2)\cos(\pi s/2)\, ds \\
&\le \left(\frac{3a}{4}\right)^2 d(\psi, \phi)\,(1 - \cos(\pi t)),
\end{aligned}$$

and $d(T^2\psi, T^2\phi) \le 2\left(\frac{3a}{4}\right)^2 d(\psi, \phi)$, so that T^2 is a contraction if $0 \le a < 4/(3\sqrt{2})$. Apply Corollary 5.10.

Sol. 6.7:   "$\mathbb{Q}$ is not a G_δ." If $\mathbb{Q}$ was a G_δ, it would be a countable intersection of open dense sets in $\mathbb{R}$. Thus $\mathbb{R} \setminus \mathbb{Q}$ would be a countable union of closed sets with empty interior, but this would imply that $\mathbb{R} = \cup_{t \in \mathbb{Q}}\{t\} \cup (\mathbb{R} \setminus \mathbb{Q})$ would also be a countable union of closed set with empty interior, a contradiction since $\mathbb{R}$ is a Baire space. Therefore $\mathbb{Q}$ is not a G_δ.

"The set of points of continuity of $f : \mathbb{R} \to \mathbb{R}$ is a G_δ." Indeed, for each $n \in \mathbb{N}$, let A_n be the set of points $t \in \mathbb{R}$ for which there is an open set V_t, with $t \in V_t$, so that $|f(s) - f(r)| < 1/n$ for all $s, r \in V_t$. So A_n is open, since if $t \in A_n$, then $V_t \subset A_n$. Now, from the definition of continuity, it follows that $\cap_n A_n$ is the set of points of continuity of f; hence such set is a G_δ.

Sol. 6.12: Since X is separable, there exists a sequence $(\xi_n)_{n\in\mathbb{N}}$ dense in X. Thus, the set of points whose orbit is dense in X is characterized by

$$D = \bigcap_{n,k} \bigcup_m h^m(B(\xi_n, 1/k)), \qquad m \in \mathbb{Z}, \ n, k \in \mathbb{N},$$

hence it is G_δ, since h is a homeomorphism. If h is transitive, there exists $\xi \in X$ with $\mathcal{O}(\xi) = \{h^m(\xi) : m \in \mathbb{Z}\}$ dense in X; by noting that, for each $m \in \mathbb{Z}$ the orbit $\mathcal{O}(h^m(\xi))$ is also dense, it then follows that in this case D is a G_δ dense subset of X; therefore its complement, that is, the set of points whose orbits are not dense, is a meager F_σ.

Sol. 8.9: Suppose $0 \neq f \in \mathcal{N}^*$ (and $\mathcal{N} \neq \{0\}$); then the answer to the question is affirmative. Indeed:

If $\emptyset \neq A \subset \mathcal{N}$ is open, consider $a \in f(A)$; hence there exists $\xi \in A$ with $a = f(\xi)$. It will be shown that $a \in \operatorname{int} f(A)$ (the interior of $f(A)$). Pick $\eta \in \mathcal{N}$ with $f(\eta) = 1$ (this is possible since f is surjective). By continuity of the scalar multiplication, there exists $\varepsilon > 0$ so that $(\xi + t\eta) \in A$, for all $|t| < \varepsilon$. Thus, $f(\xi + t\eta) \in f(A)$, for all $|t| < \varepsilon$, and since $f(\xi + t\eta) = f(\xi) + t\, f(\eta) = a + t$, it follows that $a \in \operatorname{int} f(A)$.

Tow remarks are in order. First, this solution does not explicitly use the continuity of the linear functional f and second, if $\mathcal{N}$ is complete the result is a consequence of Open Mapping Theorem ($f \neq 0$).

Sol. 8.14: $\mathcal{F}$ is linear and since for all n one has $(\mathcal{F}\psi)_n \leq \|\psi\|_1$, then $\|\mathcal{F}\psi\|_\infty \leq \|\psi\|_1$, so that $\mathcal{F} \in B(\mathcal{B}, l^\infty)$; by taking into account Riemann-Lebesgue, one has $\mathcal{F} \in B(\mathcal{B}, c_0)$. The point in the exercise is to verify that $\mathcal{F}$ is not surjective. Suppose, then, that this operator is surjective; by Open Mapping its inverse $\mathcal{F}^{-1}$ would be bounded. If $\phi^m = \sum_{n=-m}^m e^{int}$ (which belongs to $\mathcal{B}$), then $(\mathcal{F}\phi^m)_n = 1$ if $|n| \leq m$ and zero otherwise, so that $\|\mathcal{F}\phi^m\|_\infty = 1$ for all m. However, in the following it will be argued that $\|\phi^m\|_1 \to \infty$ as $m \to \infty$, showing that $\mathcal{F}^{-1}$ is not bounded, hence $\mathcal{F}$ cannot be surjective.

To get an adequate estimate for $\|\phi^m\|_1$, note that

$$\phi^m(t) = \frac{\sin((m+1/2)t)}{\sin(t/2)}, \qquad t \neq 0,$$

and $\phi^m(0) = 2m + 1$. Thus, recalling that $|\sin(t)| \leq |t|$, an estimate similar to the one in the proof of Corollary 7.6 implies that

$$\|\phi^m\|_1 \geq \frac{8}{\pi} \sum_{n=1}^m \frac{1}{n} \to \infty \quad \text{as} \quad m \to \infty.$$

Sol. 9.13: $T \in B(\mathcal{B}_1, \mathcal{B}_2)$ has closed graph. Since it is invertible $T^{-1} : \mathcal{B}_2 \to \mathcal{B}_1$, then $\mathcal{G}(T^{-1}) = H\mathcal{G}(T)$, with $H : \mathcal{B}_1 \times \mathcal{B}_2 \to \mathcal{B}_2 \times \mathcal{B}_1$ being the homeomorphism (isometric) $H(\xi_1, \xi_2) = (\xi_2, \xi_1)$; therefore $\mathcal{G}(T^{-1})$ is also closed. By the Closed Graph Theorem, T^{-1} is bounded.

Sol. 9.14:   For each $n \in \mathbb{N}$, define $\psi_n(t) = n/(n+t^s)$; then $\|\psi_n\| = 1$, and for all $t \geq 0$, $|\psi_n(t)| \leq n/(1+t^s)$, that is, $\psi_n \in \text{dom } T$. Since $\|(T\psi_n)(t)\| = \|t^s n/(n+t^s)\|_\infty = n$, it is found that T is not bounded.

To see that T is closed, let $\psi_n \to \psi$ and $T\psi_n = t^s \psi_n \to \phi$, both with uniform convergence. Since $1 + t^s \geq 1$, then for all $t \geq 0$,

$$\left| \psi_n(t) - \frac{\psi(t) + \phi(t)}{1 + t^s} \right| \leq |(1 + t^s)\psi_n(t) - (\psi(t) + \phi(t))|,$$

and since the right-hand side of this inequality vanishes uniformly as $n \to \infty$,

$$\psi(t) = \lim_{n \to \infty} \psi_n(t) = \frac{\psi(t) + \phi(t)}{1 + t^s}, \quad \text{in } C[0, \infty).$$

Therefore $|\psi(t)| \leq (\|\psi\| + \|\phi\|)/(1 + t^s)$, that is, $\psi \in \text{dom } T$, and $\phi(t) = t^s \psi(t) = (T\psi)(t)$, and so T is closed.

Sol. 9.15:   By the Closed Graph Theorem, it is enough to show that T is closed in order to conclude that it is bounded, which, by its turn, is reduced to show that if $\xi^j \to 0$ and $T\xi^j \to \eta$ in $\mathcal{B}$ (as $j \to \infty$), then $\eta = 0$. Let $\{e_j\}$ be the canonical basis of $\mathcal{B} = l^p(\mathbb{N})$. For each pair $j, k \in \mathbb{N}$, define the linear functional $f^{j,k} : \mathcal{B} \to \mathbb{F}$, $f^{j,k}(\xi) = \xi_j (Te_j)_k$, which is bounded since $|f^{j,k}(\xi)| \leq \|Te_j\| \|\xi\|$.

Suppose that $\xi^j = (\xi_1^j, \xi_2^j, \cdots) \to 0$ and $T\xi^j \to \eta$ in $\mathcal{B}$. For each pair j, k denote by $\zeta^{j,k} = (\xi_{k+1}^j, \xi_{k+2}^j, \cdots)$; thus

$$\xi^j = \sum_{m=1}^{k} \xi_m^j e_m + S_r^k(\zeta^{j,k}),$$

and since T and S_r commute, then

$$[T(S_r^k(\zeta^{j,k}))]_k = [S_r^k(T(\zeta^{j,k}))]_k = 0.$$

Therefore,

$$(T\xi^j)_k = \left[\sum_{m=1}^{k} \xi_m^j T(e_m) \right]_k = \sum_{m=1}^{k} f^{m,k}(\xi^j).$$

As such functionals are continuous and $\xi^j \to 0$, then each component

$$\eta_k = \lim_{j \to \infty} (T\xi^j)_k = \sum_{m=1}^{k} f^{m,k}(0) = 0,$$

so that $\eta = 0$ and T is closed.

Sol. 12.10:   Set $F = \bigcap\{N(f) : f \in \mathcal{N}^* \text{ and } E \subset N(f)\}$. Since each f is continuous, then F is closed, and by construction, $E \subset F \Rightarrow \overline{E} \subset F$. If $\xi_0 \notin \overline{E}$, then by Hahn-Banach, there exists $f \in \mathcal{N}^*$, $\|f\| = 1$, and $f(\xi_0) = d(\xi_0, \overline{E}) > 0$, that is, $\xi_0 \notin F$. Therefore $F \subset \overline{E}$ and $\overline{E} = F$.

Sol. 12.11: **(a)** Informally: if $\mathcal{B} = \mathcal{B}^{**}$, then $\mathcal{B}^* = \mathcal{B}^{***}$.

In details: It is convenient to distinguish the notation of the two canonical mappings $\hat{\ }: \mathcal{B} \to \mathcal{B}^{**}$ and $\check{\ }: \mathcal{B}^* \to \mathcal{B}^{***}$. Since $\mathcal{B}$ is reflexive, for each $g \in \mathcal{B}^{**}$ there exists $\xi_g \in \mathcal{B}$ with $\hat{\xi}_g = g$. Thus, if $h \in \mathcal{B}^{***}$, for all $g \in \mathcal{B}^{**}$ one has

$$h(g) = h(\hat{\xi}_g) = (h \circ \hat{\ })(\xi_g) = u(\xi_g),$$

with $u = (h \circ \hat{\ }) \in \mathcal{B}^*$. Hence

$$h(g) = u(\xi_g) = \hat{\xi}_g(u) = g(u) = \check{u}(g),$$

so that $h = \check{u}$ and $\check{\ }$ is surjective, that is, $\mathcal{B}^*$ is reflexive.

(b) The same notations for the canonical mappings as in part **(a)** will be used. If $\mathcal{B}$ is reflexive, then, by **(a)**, it follows that $\mathcal{B}^*$ is reflexive.

Suppose now that $\mathcal{B}^*$ is reflexive. If $\mathcal{B}$ is not reflexive, there exists $g \in \mathcal{B}^{**} \setminus \hat{\mathcal{B}}$. Then, since $\mathcal{B}$ is complete, $\hat{\mathcal{B}}$ is a proper closed subspace of $\mathcal{B}^{**}$, and there exists $h \in \mathcal{B}^{***}$ with $h(g) \neq 0$ and $h|_{\hat{\mathcal{B}}} = 0$ (Proposition 12.2). Since $\mathcal{B}^*$ is reflexive, there exists $f \in \mathcal{B}^*$ with $h = \check{f}$; thus

$$0 \neq h(g) = \check{f}(g) = g(f) \Rightarrow f \neq 0.$$

On the other hand, for all $\xi \in \mathcal{B}$,

$$0 = h(\hat{\xi}) = \check{f}(\hat{\xi}) = \hat{\xi}(f) = f(\xi) \Rightarrow f = 0.$$

This contradiction shows that $\mathcal{B}$ is reflexive.

Sol. 13.2: $g \in N(T^a) \Leftrightarrow T^a(g) = 0 \Leftrightarrow (T^a g)(\xi) = 0, \forall \xi \in \mathcal{N}_1 \Leftrightarrow g(T\xi) = 0, \forall \xi \in \mathcal{N}_1$. Since T^a is linear, it is injective if and only if $N(T^a) = \{0\}$, and since g is a continuous functional, one has

$$[g(T\xi) = 0, \forall \xi \in \mathcal{N}_1 \Rightarrow g = 0] \iff [\text{rng } T \text{ is dense in } \mathcal{N}_2],$$

that is, T^a is injective if and only if rng T is dense in $\mathcal{N}_2$.

Sol. 13.5: The fact that $N(T^a) = (\text{rng } T)^0$ is in the solution to Exercise 13.2. For the other relation: $\xi \in N(T) \Leftrightarrow T(\xi) = 0 \Leftrightarrow g(T\xi) = 0, \forall g \in \mathcal{N}_2^* \Leftrightarrow (T^a g)(\xi) = 0, \forall g \in \mathcal{N}_2^* \Leftrightarrow \xi \in (\text{rng } T^a)^\dagger$, that is, $N(T) = (\text{rng } T^a)^\dagger$.

Sol. 13.8: Write $c_0 \ni \xi = \sum_{j\geq 1} \xi_j e_j$. If $f \in c_0^*$, then $f(\xi) = \sum_{j\geq 1} \xi_j \alpha_j$, with $\alpha_j = f(e_j)$. By picking $\xi^n \in c_0$ with entries $\xi_j^n = \overline{\alpha_j}/|\alpha_j|$ (or zero if $\alpha_j = 0$) if $1 \leq j \leq n$ and zero if $j > n$, it follows that $\|\xi^n\| \leq 1$ and $\|f\| \geq |f(\xi^n)| = \sum_{j=1}^n |\alpha_j|$, and since this holds for all n one has $\alpha = (\alpha_1, \alpha_2, \alpha_3, \cdots) \in l^1(\mathbb{N})$ and $\|\alpha\|_1 \leq \|f\|$.

Note that the above linear mapping $c_0^* \ni f \mapsto \alpha \in l^1(\mathbb{N})$ is injective, since $\alpha = 0$ if and only if $f = 0$. Now it will be shown that such mapping is onto and isometric. Indeed, if $\alpha \in l^1$ then define $f : c_0 \to \mathbb{F}$ by $f(\xi) = \sum_{j \geq 1} \xi_j \alpha_j$, hence

$$|f(\xi)| \leq \|\xi\|_\infty \|\alpha\|_1, \quad \forall \xi \in c_0,$$

so that $f \in c_0^*$ and $\|f\| \leq \|\alpha\|_1$. It also follows that the above mapping satisfies $\|f\| = \|\alpha\|_1$.

Sol. 14.9: Since $\delta_t : C[a, b] \to \mathbb{F}$, given by $\delta_t(\phi) = \phi(t)$, $\phi \in C[a, b]$, is an element of $C[a, b]^*$ for all $t \in [a, b]$, it follows that $\psi_n(t) = \delta_t(\psi_n) \to \delta_t(\psi) = \psi(t)$.

Sol. 14.10: By Riesz-Markov, each $f \in C[-1, 1]^*$ is represented by a finite measure μ (Borelian). Thus, for the given sequence (ψ_n) in $C[-1, 1]$, one has $f(\psi_n) = \int_{[-1,1]} \psi_n \, d\mu \to \psi(0)\mu(\{0\})$ by the Dominated Convergence Theorem, hence $f(\psi_n)$ is convergent for all $f \in C[-1, 1]^*$. By Exercise 14.9, if ψ_n converges weakly to some φ, then it converges pointwise to φ, but φ cannot be represented by a continuous function.

Sol. 14.13: The statement of the exercise is equivalent to $\xi \in E = \overline{\text{Lin}((\xi_n))}$. If $\xi \notin E$, then $\xi \neq 0$, and by Hahn-Banach corollaries, there exists $f \in \mathcal{N}^*$ with $f(\xi) \neq 0$ and $f|_E = 0$, which is a contradiction with $f(\xi_n) \to f(\xi)$. Therefore, $\xi \in E$.

Sol. 15.6: It will be discussed only the case of weak topology; the other case is similar. Since $\dim \mathcal{N} = \infty$, note that $N(f) \neq \{0\}$, $\forall f \in \mathcal{N}^*$. Let $\{f_1, \cdots, f_n\}$ in $\mathcal{N}^*$ be linearly independent (it is sufficient to deal with this case), and consider the system of equations $f_j(\xi) = 0$, $1 \leq j \leq n$, whose solution is the vector space $U = \cap_{j=1}^n N(f_j)$; $U \neq \{0\}$, since if not $\{f_1, \cdots, f_n\}$ would be a basis of $\mathcal{N}^*$, since for any $f \in \mathcal{N}^*$ one would get $U \subset N(f)$, and, by Proposition 15.9, this functional would be linearly generated by $\{f_1, \cdots, f_n\}$. This concludes the first part of the exercise.

A general set of an open basis of $\tau(\mathcal{N}, \mathcal{N}^*)$ has the form $V(\xi; f_1, \cdots, f_n; \varepsilon) = \{\eta \in \mathcal{N} : \max_{1 \leq j \leq n} |f_j(\xi) - f_j(\eta)| < \varepsilon\}$, which contains $\xi + U$, with $U = \cap_{j=1}^n N(f_j)$, since for all $\zeta \in U$

$$\max_{1 \leq j \leq n} |f_j(\xi) - f_j(\xi + \zeta)| = 0 < \varepsilon.$$

Thus, every nonempty open set of the weak topology ($\dim \mathcal{N} = \infty$) contains elements of arbitrarily large norm.

The inverse image, under the norm mapping, of the open set $(-\infty, 1)$ in $\mathbb{R}$ is $B_\mathcal{N}(0; 1)$; but since this set is not bounded, it is not an element of $\tau(\mathcal{N}, \mathcal{N}^*)$, and it follows that the norm is not a continuous mapping in the weak topology.

Sol. 15.7: If $\tau(\mathcal{N}, \mathcal{N}^*)$ was generated by a norm $\|| \cdot \||$, then the open ball under this norm $B_{\||\cdot\||}(0; 1)$ would be an open set containing ($\dim \mathcal{N} = \infty$) a nontrivial vector subspace U (see the solution to Exercise 15.6); but this is impossible, since if $\xi \in B_{\|\|\cdot\|\|}(0; 1) \cap U$, with $\|\|\xi\|\| = 1/2$, then $4\xi \in B_{\|\|\cdot\|\|}(0; 1)$, which may not hold since $\|\|4\xi\|\| = 2$.

Sol. 15.9: The sets of an open basis of $\tau(X, Y)$ can be chosen as

$$U(\eta; f_1, \cdots, f_n; \varepsilon) = \left\{ \xi \in X : \max_{1 \leq j \leq n} |f_j(\xi) - f_j(\eta)| < \varepsilon \right\},$$

with $\{f_1, \cdots, f_n\} \subset Y$.

(a) If $f \in Z^*$, then $\{\xi \in X : |f(\xi)| < 1\} = |f|^{-1}(-\infty, 1)$ is open and contains the null vector, hence it contains $U := U(0; f_1, \cdots, f_n; \varepsilon)$ for some $\varepsilon > 0$ and some set $\{f_1, \cdots, f_n\} \subset Y$. If $\xi \in X$, denote $M = M(\xi) = \max_{1 \leq j \leq n} |f_j(\xi)|$; if $M \neq 0$ one has that

$$\frac{\varepsilon \xi}{2M} \in U \Rightarrow \left| f\left(\frac{\varepsilon \xi}{2M}\right) \right| < 1 \Rightarrow |f(\xi)| < \frac{2}{\varepsilon} M.$$

If $M(\xi) = 0$, then for all $t > 0$, $f_j(t\xi) = 0$, for all j; hence $|f(t\xi)| < 1$ and so $|f(\xi)| < 1/t$. By taking $t \to \infty$, it is found that $f(\xi) = 0$ and the above inequality also holds.

(b) If $f \in Z^*$, then (a) implies that $\cap_{j=1}^n N(f_j) \subset N(f)$ for some $\{f_1, \cdots, f_n\} \subset Y$ and, by Proposition 15.9, there are scalars $a_j, 1 \leq j \leq n$, so that $f = \sum_{j=1}^n a_j f_j$, and so $f \in Y$.

Sol. 16.1: Let $(f_j) \subset S$. The metric convergence $f_j \to f$ is equivalent to $\hat{\xi}_n(f_j) \to \hat{\xi}_n(f)$ for a dense set $(\hat{\xi}_n)$ in $\mathcal{N}^*$ (see the proof of Proposition 16.2), and since S is compact, it is also bounded in this metric; now apply Proposition 15.3.

Sol. 16.2: Since (f_n) is bounded, there is $r > 0$ so that this convergence occurs in the compact set $\overline{B}_{\mathcal{N}^*}(0; r)$ in the topology $\tau(\mathcal{N}^*, \hat{\mathcal{N}})$. Since $\mathcal{N}$ is separable, the corresponding induced topology in this ball is metrizable (Proposition 16.3), and hence, sequencially compact. By Exercise 16.1, this metric convergence is equivalent to the weak* convergence of sequences.

Sol. 16.3: Since $\mathcal{B}$ is reflexive, one has $\mathcal{B}^{**} = \hat{\mathcal{B}}$; since $\mathcal{B}$ is separable, it follows that $\hat{\mathcal{B}}$ is separable, so $\mathcal{B}^{**}$ is separable. By Proposition 12.4, $\mathcal{B}^*$ is separable.

Sol. 16.7: $\delta_n \in l^2(\mathbb{N})^* = l^2(\mathbb{N})$, $\delta_n(\xi_1, \xi_2, \cdots) = \xi_n$. For each $\xi \in l^2(\mathbb{N})$ one has $\hat{\xi}(\delta_n) = \delta_n(\xi) = \xi_n$ which vanishes as $n \to \infty$, hence $\delta_n \xrightarrow{w^*} 0$. On the other hand, since δ_n is "represented" by e_n (element of the canonical basis) in $l^2(\mathbb{N})$, one has $\|\delta_n - \delta_k\|_2 = \sqrt{2}$, for all $n \neq k$, and so (δ_n) has no Cauchy subsequence in $l^2(\mathbb{N})$.

Sol. 17.16: If $S(\xi) = S(\eta)$, then $0 = \langle S(\xi) - S(\eta), S(\xi) - S(\eta) \rangle = \langle S(\xi), S(\xi) \rangle - \langle S(\xi), S(\eta) \rangle - \langle S(\eta), S(\xi) \rangle + \langle S(\eta), S(\eta) \rangle = \langle \xi, \xi \rangle - \langle \xi, \eta \rangle - \langle \eta, \xi \rangle + \langle \eta, \eta \rangle = \|\xi - \eta\|^2$, therefore S is injective and $S^{-1} : \mathcal{H} \to \mathcal{H}$ exists.

If $S^{-1}(\xi_1) = \xi$ and $S^{-1}(\eta_1) = \eta$, since $\langle S(\xi), S(\eta) \rangle = \langle \xi, \eta \rangle$ then $\langle \xi_1, \eta_1 \rangle = \langle S^{-1}(\xi_1), S^{-1}(\eta_1) \rangle$; since S is bijective, such relation holds for every vector in the space. In this relation, if $\xi_1 = S(\xi_2)$, then $\langle S(\xi_2), \eta_1 \rangle = \langle \xi_2, S^{-1}(\eta_1) \rangle$, again for all vectors of $\mathcal{H}$.

Now, for all $\eta, \xi, \zeta \in \mathcal{H}$ and $a, b \in \mathbb{F}$, one has $\langle S(a\xi+b\eta), \zeta \rangle = \langle a\xi+b\eta, S^{-1}(\zeta) \rangle = \bar{a}\langle \xi, S^{-1}(\zeta) \rangle + \bar{b}\langle \eta, S^{-1}(\zeta) \rangle = \bar{a}\langle S(\xi), \zeta \rangle + \bar{b}\langle S(\eta), \zeta \rangle = \langle aS(\xi), \zeta \rangle + \langle bS(\eta), \zeta \rangle = \langle aS(\xi) + bS(\eta), \zeta \rangle$, showing that $S(a\xi + b\eta) = aS(\xi) + bS(\eta)$, that is, S is linear.

Sol. 18.14: Note that since $\|S(\xi) - S(\eta)\| = \|\xi - \eta\|$, by picking $\eta = 0$ one gets $\|S(\xi) - S(0)\| = \|\xi\|$, and so $V(\xi) = S(\xi) - S(0)$ is an isometry on $\mathcal{H}$. By polarization (real)

$$\langle V(\xi), V(\eta) \rangle = \frac{1}{4} \left(\|V(\xi) + V(\eta)\|^2 - \|V(\xi) - V(\eta)\|^2 \right);$$

writing out this expression and using that V is an isometry, it is found that $\langle V(\xi), V(\eta) \rangle = \langle \xi, \eta \rangle$. As in the solution to Exercise 17.16, it follows that V is linear.

Sol. 18.15: If E is invariant under T, then for all $\xi \in \mathcal{H}$, $TP_E\xi \in E$, and so $P_E T P_E \xi = T P_E \xi$ and $T P_E = P_E T P_E$. Now, if such relation holds, then for $\eta \in E$, $T\eta = T P_E \eta = P_E T P_E \eta$, which belongs to E. Therefore, $T(E) \subset E$.

Sol. 19.14: If $T_b\eta = 0$ one has $0 = \|T_b\eta\| \geq c\|\eta\|$ and so $\eta = 0$; thus T_b is invertible. If $\eta = T_b^{-1}\xi$, then $c\|T_b^{-1}\xi\| \leq \|\xi\|$ and so T_b^{-1} is bounded. Now if $\zeta \in \overline{\text{rng } T_b}$, there is $T_b\eta_n \to \zeta$, and, by the inequality in the hypothesis, it follows that η_n is Cauchy (since $T_b\eta_n$ is Cauchy), so convergent $\eta_n \to \xi$. Since T is continuous, $T\xi = \zeta$ and $\zeta \in \text{rng } T_b$; hence rng T_b is closed.

Sol. 20.4: For $T \in B(\mathcal{H})$ define $T_R = (T + T^*)/2$ and $T_I = (T - T^*)/(2i)$; it is immediate that $T = T_R + iT_I$ and $T^* = T_R - iT_I$. If T_R commutes with T_I, then T commutes with its adjoint, hence T is normal. Now, if T commutes with T^*, then from this decomposition it is found that

$$-i(T_R T_I - T_I T_R) = i(T_R T_I - T_I T_R),$$

and so $(T_R T_I - T_I T_R) = 0$. Using such relations in $TT^* = \mathbf{1} = T^*T$ and equating the respective real and imaginary parts, the characterization of unitary operators is found.

Sol. 20.9: **(a)** Combine the relation $\|T\xi\|^2 - \|T^*\xi\|^2 = \langle T\xi, T\xi \rangle - \langle T^*\xi, T^*\xi \rangle = \langle \xi, (T^*T - TT^*)\xi \rangle$ with Proposition 19.11.

(d) From $(\mathbf{1} - T)\xi = \xi - T\xi$, one has $\xi \in N(\mathbf{1} - T)$ if and only if $T\xi = \xi$; thus, rng $T \supset N(\mathbf{1} - T)$. By using that $T^2 = T$, if $\xi \in \text{rng } T$, then $\xi = T\eta$ and $T\xi = T^2\eta = T\eta = \xi$, showing that, indeed, rng $T = N(\mathbf{1} - T)$, hence it is a closed set. Since $(\mathbf{1} - T)^2 = \mathbf{1} - T$, in a similar way one concludes that rng $(\mathbf{1} - T) = N(T)$.

Since T is normal, $\|T\xi\| = \|T^*\xi\|$ and $N(T) = N(T^*)$. Combining such relations with $N(T^*) = (\text{rng } T)^\perp$, one gets rng $(\mathbf{1} - T) = (\text{rng } T)^\perp$ and rng $T = (\text{rng } (\mathbf{1} - T))^\perp$.

Thus,

$$\begin{aligned} \langle T\xi, \eta \rangle &= \langle T\xi, T\eta + (1-T)\eta \rangle = \langle T\xi, T\eta \rangle \\ &= \langle \xi - (1-T)\xi, T\eta \rangle = \langle \xi, T\eta \rangle, \end{aligned}$$

and T is self-adjoint, hence an orthogonal projection operator by Theorem 20.11.

Sol. 21.2: From the definition of orthonormal basis, it is enough to combine Proposition 3.4 (that is, the space $\mathcal{H}$ is separable if and only if it has a total [linearly independent] countable set) with the Gram-Schmidt process.

Sol. 21.5: **a)** Let $(\xi_j)_j$ be an orthonormal sequence in $\mathcal{H}$; then $\|\xi_j - \xi_k\|^2 = 2$ if $j \neq k$. Hence, it has no Cauchy subsequence. By Bessel Inequality, for all $\xi \in \mathcal{H}$ the series $\sum_j |\langle \xi_j, \xi \rangle|^2$ is convergent, so $\langle \xi_j, \xi \rangle \to 0$ if $j \to \infty$; it follows that $\xi_j \xrightarrow{w} 0$.

Sol. 21.10: Let $f \in \mathcal{H}^*$, $\{\xi_\alpha\}_{\alpha \in J}$ be an orthonormal basis of $\mathcal{H}$ and $a_\alpha = f(\xi_\alpha)$. If J_f denotes the set of indices $\alpha \in J$ so that $a_\alpha \neq 0$, it will initially be shown that J_f is countable. Consider a finite set $\xi_1, \cdots, \xi_n$ and $\delta^2 = \sum_{j=1}^n |a_j|^2$; if $\delta > 0$, introduce $\xi := 1/\delta \sum_{j=1}^n \overline{a_j} \xi_j$ and, by Pythagoras, $\|\xi\| = 1$ and $f(\xi) = \delta$; hence $\|f\| \geq |f(\xi)| = \delta$ and at most $n-1$ of such ξ_α's can satisfy $|a_\alpha| > \|f\|/\sqrt{n}$. It then follows that J_f is countable.

Hence one may restrict the index j to J_f and, for simplicity, assume that $J_f = \mathbb{N}$ (if J_f is finite, the argument below is easily adapted); since $\sum_{j=1}^n |a_j|^2 \leq \|f\|^2$ for all n, it follows that $\sum_{j=1}^n \overline{a_j} \xi_j$ is a Cauchy sequence and so it converges to $\eta = \sum_j^\infty \overline{a_j} \xi_j$. Now, for $\xi = \sum_\alpha \langle \xi_\alpha, \xi \rangle \xi_\alpha$ (sum over a countable set), it follows that $f(\xi) = \langle \eta, \xi \rangle$, that is, $f = f_\eta$. Therefore γ is onto.

Sol. 22.3: Since $t \mapsto \langle \xi, \psi(t) \rangle$ is measurable for all $\xi \in \mathcal{H}$, then if (ψ_j) is a (countable) orthonormal basis of $\mathcal{H}$, by Parseval, one has

$$\|\psi(t)\|^2 = \sum_j |\langle \xi_j, \psi(t) \rangle|^2,$$

and so $t \mapsto \|\psi(t)\|^2$ is limit of measurable mappings, hence it is also mensurable. By polarization, it follows that $t \mapsto \langle \phi(t), \psi(t) \rangle$ is mensurable.

Sol. 22.9: Set $E = \{\xi \in \mathcal{H} : U\xi = \xi\}$ and $F = \{\xi \in \mathcal{H} : U^*\xi = \xi\}$; it is easy to check that both sets are closed vector subspaces. If $\xi \in E$, then $U\xi = \xi$, $\xi = U^*U\xi = U^*\xi$, and $E \subset F$. Similarly one gets $F \subset E$; hence, $E = F$.

Since $E = N(1-U)$ and $(1-U)$ is normal, then by Exercise 20.9, one finds that $E = (\text{rng }(1-U))^\perp$; thus, $\mathcal{H} = E \oplus \overline{\text{rng }(1-U)}$.

If $\xi \in E$, then $U^j\xi = \xi$, for all $j \in \mathbb{Z}$, and $T_n\xi \to \xi$. If $\xi \in \text{rng }(1-U)$, then $\xi = (1-U)\eta$ and $U^j\xi = (U^j - U^{j+1})\eta$, so that

$$\frac{1}{n+1} \sum_{j=0}^n U^j\xi = \frac{1}{n+1}(\eta - U^{n+1}\eta) \to 0, \quad n \to \infty.$$

Now, if $\xi \in E^\perp = \overline{\text{rng} \, (1-U)}$, it is possible to approximate this vector by a sequence $\xi_k = (1-U)\eta_k$ in rng $(1-U)$, and by the triangle inequality, the limit above also vanishes for such vectors. Since any vector $\xi \in \mathcal{H}$ can be decomposed in the form $\xi = P_E\xi + P_{E^\perp}\xi$, the result follows.

Sol. 22.13: (a) Since $T(t)$ is continuous, then the mapping $t \mapsto \langle \eta, T(t)\xi \rangle$ is mensurable.

(b) Since for each $\xi \in \mathcal{H}$, $T(t)\xi$ converges as $t \to \infty$, then by Banach-Steinhaus, $M = \sup_{t \in [0,\infty)} \|T(t)\| < \infty$; thus,

$$\int_0^\infty \|e^{-\delta t}T(t)\| \, \mathrm{d}t \leq M \int_0^\infty e^{-\delta t} \, \mathrm{d}t = \frac{M}{\delta},$$

and $\int_0^\infty e^{-\delta t}T(t)\xi \, \mathrm{d}t$ is well defined (Proposition 22.8).

Now, for each $\varepsilon > 0$ there exists $s \in \mathbb{R}$ with $\|(S - T(t))\xi\| < \varepsilon$ if $t > s$. Thus, the integral $\delta \int_0^\infty e^{-\delta t}T(t)\xi \, \mathrm{d}t$ is equal to

$$\delta \int_0^s e^{-\delta t}T(t)\xi \, \mathrm{d}t + \delta \int_s^\infty e^{-\delta t}(T(t) - S)\xi \, \mathrm{d}t + \delta \int_s^\infty e^{-\delta t}S\xi \, \mathrm{d}t.$$

As $\delta \to 0^+$, the first integral vanishes. The second one is

$$\leq \delta \int_s^\infty e^{-\delta t}\|(T(t) - S)\xi\| \, \mathrm{d}t \; \leq \; \varepsilon\delta \int_s^\infty e^{-\delta t} \, \mathrm{d}t \leq \varepsilon;$$

whereas the third one

$$= \left(\delta \int_s^\infty e^{-\delta t} \, \mathrm{d}t \right) S\xi = e^{-\delta t}S\xi,$$

and so it converges to $S\xi$ as $\delta \to 0^+$. Since this holds for all $\varepsilon > 0$, the proposed result follows.

Sol. 24.3: It is a straightforward consequence of Theorem 2.2.

Sol. 24.6: (a) Since $\mathcal{N}_1 = \bigcup_{j=1}^\infty B(0; j)$, then for $T : \mathcal{N}_1 \to \mathcal{N}_2$, one has rng $T = \bigcup_{j=1}^\infty T(B(0; j))$. In order to conclude the exercise, it is sufficient to show that for each $j \in \mathbb{N}$ the set $TB(0; j)$ has a countable dense subset. If T is compact, $TB(0; j)$ is totally bounded; thus, for each $m \in \mathbb{N}$, it can be covered by a finite number of open balls of radii $1/m$, centered at points of $TB(0; j)$. The union of the centers of such open balls, for all $m \in \mathbb{N}$, is a dense countable subset of $TB(0; j)$.

Sol. 24.12: Note that, in the proof of Proposition 12.6, the operators T_j have rank one, so they are compact. Hence, that proof only makes use of the fact that $B(\mathcal{N}_1, \mathcal{N}_2)$ is complete to conclude that $\mathcal{N}_2$ is complete. Summing up, just replace $B(\mathcal{N}_1, \mathcal{N}_2)$ by $B_0(\mathcal{N}_1, \mathcal{N}_2)$ in that proof.

Sol. 24.13: Consider, to make the notation easier, $L^2[0, 2\pi]$; then the sequence $\psi_n(t) = e^{int}/\sqrt{2\pi}$, $n \in \mathbb{N}$, is orthonormal and converges weakly to zero. Since

$\|\mathcal{M}_\phi \psi_n\|_2 = \|\phi(t)\|_2 \neq 0$, for all n, $\mathcal{M}_\phi \psi_n$ does not converge to zero and this operator is not compact by Proposition 24.8.

Sol. 25.3: For $\eta \in \mathcal{E}$ one has $T\eta \in \mathcal{N}$ and $\widehat{T\eta} \in \mathcal{N}^{**}$. Thus, for all $g \in \mathcal{N}^*$, $(T^a g) \in \mathcal{E}^*$ and since

$$\widehat{T\eta}(g) = g(T\eta) = (T^a g)(\eta) = \breve{\eta}(T^a g) = (T^{aa}\breve{\eta})(g),$$

one concludes that $\widehat{T\eta} = T^{aa}\breve{\eta}$.

Sol. 25.4: Let T be a finite rank operator and $\{\xi_1, \cdots, \xi_n\}$ an orthonormal basis of rng T. Thus, for every $\xi \in \mathcal{H}$ one has $T\xi = \sum_{j=1}^n a_j \xi_j$ and so $a_j = \langle \xi_j, T\xi \rangle = \langle T^*\xi_j, \xi \rangle$; by denoting $\eta_j = T^*\xi_j$, the general form of finite rank operators is found

$$T\cdot = \sum_{j=1}^n \langle \eta_j, \cdot \rangle \xi_j.$$

From this expression it follows that $T^* \cdot = \sum_{j=1}^n \langle \xi_j, \cdot \rangle \eta_j$.

Suppose that $\{\eta_1, \cdots, \eta_n\}$ is linearly dependent; it is then possible to assume that $\eta_n = \sum_{j=1}^{n-1} c_j \eta_j$. Thus,

$$T\xi = \sum_{j=1}^n \langle \eta_j, \xi \rangle \xi_j = \sum_{j=1}^{n-1} \langle \eta_j, \xi \rangle (\xi_j + \bar{a}_j \xi_n),$$

and $\dim \operatorname{rng} T \leq (n-1)$; this contradiction with with $\dim \operatorname{rng} T = n$ shows that $\{\eta_1, \cdots, \eta_n\}$ is linearly independent and so $\dim \operatorname{rng} T^* = n$.

Sol. 25.8: If $T \in \mathrm{B}(\mathrm{C}(X))$, then for any $\{\xi_1, \cdots, \xi_n\} \subset X$ and $\{f_1, \cdots, f_n\} \subset \mathrm{C}(X)$, the operator $(T_n \psi) = \sum_{j=1}^n (T\psi)(\xi_j) f_j$, $\psi \in \mathrm{C}(X)$, belongs to $\mathrm{B_f}(\mathrm{C}(X))$.

If T is compact, then $T\overline{B}(0;1)$ is bounded and equicontinuous, thus for any $\varepsilon > 0$ and $\xi \in X$ there exists $B(\xi; r_\xi)$ so that, for all η in this ball, $|(T\psi)(\xi) - (T\psi)(\eta)| < \varepsilon$, for all $\psi \in \overline{B}(0;1)$. Since X is compact, it is covered by a finite number of such balls $X = \cup_{j=1}^n B(\xi_j; r_j)$. Taking $\{f_1, \cdots, f_n\}$ as the partition of unity associated with such balls (that is, $0 \leq f_j \leq 1$, $f_j = 0$ in the complement of $B(\xi_j; r_j)$, and for all $\xi \in X$, $\sum_{j=1}^n f_j(\xi) = 1$) one has, for $\psi \in \overline{B}(0;1)$,

$$\begin{aligned}
|(T_n \psi)(\xi) - (T\psi)(\xi)| &= \left| \sum_{j=1}^n (T\psi)(\xi_j) f_j(\xi) - (T\psi)(\xi) f_j(\xi) \right| \\
&\leq \sum_{j=1}^n |(T\psi)(\xi_j) - (T\psi)(\xi)| f_j(\xi).
\end{aligned}$$

Recall that if $f_j(\xi) \neq 0$, then necessarily $\xi \in B(\xi_j; r_j)$; then each term in the latter sum vanishes, with the possible exception of only one of them, which is less than ε;

therefore, $\|T_n\psi - T\psi\| < \varepsilon$, for all $\psi \in \overline{B}(0;1)$, hence $\|T_n - T\| \leq \varepsilon$ and $B_f(C(X))$ is dense in $B_0(C(X))$.

Sol. 27.6: By Theorem 27.6, if $\lambda_0 \in \rho(T)$ and $\|R_{\lambda_0}(T)\|\,|\lambda - \lambda_0| < 1$, then $\lambda \in \rho(T)$. Thus, if $\lambda \in \sigma(T)$, necessarily $\|R_{\lambda_0}(T)\|\,|\lambda - \lambda_0| \geq 1$, that is,

$$\|R_{\lambda_0}(T)\| \geq \frac{1}{|\lambda - \lambda_0|}, \quad \forall \lambda \in \sigma(T),$$

in particular, this completes the solution to the exercise (recall that $\sigma(T) \neq \emptyset$).

Sol. 28.1: Put $S = T_1 T_2 \cdots T_n$. If each T_j is invertible in $B(\mathcal{B})$, it is clear that $S^{-1} = T_n^{-1} \cdots T_1^{-1}$ exists and is bounded. Now suppose that $S^{-1} \in B(\mathcal{B})$; then the operators

$$S^{-1}T_2 T_3 \cdots T_n, \qquad T_2 T_3 \cdots T_n S^{-1}$$

are bounded and are the left inverse and right inverse of T_1, respectively (it is enough to perform the products with T_1 to check this). Similarly one works with T_j, $j \geq 2$. However, if an element u has a left inverse u_l and also a right inverse u_r, such elements coincide. Indeed, if one has $uu_r = \mathbf{1}$ and $u_l u = \mathbf{1}$, by properly multiplying such relations by u_l and u_r it is found that

$$u_l u u_r = u_r, \quad u_l u u_r = u_l,$$

and so $u_l = u_r$.

Sol. 28.14: Since $(T^n)^* = (T^*)^n$, it is clear that T^n is also normal if T is. In this case, since $\sigma(T^n) = \sigma(T)^n$,

$$\begin{aligned}\|T^n\| = r_\sigma(T^n) &= \sup\{|\lambda| : \lambda \in \sigma(T^n)\} = \sup\{|\lambda^n| : \lambda \in \sigma(T)\} \\ &= (\sup\{|\lambda| : \lambda \in \sigma(T)\})^n = r_\sigma(T)^n = \|T\|^n.\end{aligned}$$

Sol. 29.5: If there exists such (ξ_n), then λ cannot belong to $\rho(S)$, since one would get $1 = \|\xi_n\| = \|R_\lambda(S)S_\lambda\xi_n\| \leq \|R_\lambda(S)\|\,\|S_\lambda\xi_n\| \to 0$ as $n \to \infty$.

If λ is a boundary point of $\sigma(S)$ (then it belongs to the spectrum, since $\sigma(S)$ is closed), then there exists $(\lambda_n) \subset \rho(S)$, $\lambda_n \to \lambda$. By Corollary 27.10, $\|R_{\lambda_n}(S)\| \geq 1/d(\lambda_n, \sigma(S))$, and it is possible to pick a sequence $(\eta_n) \subset \mathcal{H}$, without null elements, so that

$$\left\|R_{\lambda_n}(S)\frac{\eta_n}{\|\eta_n\|}\right\| \to \infty, \qquad n \to \infty,$$

and simultaneously with $\xi_n = R_{\lambda_n}(S)\eta_n$ normalized. Thus, $1 = \|\xi_n\| = \|R_{\lambda_n}(S)\frac{\eta_n}{\|\eta_n\|}\|\,\|\eta_n\|$, and so $\|\eta_n\| \to 0$; hence,

$$\|S_\lambda\xi_n\| = \|\eta_n + (\lambda_n - \lambda)\xi_n\| \leq \|\eta_n\| + |\lambda_n - \lambda| \to 0, \quad n \to \infty.$$

Therefore, (ξ_n) is the searched sequence.

Sol. 29.10: **(d)** Let $S = \mathbf{1} + \lambda T^2$ ($\lambda \geq 0$). For all $\xi \in \mathcal{H}$ one has

$$\|S\xi\|^2 = \|\xi\|^2 + \lambda^2\|T^2\xi\|^2 + 2\lambda\langle T\xi, T\xi\rangle \geq \|\xi\|^2.$$

By Theorem 29.1, it is follows that $0 \notin \sigma(S)$, hence $(\mathbf{1}+\lambda T^2)^{-1} \in B(\mathcal{H})$ for all $\lambda \geq 0$.

(e) It is similar to part **(d)**, since the important fact was that T^2 is positive, which also holds for T^*T and TT^*.

Sol. 29.13: If there exists $\varepsilon_0 > 0$ with $\mu(\phi^{-1}B(\lambda;\varepsilon_0)) = 0$, then $Q = \mathcal{M}_{1/(\phi-\lambda)}$ satisfies [denoting $\Omega' = \Omega \setminus \phi^{-1}B(\lambda;\varepsilon_0)$]

$$\begin{aligned}
\|Q\psi\|_2^2 &= \int_\Omega \left|\frac{1}{\phi(t) - \lambda}\right|^2 |\psi(t)|^2 \mathrm{d}\mu(t) = \int_{\Omega'} \left|\frac{1}{\phi(t) - \lambda}\right|^2 |\psi(t)|^2 \, \mathrm{d}\mu(t) \\
&\leq \int_{\Omega'} \frac{1}{\varepsilon_0^2} |\psi(t)|^2 \, \mathrm{d}\mu(t) \leq \frac{1}{\varepsilon_0^2} \|\psi\|_2^2, \qquad \forall \psi \in \mathrm{L}_\mu^2(\Omega),
\end{aligned}$$

which implies that Q is bounded with $\|Q\| \leq 1/\varepsilon_0$, and since $Q\mathcal{M}_{(\phi-\lambda)} = \mathcal{M}_{(\phi-\lambda)}Q = \mathbf{1}$, then $\lambda \in \rho(\mathcal{M}_\phi)$.

Now, if $\lambda \in \rho(\mathcal{M}_\phi)$, there exists a (nonzero) bounded operator R so that $R\mathcal{M}_{(\phi-\lambda)} = \mathcal{M}_{(\phi-\lambda)}R = \mathbf{1}$; thus, for all $\psi \in \mathrm{L}_\mu^2(\Omega)$ one has $\|\psi\|^2 \leq \|R\|^2\|\mathcal{M}_{(\phi-\lambda)}\psi\|^2$, that is,

$$0 \leq \int_\Omega \left(|\phi(t) - \lambda|^2 - \frac{1}{\|R\|^2}\right) |\psi(t)|^2 \, \mathrm{d}\mu(t),$$

therefore $1/\|R\| \leq |\phi(t) - \lambda|$ μ-a.e.; in other words, this result says that $\mu(\phi^{-1}B(\lambda; 1/\|R\|)) = 0$.

Sol. 30.2: *(iv)* Let $T \in B_0(\mathcal{B})$; if zero does not belong to the spectrum of T, then $T^{-1} = R_0(T) \in B(\mathcal{B})$ and $\mathbf{1} = TT^{-1}$ is a compact operator; then $\overline{B}(0;1) = \mathbf{1}(\overline{B}(0;1))$ is a compact set, and therefore, $\dim \mathcal{B} < \infty$.

Sol. 30.22: Since S is self-adjoint, its spectrum is real; then it is possible to restrict the argument to $\lambda \in \mathbb{R}$. By picking a sequence $z_n \to \lambda$, with $\mathrm{Im}\, z_n \neq 0$, one has

$$\|AR_{z_n}(T)A^* - AR_\lambda(T)A^*\| \leq \|A\|^2 \|R_{z_n}(T) - R_\lambda(T)\| \to 0,$$

which vanishes by continuity of the resolvent. Hence, by Theorem 24.10, $AR_\lambda(T)A^*$ is compact for all $\lambda \in \rho(T)$; note that it is also self-adjoint for real λ.

By hypothesis, $(\mathbf{1}+AR_\lambda(T)A^*)$ has no inverse in $B(\mathcal{H})$, then $-1 \in \sigma(AR_\lambda(T)A^*)$ and by compactness of this operator, it is an eigenvalue. Thus, there exists $0 \neq \xi \in \mathcal{H}$ so that $-\xi = AR_\lambda(T)A^*\xi = A\eta$, with $0 \neq \eta = R_\lambda(T)A^*\xi$. Now,

$$S_\lambda\eta = T_\lambda\eta + A^*A\eta = T_\lambda R_\lambda(T)A^*\xi + A^*(-\xi) = 0,$$

which implies that η is an eigenvector of S associated with the eigenvalue λ.

Sol. 30.23: If $\lambda \in \sigma(T + A) \cap \rho(T)$, then from the relation

$$T + A - \lambda\mathbf{1} = (T - \lambda\mathbf{1})[\mathbf{1} + R_\lambda(T)A],$$

it follows that -1 is an eigenvalue of the compact self-adjoint operator $R_\lambda(T)A$ (since $R_\lambda(T) \in \mathrm{B}(\mathcal{H})$ and $\lambda \in \mathbb{R}$), and so the kernel of $T + A - \lambda\mathbf{1}$ is nontrivial, that is, λ is an eigenvalue of $T + A$. Note that the commutativity between T and A was only used to conclude that $R_\lambda(T)A$ is self-adjoint, but, in fact, this solution holds even without the commutative hypothesis (it is enough to know that, except zero, the spectrum of any compact operator is constituted only of eigenvalues).

Sol. 30.24: Denote by ξ_j the eigenvectors of $T = \sum_j \lambda_j P_j$, and $S = \sum_j \sqrt{\lambda_j}\, P_j$. Let Q be a positive compact operator so that $Q^2 = T$; the goal is to conclude that $Q = S$. By the relation $QT = QQ^2 = Q^2Q = TQ$ and Lemma 30.10, one has $Q\xi_j = q_j\xi_j$, for all indices j (q_j are the eigenvalues of Q), and this implies that $QS = SQ$. Since $(S - Q)$ is self-adjoint, both operators $R_1 = (S - Q)S(S - Q)$ and $R_2 = (S - Q)Q(S - Q)$ are positive, and

$$R_1 + R_2 = (S - Q)(S^2 - Q^2) = (S - Q)(T - T) = 0;$$

hence, by Proposition 19.11, $R_1 = R_2 = 0$. Now

$$(S - Q)^4 = (S - Q)(S - Q)^3 = (S - Q)(R_1 - R_2) = 0,$$

and by Proposition 20.7, it is found that

$$\|S - Q\|^4 = \|(S - Q)^4\| = 0.$$

Therefore $Q = S$.

Sol. 30.26: Suppose that $R^2 = S_l$. From this relation one has:

1. $\mathrm{N}(R) \subset \mathrm{N}(S_l) = \mathrm{Lin}(\{e_1\})$, with $e_1 = (1, 0, 0, 0, \cdots)$.

2. R is onto since S_l is onto.

3. R is not injective since S_l is not. Thus, $\mathrm{N}(R) \neq 0$ and so it is unidimensional and equal to $\mathrm{Lin}(\{e_1\})$.

Pick $\eta \in l^2$ so that $R\eta = e_1$. Since $R^2\eta = Re_1 = 0$, it follows that $\eta = \lambda e_1 = \lambda R\eta$ (for some scalar λ), and so $e_1 = R\eta = \lambda R^2\eta = 0$, and the contradiction with $e_1 \neq 0$ completes the solution.

Bibliography

[Alaog]   Alaoglu, L.: Weak Topologies of Normed Linear Spaces, Ann. Math. **41**, 252–267 (1940)

[ArSmi]   Aronszajn, N., Smith, K. T.: Invariant subspaces of completely continuous operators, Ann. Math. **60**, 345–350, (1954)

[BachNa]   Bachman, G., Narici, L.: Functional Analysis. Academic Press, New York (1966)

[Banach]   Banach, S.: Theorie des Opérations Linéaires. Mathematical Monograph vol. **1**, Warsaw (1932)

[BirKre]   Birkhoff, G., Kreyszig, E.: The establishment of functional analysis, Historia Math. **11**, 258–321 (1984)

[BlExHa]   Blank, J., Exner, P., Havlíček, M.: Hilbert Space Operators in Quantum Physics, Second Edition. Springer-Verlag, New York (2008)

[BohSob]   Bohnenblust, H. F., Sobczyk, A.: Extensions of functionals on complex linear spaces, Bull. Amer. Math. Soc. **44**, 91–93 (1938)

[Brezis]   Brezis, H.: Analyse Fonctionnelle. Masson, Paris (1983)

[Buskes]   Buskes, G.: The Hahn-Banach theorem surveyed, Dissertationes Math. **327**, (1993)

[Carles]   Carleson, L.: On the convergence and growth of partial sums of Fourier series, Acta Math. **116**, 135–157 (1966)

[Clarks]   Clarkson, J. A.: Uniformly convex spaces, Trans. Amer. Math. Soc. **40**, 396–414 (1936)

[Cohen1]   Cohen, P. J.: The independence of the continuum hypothesis, Proc. Nat. Acad. Sci. U.S.A. **50**, 1143–1148 (1963)

[Cohen2]   Cohen, P. J.: The independence of the continuum hypothesis. II, Proc. Nat. Acad. Sci. U.S.A. **51**, 105–110 (1964)

[Conway]   Conway, J.: A Course in Functional Analysis. Springer-Verlag, New York (1985)

[Davie1]   Davie, A. M.: The approximation problem for Banach spaces, Bull. London Math. Soc. **5**, 261–266 (1973)

[Davie2]   Davie, A. M.: The Banach approximation problem. Collection of articles dedicated to G. G. Lorentz on the occasion of his sixty-fifth birthday, IV. J. Approx. Theory **13**, 392–394 (1975)

[Day]   Day, M. M.: The Spaces L^p with $0 < p < 1$, Bull. Amer. Math. Soc. **46**, 816–823 (1940)

[de Oliv]   de Oliveira, C. R.: Intermediate Spectral Theory and Quantum Dynamics. Birkhäuser, Basel (2009)

[Dieud]   Dieudonné, J.: History of Functional Analysis. North-Holland, Oxford (1981)

[DunSch]   Dunford, N., Schwartz, J.: Linear Operators I and II. Interscience, New York (1958,1963)

[Enflo]   Enflo, P.: A counterexample to the approximation problem in Banach spaces, Acta Math. **130**, 309–317 (1973)

[Foguel]   Foguel, S. R.: On a theorem of A. E. Taylor, Proc. Amer. Math. Soc. **9**, 325 (1958)

[Halmos]   Halmos, P. R.: Schauder bases, Amer. Math. Monthly **85**, 256–257 (1978)

[HewStr]   Hewitt, E., Stromberg, K.: Real and Abstract Analysis. GTM 25, Springer-Verlag, New York, (1975)

[HirLac]   Hirsch, F., Lacombe, G.: Elements of Functional Analysis. Springer-Verlag, Berlin (1999)

[Hochst]   Hochstadt, H.: Edward Helly, father of the Hahn-Banach theorem, Math. Intell. **2**, 123–125 (1979)

[Holsch]   Holschneider, M.: Wavelets: An Analysis Tool. Oxford University Press, New York, (1999)

[James]   James, R. C.: Characterizations of reflexivity, Studia Mathematica **23**, 205–216 (1964)

[Kato]   Kato, T.: Perturbation Theory for Linear Operators. Springer-Verlag, Berlin (1966)

[Kreysz]   Kreyszig, E.: Introductory Functional Analysis with Applications. John Wiley & Sons, New York (1978)

[KryBog]   Krylov, N., Bogolioubov, N.: La théorie générale de la mesure dans son application à l'étude des systèmes dynamiques de la mécanique non linéaire, Ann. of Math. (2) **38**, 65–113 (1937)

[Kubr]   Kubrusly, C. S.: Elements of Operator Theory. Birkhäuser, Boston (2001)

[LaSalle] La Salle, J. P.: Pseudo-normed linear spaces, Duke Math. J. **8**, 131–135 (1941)

[Last]   Last, Y.: Quantum dynamics and decompositions of singular continuous spectra, J. Funct. Anal. **142**, 406–445 (1996)

[LevSar]   Levitan, B. M., Sargsjan, L. S.: Introduction to Spectral Theory. Transl. Math. Monographs, AMS, vol. **39**, (1975)

[LinLin]   Lin, S.-Y. T., Lin, Y.-F.: Set Theory: An Intuitive Approach. Houghton Mifflin Company, Boston (1974)

[LusSob]   Lusternik, L., Sobolev, V.: Précis d'Analyse Fonctionnelle. Mir, Moscow (1989)

[Oberlin] Oberlin, D. M.: A measure-theoretic proof of a theorem on reflexivity, Proc. Amer. Math. Soc. **41**, 325–326 (1973)

[Phelps]   Phelps, R. R.: Uniqueness of the Hahn-Banach extensions and unique best approximation, Trans. Amer. Math. Soc. **95**, 238–255 (1960)

[Ptak]   Pták, V.: The principle of uniform boundedness and the closed graph theorem, Czech. Math. J. **12**, 523–528 (1962)

[RadRos]   Radjavi, H., Rosenthal, P.: Invariant Subspaces. Springer-Verlag, New York (1973)

[ReedSi]   Reed, M., Simon, B.: Methods of Modern Mathematical Physics I. Functional Analysis, Second Edition. Academic Press, New York (1980)

[RiesSz]   Riesz, F., Sz.-Nagy, B.: Functional Analysis. F. Ungar Publ. Co., New York (1955)

[Rogers]   Rogers, C. A.: Hausdorff Measures. Cambridge Univ. Press, London (1970)

[Royden]   Royden, H. L.: Real Analysis. Macmillan Company, London (1968)

[Rudin1]   Rudin, W.: Functional Analysis. McGraw-Hill, New York (1973)

[Rudin2]   Rudin, W.: Real and Complex Analysis, Second Edition. McGraw-Hill, New York (1974)

[Saidi]   Saidi, F. B.: An extension of the notion of orthogonality to Banach spaces, J. Math. Anal. Appl. **267**, 29–47 (2002)

[Schatt]   Schatten, R.: Norm Ideals of Completely Continuous Operators. Springer-Verlag, Berlin (1960)

[Schecht] Schechter, M.: Principles of Functional Analysis. Academic Press, New York (1971)

[Schw]   Schwartz, J.: A note on the space L_p^*, Proc. Amer. Math. Soc. **2**, 270–275 (1951)

[Simm]   Simmons, G. F.: Introduction to Topology and Modern Analysis. McGraw-Hill, London (1963)

[Steen]   Steen, L. A.: Highlights in the history of spectral theory, Amer. Math. Monthly **80**, 359–381 (1973)

[Stone]   Stone, M.: Linear Transformations in Hilbert Spaces and Their Applications to Analysis. Amer. Math. Soc. Colloq. Publ., vol. **15**, New York (1932)

[Swartz] Swartz, C.: The evolution of the uniform boundedness principle, Math. Chronicle **19**, 1–18 (1990). Addendum: Math. Chronicle **20**, 157–159 (1991)

[Taylor] Taylor, A. E.: Notes on the history of the uses of analyticity in operator theory, Amer. Math. Monthly **78**, 331–342 (1971)

[Trenon] Trénoguine, V.: Analyse Fonctionnelle. Mir, Moscow (1985)

[Yosida] Yosida, K.: Functional Analysis, Second Edition. Springer-Verlag, Berlin (1968)

Index

$(\Omega, \mathcal{A}, \mu)$, 3
$B_X(\xi_0; r)$, 2
F_σ, 35
G_δ, 35
$M(X)$, 24
P_E, 102, 114
$R_\lambda(T)$, 154
S-bounded, 52
$S_X(\xi_0; r)$, 2
T^*, 107
T^a, 72, 107
T_λ, 155
$B(\mathcal{N}_1, \mathcal{N}_2)$, $B(\mathcal{N})$, 16, 19
$B_0(\mathcal{N}_1, \mathcal{N}_2)$, $B_0(\mathcal{N})$, 137
$B_f(\mathcal{N}_1, \mathcal{N}_2)$, $B_f(\mathcal{N})$, 137
$C(\Omega)$, 2
$\mathbb{F}$, 1, 95
$\mathbb{F}^n$, 95
$\mathcal{G}(T)$, 50
$\mathbf{1}$, 15
Im z, 1
$L^p(\Omega)$, 3
$L^p[a, b]$, 3
L^p_μ, 3
Re z, 1
$\mathcal{B}$, 2
χ_A, 6
dim X, 3
dom T, 14
$\hat{}$, 69
$\mathcal{H}$, 95
$\mathcal{P}(X)$, 55
$N(T)$, 15
s $\cdot$ lim, 77, 78
w $\cdot$ lim, 77, 78
$\mathcal{N}$, 2
$\mathcal{N}^*$, 22
$\ominus$, 101
$\oplus$, 100

$\overline{B}_X(\xi_0; r)$, 2
$\overline{\mathrm{Lin}(A)}$, 18
$\perp$, 95
$\prec$, 55
$\rho(T)$, 154
$\xrightarrow{s}$, 77, 78
$\sigma(T)$, 154
$\sigma_c(T)$, 163
$\sigma_p(T)$, 162
$\sigma_r(\mathcal{M}_\phi)$, 163
$\sqrt{T}$, 176
$\tau(\mathcal{N}, \mathcal{N}^*)$, 83
$\tau(\mathcal{N}^*, \hat{\mathcal{N}})$, 83
$\xrightarrow{w}$, 77, 78
$\xrightarrow{w^*}$, 82
$\rightarrow$, 77, 78
c, 6, 18
c_0, 6, 18, 49, 76, 79, 147
l^∞, 3
l^p, 3
HS$(\mathcal{H}_1, \mathcal{H}_2)$, 148
Lin(A), 3
rng T, 15
$\mathcal{M}_\phi$, 15, 20, 111, 142, 163, 164

absolutely summable, 8, 47
adjoint operator, 72
Alaoglu, 85, 89, 91
algebraic dimension, 3
algebraic dual, 24
annihilator, 71, 76
antilinear, 93
Axiom of Choice, 56, 129

Baire space, 35
Baire theorem, 35, 40, 47
ball, 2
Banach adjoint operator, 72
Banach space, 2

For Product Safety Concerns and Information please contact our EU
representative GPSR@taylorandfrancis.com
Taylor & Francis Verlag GmbH, Kaufingerstraße 24, 80331 München, Germany

www.ingramcontent.com/pod-product-compliance
Lightning Source LLC
Chambersburg PA
CBHW061418210326
41598CB00035B/6258

* 9 781041 106500 *